Photocatalytic Semiconductors

Aracely Hernández-Ramírez
Iliana Medina-Ramírez
Editors

Photocatalytic Semiconductors

Synthesis, Characterization, and Environmental Applications

Springer

Editors
Aracely Hernández-Ramírez
Facultad de Ciencias Químicas
Universidad Autónoma de Nuevo León
Cd. Universitaria. C.P. 66450
San Nicolás de los Garza, N.L.
México

Iliana Medina-Ramírez
Departamento de Química
Universidad Autónoma de Aguascalientes
Ciudad Universitaria. C.P. 20131
Aguascalientes, Ags.
México

ISBN 978-3-319-10998-5 ISBN 978-3-319-10999-2 (eBook)
DOI 10.1007/978-3-319-10999-2
Springer Cham Heidelberg New York Dordrecht London

Library of Congress Control Number: 2014955057

Printed on acid-free paper

Springer is part of Springer Science+Business Media (www.springer.com)

Preface

The frequent incidence of hazardous chemicals in wastewater produced by anthropogenic and industrial activities is of great concern because these pollutants contaminate lakes, rivers, and underground aquifers; furthermore, currently more pollutants, including traces of contaminants ranging from pharmaceutical drugs, hormones, and sunscreen to pesticides and dyes, are being detected at smaller concentrations in freshwater bodies. In addition, many of these contaminants are recalcitrant compounds, which cannot be degraded by the conventional methods of wastewater treatment; thus, many treated effluents that are considered "safe" for disposal still contain several toxic (bioactive) pollutants. In general, these compounds are undetectable when ingested or absorbed by living organisms and are subsequently accumulated, causing adverse health effects. Therefore, considerable efforts have been devoted to the development of a suitable environmentally friendly, clean purification process that can destroy these recalcitrant organic contaminants from wastewater to reduce the risk of pollution and toxicity.

Advanced oxidation processes (AOPs) have been proposed as alternative methods for the elimination of many toxic organic compounds in wastewater. The principle of AOPs is to produce hydroxyl radicals in water, a very powerful oxidant capable of oxidizing a wide range of organic compounds with no selectivity.

Among these AOPs, heterogeneous photocatalysis employing semiconductor materials has demonstrated its efficiency in degrading a wide range of indistinct refractory organics into readily biodegradable compounds and eventually mineralizing them to innocuous carbon dioxide and water. In this process, a semiconductor is activated with UV-Vis radiation, and a photoexcited electron is promoted from the valence band to the conduction band, forming an electron/hole pair (e^-/h^+). The photogenerated pair is able to reduce and/or oxidize a compound adsorbed on the photocatalyst surface.

Although heterogeneous photocatalysis has been actively investigated as a promising antibacterial, self-cleaning, and deodorization system, the applications of such photocatalytic process are mostly needed for the purification of water to remove pollutants and bacteria, since this will allow for the safe reuse of this scarce liquid.

However, the key point of the heterogeneous photocatalysis is the material to be used as a catalyst which must be a semiconductor with photocatalytic properties. A photocatalyst is defined as a substance that is able to produce, by the absorption of light quanta, chemical transformations of the reaction participants, repeatedly coming with them into intermediate chemical interactions and regenerating its chemical composition after each cycle of such interactions. The physicochemical properties of the material are crucial for a good performance, and these are usually established according to the photocatalyst nature (composition, size, shape, morphology) and the source of the material.

In the first part of this book, the fundamental principles that govern the physicochemical properties of semiconducting photocatalytic materials are presented (Chaps. 1 and 2) with special emphasis on titanium dioxide, since up to now, this material is the most widely used for photocatalytic applications. However, being aware of the numerous photocatalytic materials with activity under visible light that have been proposed lately, in Chap. 2 some general aspects related to these materials are discussed.

Knowledge of the different synthetic methods employed for the preparation of photocatalytic materials is considered essential, since it is well known that depending upon the production procedure, it can be easily controlled for the development of certain properties in the photoactive material, or facilitate the formation of powders or thin layers with the required characteristics that improve the performance of the catalyst in the photocatalytic process. Therefore, Chap. 3 is devoted to describe a variety of techniques which can be used for photocatalytic semiconductor preparation. Also, some recent examples of the material's preparation exploring the different techniques are included.

There are three key steps in the development of photocatalytic materials: material's preparation, property characterization, and material's activity evaluation. Characterization can be divided in two main categories, structure analysis and property measurements. Structure analysis can be conducted by means of microscopy and spectroscopy techniques, while property characterization can be carried out using electrochemical techniques. In Chaps. 4 and 5 the most common characterization techniques are briefly described including the fundamentals and their applications.

Chapters 6–8 are related to articles dealing with the application of photocatalytic semiconductors on water treatment and disinfection. Finally, in the last part of the book (Chap. 9), some perspectives for photocatalytic materials from the economic and toxicological point of view are discussed.

We hope that our book will be useful to students and researchers who are currently working on the development and/or improvement of photocatalytic materials. We also would like that this book would serve as a general introduction for people just entering the field of photocatalytic materials with environmental applications.

Aracely Hernández-Ramírez
Iliana Medina-Ramírez

Acknowledgments

We would like to express our sincere gratitude to all our colleagues who contributed as authors for the realization of this book. It is also important to express our acknowledgment to all researchers who allowed us to enrich the contents of our book consenting the use of some of their figures. We are also grateful to all the reviewers for their valuable time and observations. We are indebted to our publisher Merry Stuber for taking personal interest in completing this manuscript in time and in a very professional form and Springer for giving us the opportunity to publish this work. Last but not least we acknowledge financial support to the CONACYT-UANL-193883 project.

Contents

Chapter 1
Semiconducting Materials

Aracely Hernández-Ramírez and Iliana Medina-Ramírez

Abstract Semiconducting materials are of great technological importance in electronics industry and environmental remediation due to their ability to generate charge carriers when activated with certain energy. The favorable combination of electronic structure, light absorption properties, charge transport characteristics, and excited lifetimes of some semiconductors has made possible its application as photocatalyst. Although many publications define the fundamental aspects of the semiconductors from different points of view, in this chapter the definition of photocatalytic semiconductor, its structural characteristics, and the requirements to be used as photocatalyst are briefly recalled. In addition, the classification of photocatalytic materials and the mechanism of the photocatalytic process are also reviewed. The advantages and disadvantages of the most used photocatalyst TiO_2 and the second-generation TiO_2-based materials are also addressed.

1.1 Fundamentals

1.1.1 Band Structures

A semiconductor has band structure, roughly characterized as a series of energetically closed spaced energy levels associated with covalent bonding between atoms composing the crystallite (the valence band) and a second series of spatially diffuse, energetically similar levels lying at higher energy and associated with conduction in the macromolecular crystallite (the conduction band). The energy difference between the conduction band and the valence band is called the band gap.

The behavior of semiconductor materials has been explained by the band theory of solids. When a large number of atoms are linked to form a solid, their external

A. Hernández-Ramírez (✉)
Universidad Autónoma de Nuevo León, Facultad de Ciencias Químicas, Cd. Universitaria, c.p. 66450. San Nicolás de los Garza, NL, Mexico
e-mail: aracely.hernandezrm@uanl.edu.mx

I. Medina-Ramírez
Departamento de Química, Universidad Autónoma de Aguascalientes, Av. Universidad 940, Ciudad Universitaria, 20131 Aguascalientes, Ags., Mexico
e-mail: iemedina@correo.uaa.mx

A. Hernández-Ramírez, I. Medina-Ramírez (eds.), *Photocatalytic Semiconductors*,
DOI 10.1007/978-3-319-10999-2_1

orbitals begin to overlap, and then a large number of levels are formed with a close spacing so that it can be considered as a continuous band of energy levels. The gap of an energy band depends only on the interaction of neighbors, while the number of levels within the band depends on the total number of particles interacting (and therefore the number of atoms in a crystal). In general, a solid has a considerable number of allowed energy bands from different atomic energy levels. The energy gap (E_g) between the allowed energy bands is called forbidden energy band because the electrons are not allowed to stay there (Serway and Jewett 2014).

Based in the band theory, the differences between metals, semiconductors, and insulators depend on the band structure of each one, the filling of the bands, and the extent of the energy gap between full and empty bands.

Another important aspect in the band theory is the Fermi level; at $T = 0$ K, electrons occupy the individual molecular orbitals of the bands in accordance with the building-up principle. The energy levels that the electrons occupy are quantized, and the levels are filled from the bottom with two electrons per level. The highest occupied energy level in a solid at absolute zero is known as Fermi level.

For metals, Fermi energy (E_F) lies within the conduction band while, for semiconductors and insulators, falls into the band gap energy. In a semiconductor the Fermi level is located near the middle of the gap and E_g is small, and appreciable numbers of electrons are thermally excited from the valence band to the conduction band. Because of the many empty levels above the thermally filled levels in the conduction band, a small applied potential difference (or photonic energy) can easily raise the electrons in the conduction band into available energy states, giving as a result the formation of charge carriers.

Charge carriers in a semiconductor can be negative, positive, or both. When an electron moves from the valence band VB into the conduction band CB, it leaves behind a vacant site, the hole. This hole behaves as a charge carrier because a free electron from a nearby site can transfer into the hole; as electron leaves a site, a new hole is created; thus the net effect can be viewed as the hole which migrates through the material. Therefore, the hole acts as a particle with a positive charge. Electrons that are promoted into the conduction band are the negative charge carriers, and the positive holes move in the opposite direction of the electrons in the presence of an electric field (Serway and Jewett 2014).

In solids, the absorption of a photon can produce excitation of electrons depending on the band structure of the semiconductor. When the lowest energy excitation of an electron from the valence band to the conduction band involves no change in momentum k, the semiconductor has a *direct band gap*, and the absorption probability is high for these transitions (i.e., GaS, ZnO, and CdTe), direct band gap materials provide more efficient absorption and emission of light (Jacobsson and Edvinsson 2012). An *indirect band gap* is presented in a material where the k at the valence band maximum is different from the k at the conduction band minimum. For photons to be absorbed, momentum conservation requires coupling to a lattice vibration (a phonon) to compensate for the change in the wavevector during the transition, and then the required input photon energy is lower. GaP, TiO_2, and CdS are examples of semiconductors with an indirect band gap (Serpone and Pellizetti 1989).

1.1.2 Intrinsic and Extrinsic Semiconductors

There are two classes of semiconductors: elemental semiconductors like silicon or germanium which are widely used in the electronic industry and semiconductor compounds that are, for example, metal oxides or chalcogenides, among which are the semiconductors with photocatalytic properties.

A pure semiconductor crystal containing only one element or one compound is an *intrinsic* semiconductor which at 0 K behaves as insulator. In these semiconductors, there are equal numbers of conduction electrons and holes, and the Fermi level is located midway between the valence and conduction bands. An *extrinsic* semiconductor is a substance that is a semiconductor by addition of impurities. When impurities are added to semiconductors, the band structure is modified; this process is called doping. When a semiconductor is doped with acceptor atoms is known as a *p*-type semiconductor because these atoms can be reduced taking electrons from the valence band and increasing the density of holes, then the majority carriers are the positively charged holes. Semiconductors doped with donor impurities, which provide electrons to the conduction band, are *n*-type semiconductors because the majority carriers are the electrons. The presence of acceptor impurities in a semiconductor originates the Fermi level shift closer to the valence band, while the donor impurities cause that Fermi level shift near to the conduction band (Fig. 1.1).

In compound semiconductors, a deviation from stoichiometry generates donors or acceptors depending on whether it is the cation or the anion which is in excess (Pankove 1971). Several *d*-metal oxides, including ZnO, TiO_2, and Fe_2O_3, are *n*-type semiconductors. In their case, the property is due to small variations in stoichiometry and a small deficit of O atoms (Atkins et al. 2010). This deficiency implies the presence of anionic vacancies around in which the deficit of negative charge is compensated by a decrease of the positive charge of the surrounding cations.

p-type semiconduction is observed for some low oxidation number *d*-metal chalcogenides, including Cu_2O, FeO, and FeS. In these compounds, the loss of electrons can occur through a process equivalent to the oxidation of some of the metal atoms, with the result that holes appear in the predominantly metal band. *n*-type semiconductivity, however, tends to occur for oxides of metals in higher oxidation states, as the metal can be reduced to a lower oxidation state by occupation of a conduction band formed from the metal orbitals (Atkins et al. 2010).

In the case of TiO_2, which should formally be written TiO_{2-x}, the oxygen vacancies are compensated by adopting the 3+ oxidation state with an equivalent number of atoms of titanium. These Ti^{3+} ions act as donors of electrons, and the material is an *n*-type semiconductor.

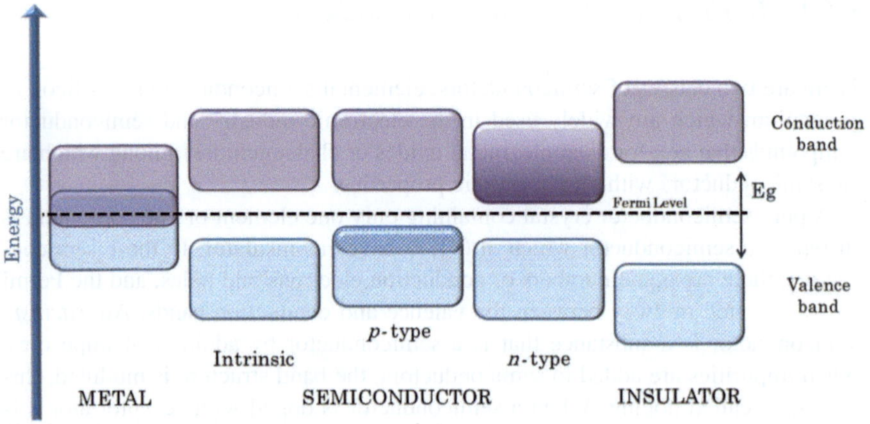

Fig. 1.1 Electronic band structures of metal, semiconductor, and insulator indicating the Fermi level

1.1.3 Band Edge Positions and Band Gaps

Based on the band theory, the main difference among metals, semiconductors, and insulators is the extent of the band gap energy; thus, the band gap of semiconductors (between 1 and 4 eV) lies in the appropriate range for a semiconductor photocatalyst (Fig. 1.1).

Besides band gaps and charge carrier mobility, the band edge positions are especially important to the photocatalytic activity of semiconductors because they are crucial for which catalytic reactions could occur at the surface.

In an electronic energy level diagram of a semiconductor (Fig. 1.2), the upper edge of the valence band E_{VB} corresponds to the energy that determines the oxidizing ability of holes, and the energy level at the bottom of the conduction band E_{CB} is the energy of electrons for reduction processes (Litter 1999). The energies of electrons in a solid are referred to the vacuum level, which is taken as the reference of zero energy value.

The energy binding of the electrons inside the semiconductor corresponds to the electrochemical potential of the electrons, $\widetilde{\mu}_e$. The Fermi energy (E_F) is equal to the electrochemical potential of the semiconductor:

$$E_F = \widetilde{\mu}_e \tag{1.1}$$

The work function of the semiconductor, Φ_{SC}, is defined as the difference between the Fermi level and the electrostatic potential, $e\Psi$, right out of the semiconductor surface:

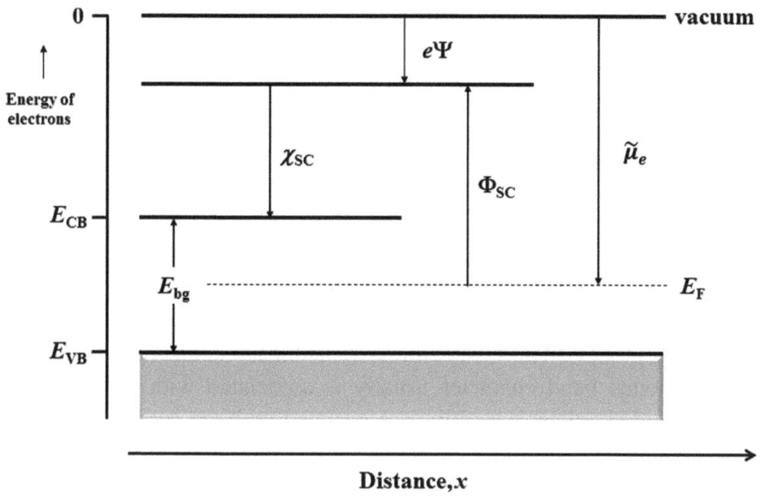

Fig. 1.2 The electronic energy band diagram for a semiconductor (Kondarides 2010)

$$\Phi_{SC} = -\widetilde{\mu}_e - e\Psi \tag{1.2}$$

The work function, which incorporates the contribution of a term related to the bulk and a term related to the surface, represents the work needed to move an electron from the solid and place it at a distance x from the surface where the classical image-force potential, $-e^2/4x$, is just negligible. A distance of 10^{-4} cm is generally sufficient to meet this requirement (Kondarides 2010).

The electron affinity, χ_{SC}, is defined as the difference between the edge of the conduction band (E_{CB}) and the electrostatic potential just outside the semiconductor:

$$\chi_{SC} = -E_{CB} - e\Psi \tag{1.3}$$

Electron affinity is an important surface parameter for estimating the band bending that occurs at the interface of two materials.

The determination of the electronic properties of the most common semiconductor photocatalysts such as band edge positions and band gap energy has been studied by different spectroscopy and electrochemical methods which will be discussed in Chaps. 4 and 5.

1.1.4 Semiconductors for Photocatalysis

According to the fundamental aspects previously revised in this chapter, many binary compounds are classified as semiconductors, but not all are suitable for photocatalytic applications.

For example, the transition metal dichalcogenides have an attractive structure because the valence band and conduction band are composed of hybridized *d*-orbitals from the metal atoms. Consequently, an electron excitation from the valence band d_{z2} orbital to the conduction band d_{xy}, d_{x2-y2} orbitals does not affect interatomic bonding and the compounds are generally very resistant to photocorrosion process. This is in contrast to the binary semiconductor compounds, where the valence band character usually is associated with the anion and the conduction band character generally is associated with the cation. For these compounds, a band to band transition weakens the interatomic bond, so photoexcitation promotes decomposition and dissolution of the solid when used in solution (Serpone and Pellizetti 1989). For photovoltaic applications, silicon and gallium arsenides have been found to have great practical potential; nevertheless they are not suitable for photocatalytic applications due to lack of chemical stability.

Therefore, an appropriate semiconductor for heterogeneous photocatalysis must fulfill with the following characteristics: appropriate band gap (in the range of 1.7–3.2 eV), efficient light absorption, high carrier mobility, precise band edge positions that straddle the water redox potentials, and nontoxic and chemically stable.

Titanium dioxide (TiO_2) in anatase form and zinc oxide (ZnO) in wurtzite phase are the most used metal oxides as photocatalysts due to their electronic band structure (Jacobsson and Edvinsson 2012; Wang and Lewis 2006) and the combination of the properties above mentioned. The band gap (3.2 eV) and conduction band edge position of these semiconductors are very similar ($E_{CB} = -0.51$ V at pH 5–7 vs. NHE) (Park et al. 2013; Soga 2006). Both are highly efficient photocatalysts; however, although ZnO is sometimes preferred over TiO_2 for degradation of various organic pollutants due to its high quantum efficiency (Farbod and Jafarpoor 2012; Rehman et al. 2009), it is not stable in acidic aqueous suspensions.

1.1.5 Mechanism of the Photocatalytic Process

The first studies that were reported with the concept of "heterogeneous photocatalysis" are those by Doerfler and Hauffe, published in 1964 when they carried out the oxidation of CO using zinc oxide as catalyst under illumination (Doerfler and Hauffe 1964).

However, the great interest in heterogeneous photocatalysis has been related to the work by Fujishima and Honda, published in 1972 (Fujishima and Honda 1972). In the aforementioned work, was described the photo-electrolysis of water under irradiation of TiO_2 electrode (with a Pt counter-electrode) which was an energy storage reaction that can be termed photogalvanic (Tobaldi et al. 2014), in fact, that work promised utilization of TiO_2-based materials for solar energy conversion and storage.

Since then, many efforts for understanding the photocatalytic process were done especially with TiO_2 catalyst for degradation of organic pollutants in water; nowadays the mechanism of the illuminated catalyst, the surface phenomena, and the generated species has been elucidated (Bahnemann 2004; Friedmann et al. 2010; Pichat 2013). Also the parameters that affect the photocatalytic process and consequently the reaction rate have been studied. Several articles already were published where a detailed explanation about the influence of solution pH, catalyst dosage, substrate concentration, temperature, photonic flux, and the reactor design has been described (Pichat 2013; Ahmed et al. 2011; Friedmann et al. 2010; Herrmann 2005, 2010).

Therefore, it has been well established that the photocatalytic process begins when photons of energy higher or equal to the band gap energy are absorbed by a semiconductor particle and an electron (e^-) from the valence band (VB) is transferred to the conduction band (CB) generating a hole (h^+) in the VB. The absorption of these photons creates within the bulk electron–hole pairs, which dissociate into free photoelectrons in the conduction band and photoholes in the valence band.

The e^- and h^+ can recombine on the surface or in the bulk of the particle releasing the energy as heat or migrate to the surface where they can react with adsorbed molecules on the surface of the particle as illustrated in Fig. 1.3.

In a fluid medium, the flat band potential, V_{fb}, of the semiconductor locates the energy of both charge carriers at the semiconductor–electrolyte interface and depends on the nature of the material and the system equilibria. Adsorbed species can be reduced by CB electrons if they have redox potentials more positive than the V_{fb} of the CB and can be oxidized by holes if they have redox potentials more negative than the V_{fb} of the VB. In presence of adsorbed water, electrons transfer from water molecule to the positive holes to produce $^{\bullet}OH$ radicals which are powerful oxidants and react with organic and toxic compounds. $^{\bullet}OH$ radicals play an important role in initiating oxidation reactions, especially for substances that adsorb weakly on the TiO_2 surface. This oxidation pathway is known as indirect oxidation to differentiate it from the direct oxidation by holes. However, the role of the $^{\bullet}OH$ radicals is probably overestimated, and some controversial aspects have been reported regarding the origin of photogenerated free $^{\bullet}OH$ radicals (Salvador 2007; Fujishima et al. 2008).

In Fig. 1.4 several semiconductors and their band edge positions in contact with aqueous media at $pH = 0$ are presented. The band levels usually shift with a change in pH (0.059 V/pH) for oxide materials. The standard potentials of these semiconductors indicate the thermodynamic limitations for the photoreactions that can be carried out with the charge carriers. It is to be recognized that E_g does not represent the driving force of the reaction, but only the system total energy. The final products of the process and its efficiency and selectivity are of course

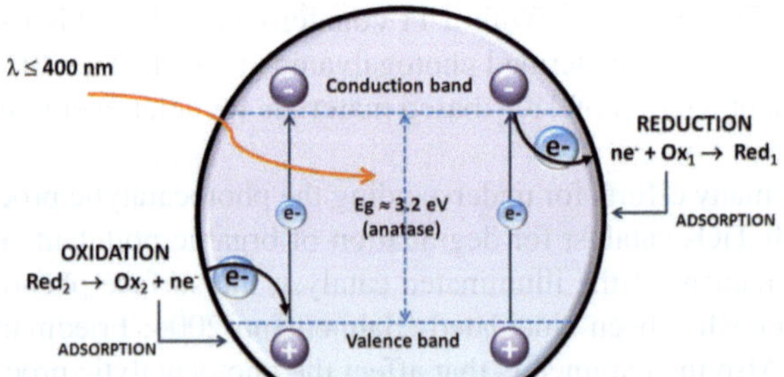

Fig. 1.3 Schematic representation of the electron transfer reactions on TiO_2 particle

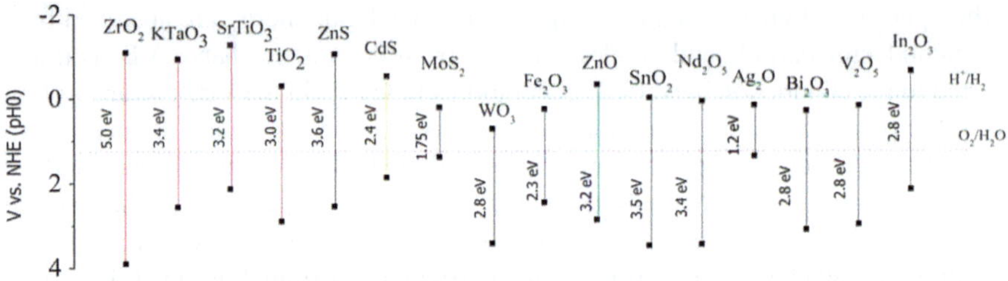

Fig. 1.4 Band edge positions of some semiconductors at pH = 0 with permission from Wang et al. (2013)

affected by several other parameters which are essentially of kinetics nature (Schiavello 1987).

In summary, the efficiency of a photocatalyst depends on the competition of different interface transfer processes involving electrons and holes and their deactivation by recombination. In this context, there are intrinsic and extrinsic parameters to the photocatalytic semiconductor that affect the kinetics and mechanisms of photocatalytic reactions in aqueous media. The crystallographic phase, the exposed crystal face, the crystallite size, and the presence of dopants, impurities, vacancies, and different surface states can be enlisted as intrinsic factors, while the surrounding environment and the photocatalytic conditions (pH of the solution, pollutant and its initial concentration, the presence of impurities in the system, light intensity, catalyst dosage, and flow rate) are the extrinsic parameters (Friedmann et al. 2010).

1.2 Photocatalytic Semiconductors: Classification

1.2.1 Metal Oxides

Simple binary oxides with a *d*-transition metal—TiO_2, ZnO, WO_3, and Fe_2O_3—have been the most studied metal oxides with photocatalytic purposes (Malato et al. 2009; Lezner et al. 2012; Wang et al. 2013; Topkaya et al. 2014). Among them, TiO_2 is the photocatalyst mainly used due to its nontoxicity, water insolubility, hydrophilicity, low cost, stability, and resistance to photocorrosion. However, several binary metal oxides have been tested in the photocatalytic degradation of various organic compounds as possible alternatives to TiO_2.

Zinc oxide is an *n*-type semiconductor whose thermodynamically stable phase is hexagonal wurtzite structure where each anion is surrounded by four cations at the corners of a tetrahedron and vice versa. This tetrahedral coordination is typical of sp^3 covalent bonding nature, but this material also has a substantial ionic character that tends to increase the band gap beyond the one expected from the covalent bonding (Morkoç and Özgür 2009).

Despite this fact, ZnO is the second most employed semiconductor after TiO_2 because of its good optoelectronic, catalytic, and photochemical properties along with its low cost and nontoxic nature. The band gap value for ZnO semiconductor depends on preparation method because different kind of structural defects can be produced and may alter the electronic structure (Asthana et al. 2011). Thereby the band gap energy reported for ZnO in various articles varies from 2.8 (Malato et al. 2009) to 3.37 eV (Asthana et al. 2011; Jacobsson and Edvinsson 2012; Rivera-Reyna et al. 2013; Di Paola et al. 2012).

The quantum efficiency of ZnO powder is also significantly larger than that of TiO_2 powder. Higher catalytic efficiencies, in many cases, have been reported for ZnO (Chakrabarti and Dutta 2004; Farbod and Jafarpoor 2012), but the occurrence of photocorrosion and the susceptibility of ZnO to facile dissolution at acidic pH values have limited its application in photocatalysis (Di Paola et al. 2012). Nevertheless the photocatalytic advantages offered by ZnO can be exploited by controlling the solution pH at neutral conditions (Delgado-Balderas et al. 2012; Kansal et al. 2011).

On the other hand, tungsten oxide (WO_3) is also an *n*-type semiconductor. The conduction band edge of WO_3 is composed of W 5d–O 2p π^* orbitals; therefore E_{CB} of WO_3 is dependent on the W–O–W bond angle. The most stable phase of WO_3 at room temperature is monoclinic γ-WO_3. At elevated temperatures, monoclinic WO_3 adopts a more symmetric crystal structure, becoming orthorhombic between 330 and 740 °C and then tetragonal between 740 °C and the melting point. The evolution of the crystal structure of WO_3 is accompanied by a reduction in E_g from 2.6 to 2.7 eV at room temperature to ~1.8 eV at 750 °C, corresponding to a color change from pale yellow to red. E_g values of different WO_3 phases decrease in the order monoclinic > orthorhombic > tetragonal ≥ cubic (Mi et al. 2012).

WO_3 has also been considered to be a photocatalytic material, because of its high stability in aqueous solution under acidic condition, non-photocorrosion, and a proper band gap for visible light absorption (2.4–2.8 eV) which makes it a suitable candidate for the photocatalytic degradation of organic pollutants under solar irradiation (Aslam et al. 2014; Hong et al. 2009). However, although WO_3 absorbs light in the visible region, its photocatalytic activity is very low.

The valence band (VB) edge of WO_3 is located at approximately 3 eV (+3.1 to 3.2 VNHE), but the lower conduction band (CB) edge value (+0.3 to 0.5 VNHE) does not provide a sufficient potential to reduce O_2 [$E^0(O_2/O_2^{-\bullet})$ – 0.33 VNHE and $E^0(O_2/HO_2^{\bullet})$ – 0.05 VNHE] (Kim et al. 2010). The inability of O_2 to scavenge CB electrons in WO_3 makes it less favorable to separate the photogenerated electrons from the hole, and thus WO_3 alone is not an efficient photocatalyst. In order to achieve good photoactivity toward organic substance degradation, a suitable co-catalyst should be used (He et al. 2013; Sayama et al. 2010; Widiyandari et al. 2012).

Iron oxide (α-Fe_2O_3, hematite) is an *n*-type semiconductor material with a narrow band gap at about 2.1 eV, which corresponds to visible region energy of the solar spectrum (Zhang et al. 2010a). It shows stability in most aqueous solution at $pH > 3$. Iron oxides with Fe^{3+} ions surrounded octahedrally by oxygen anions have complex band structures where the electron transfer occurs between 3*d* bands. Similar to WO_3, the conduction band of iron oxide (+0.3 VNHE) is located below the H^+/H_2 redox potential which means that the reduction power is not enough to reduce water.

High electron–hole recombination rate and short hole diffusion length (2–4 nm) are the major challenges for its practical application (Zhang et al. 2010b). Although Fe_2O_3 is a good candidate to absorb visible light, the drawback of this semiconductor is the low quantum yield of the process which inhibits the efficiency of the degradation reaction (Bandara et al. 2007). However, several works have reported the application of this semiconductor as an efficient photocatalyst in the degradation of some organic compounds such as orange II, methyl orange, phenols, and chlorophenols (Bakardjieva et al. 2007; Bandara et al. 2001; Sharma et al. 2014).

Other binary oxides such as Nb_2O_5, V_2O_5, Sb_2O_3, Bi_2O_3, NiO, Ta_2O_5, ZrO_2, CeO_2, Ga_2O_3, CuO, and Cu_2O have been tested as photocatalysts (Di Paola et al. 2012); some of them have wide band gaps resulting in being photoactivated only by irradiation from UV range. For example, ZrO_2 with a wide band gap energy of 5.0 eV (Sreethawong et al. 2013) and Ga_2O_3 (band gap 4.8 eV) can be photoexcited by irradiation energy of 248–260 nm. Nevertheless, depending on synthetic approaches, these metal oxides have been shown to have higher photocatalytic performance in the degradation of organic pollutants and in some cases to be superior to that exhibited by commercial P-25 TiO_2.

1.2.2 Chalcogenides: Other than Oxides

In addition to metal oxides, several chalcogenides have been investigated regarding their photocatalytic activity. Among them, metal sulfides are the most common semiconducting materials studied because of their low band gap energy (1.3–2.40 eV) which take advantage of solar energy utilization. There are various sulfide semiconductor materials possessing narrow band gaps (i.e., CdS, ZnS, Sb_2S_3, Bi_2S_3, MoS) with appropriate characteristics to be used in the photocatalytic process. The valence bands of sulfide semiconductors usually consist of S*3p* orbitals which are in a more negative position than O2*p*. Nevertheless, metal sulfides have the problem of photocorrosion, but in spite of this, they have attracted attention as photocatalysts with a visible light response. The representative chalcogenide photocatalyst is cadmium sulfide (CdS) which is active for H_2 evolution under visible light irradiation; but this material is known to be toxic in aqueous media as it undergoes photoanodic corrosion, releasing dangerous metal ions (Cd^{2+}) into the solution and then limiting CdS practical application for degradation of aqueous pollutants. In some cases the photocorrosion can be considerably suppressed in the presence of sacrificial reagents (Kudo and Miseki 2009). Anyway CdS has been used in laboratory trials for the treatment of dyes and organic pollutants in aqueous media (Giribabu et al. 2012; Yu et al. 2011)

Among other metal sulfides such as ZnS, Sb_2S_3, Bi_2S_3, and MoS, only zinc sulfide possesses a wide band gap of 3.5 eV; even so this material has the advantage to be nontoxic, exhibiting good photocatalytic activity on removal of hazardous compounds (Bai et al. 2014).

Cadmium selenide and telluride have the ability to reduce water for hydrogen production because of the negative conduction band position vs. ENH; nevertheless these chalcogenides are unable to oxidize water because the valence band redox potential is lower than 1.23 V; thereby CdSe and CdTe exhibit poor photocatalytic activity for oxidation of organic pollutants.

1.2.3 Ternary Compounds

Various metallates, oxysulfides, oxyhalides, and oxynitrides can be classified as photocatalytic ternary and quaternary semiconductors which can be activated with visible light radiation. Many of these compounds have been mainly investigated for oxygen and hydrogen evolution from water splitting; nevertheless, degradation of organic pollutants has also been carried out under visible or solar light (Hernández-Alonso et al. 2009; Kudo and Miseki 2009).

Aluminates, ferrites, niobates, tantalates, titanates, tungstates, and vanadates are some examples of metallates which have been tested as photocatalytic semiconductors in the degradation of different organic dyes, mainly methylene blue (MB), rhodamine B, and methyl orange (MO), among others. It is important to remark that dye decolorization, especially in the visible range, is an inappropriate test to demonstrate the real photocatalytic activity of a semiconductor material. If dyes are used as model compounds, attention must be taken in the analysis of experimental results; not only measurement of parent compound disappearance is sufficient, but also their mineralization by TOC determination is necessary (Herrmann 2010; Ohtani 2008).

Regarding to aluminates, $ZnAl_2O_4$ with face-centered cubic spinel structure and $AgAlO_2$ (orthorhombic structure) have band gaps of 4.11 (Sampath et al. 1999) and 3.2, respectively. These aluminates presented lower activity in the visible region due to their wide band gaps; despite its wide band gap, $ZnAl_2O_4$ was successfully used in the degradation of gaseous toluene under artificial UV irradiation (Li et al. 2011).

Ferrites $ZnFe_2O_4$ and $BiFeO_3$ which have spinel and perovskite structure, respectively, have been investigated in the degradation of organic pollutants. The photocatalytic activity of $ZnFe_2O_4$ with 1.9 eV band gap (Xie et al. 2012) was tested for degradation of phenol under solar light and artificial UV irradiation, while $BiFeO_3$ activity was investigated in the decomposition of MO irradiating with UV and visible light lamp. Both catalysts showed an efficient photocatalytic activity only under ultraviolet light (Anchieta et al. 2014; Liu et al. 2010).

Many niobates, tantalates, and titanates possess layered perovskite structure and can be activated with visible light especially for water splitting; $Ca_2Nb_2O_7$, $Ba_5Nb_4O_{15}$, $NaTaO_3$, $KTaO_3$, $AgTaO_3$, $KTaO_3$, $Ca_2Ta_2O_7$, $Sr_2Ta_2O_7$, $CaTiO_3$, and $SrTiO_3$ are some ternary compounds with very interesting conduction band edges. For example, $SrTiO_3$ and $KTaO_3$ have high negative conduction band levels (−0.3 to 0.4 V vs. NHE) which can decompose water into H_2 and O_2; moreover $SrTiO_3$ is also active for reduction of NO_3^- using water as an electron donor (Kudo and Miseki 2009).

Oxysulfides, oxyhalides, and oxynitrides can induce a significant increase in visible light absorption. As an example, tantalum oxynitride (TaON) exhibits a narrow band gap of 2.5 eV and an absorption edge that is redshifted by 170 nm compared with Ta_2O_5. Under visible light irradiation, TaON enabled the effective evolution of H_2 and O_2 from methanol and $AgNO_3$ solution, respectively (Tong et al. 2012).

1.2.4 Quaternary Compounds

Oxides, oxynitrides, oxysulfides, and oxyhalides as quaternary compounds also have been investigated for evaluation of their activity in the degradation of organic pollutants and water splitting (Di Paola et al. 2012).

Quaternary oxides $ABi_2Nb_2O_9$ ($A = $ Ca, Sr, Ba) photocatalysts with perovskite structure were prepared for both H_2 and O_2 evolution from aqueous solutions containing sacrificial reagents; the photocatalytic performance decreased in the order of $SrBi_2Nb_2O_9 > BaBi_2Nb_2O_9 > CaBi_2Nb_2O_9$ (Ismail and Bahnemann 2014).

Oxynitrides such as LaTiON and oxysulfides such as $Sm_2Ti_2S_2O_5$ can also effectively harvest visible light and exhibit higher photocatalytic activity for water splitting and organic contaminant degradation than their oxide precursors.

Among these nonmetal elements, S can often completely replace O to convert an oxide into a sulfide, because both O and S belong to group 16. A series of sulfides have been reported, including $AgInZn_7S_9$ and $ZnIn_2S_4$ in which the visible light absorption is better than their oxide analogues. These sulfides thus exhibit high photocatalytic activity for the evolution of H_2 under visible light irradiation (Tong et al. 2012).

Bi-based layered oxyhalides were tested for the degradation of methyl orange under UV and visible illumination. Bi_4NbO_8Cl ($E_g = 2.38$ eV) showed an excellent visible light efficiency and was more active than the ternary oxychloride Bi_3O_4Cl ($E_g = 2.80$ eV). Meanwhile, $Na_{0.5}Bi_{1.5}O_2Cl$ with band gap of 3.04 eV was more efficient than BiOCl ($E_g = 3.44$ eV) under visible light illumination but less active under UV light (Di Paola et al. 2012).

Table 1.1 summarizes the photocatalyst classification indicating the band gap energy and the corresponding absorption edge of various semiconductors which have been reported in different literature sources.

In summary, photocatalytic materials applied for environmental remediation or water splitting includes a wide range of semiconductor materials such as binary, ternary, and quaternary oxides. Additionally, a large number of sulfides, nitrides, and oxynitrides (CdS, Ta_3N_5, and LaTiON) have been also investigated as alternative materials to TiO_2 for visible light or solar photocatalysis. However, TiO_2 is still the most used photocatalyst showing the best photocatalytic performance, especially under UV irradiation.

1.3 Advantages and Disadvantages in the Use of Titanium Dioxide (TiO_2) as Photocatalytic Semiconductor

Titanium dioxide is a binary metal oxide and crystallizes in three different allotropic forms: anatase (tetragonal, D_{4h}^{19}-I4$_1$/amd, $a = b = 3.782$ Å, $c = 9.502$ Å), rutile (D_4^{14}-P4$_2$/mnm, $a = b = 4.584$ Å, $c = 2.953$ Å), and brookite (rhombohedrical, D_{2h}^{15}-Pbca, $a = 5.436$ Å, $b = 9.166$ Å, $c = 5.135$ Å) (Diebold 2003); nevertheless, only anatase and rutile phases have adequate structural features for application in photocatalysis. Both crystal structures are formed of TiO_6 octahedra chains which differ by the distortion of each octahedron. The octahedron in rutile is not regular, showing a slight orthorhombic distortion, whereas the octahedron in anatase is significantly distorted so that its symmetry is lower than orthorhombic. These differences in lattice structures cause different electronic band structures between the two TiO_2 crystalline forms. Hence, the difference between the 3.2 eV band gap of anatase and the 3.0 eV band gap of rutile lies mainly in the position of the CB edge, the edge for anatase being ~0.2 eV higher than that of rutile. The valence

Table 1.1 Common photocatalytic semiconductors, band gap energy, and the corresponding absorption edge

Semiconductor	Band gap (eV)	Absorption edge (nm)
Binary compounds		
Metal oxides		
Ag_2O	1.20[a]	1,033
Bi_2O_3	2.80[a]	443
CeO_2	2.94[b]	422
Cr_2O_3	3.40[c]	365
Cu_2O	2.0–2.2[b]	564–620
Fe_2O_3	2.30[c*]	539
Ga_2O_3	4.80[b]	258
In_2O_3	2.80[a]	443
Nb_2O_5	3.40[b]	365
Nd_2O_5	3.40[a]	365
SnO_2	3.50[a]	354
Ta_2O_5	3.00[b]	413
TiO_2	3.00[c*]	413
V_2O_5	2.20[a]	564
WO_3	2.80[c*]	443
ZnO	3.20[a]	388
ZrO_2	5.00[a]	248
Chalcogenides		
Bi_2S_3	1.3[g]	953
CdS	2.40[c*]	517
CdSe	1.70[c*]	729
CdTe	1.40[c*]	886
MoS_2	1.75[c*]	709
Sb_2S_3	1.55[b]	800
ZnS	3.60[c*]	344
Other binary compounds		
β-Ge_3N_4	3.80[d]	326
GaAs	1.40[c**]	886
GaP	2.25[c**]	551
SiC	3.00[c**]	413
Ta_3N_5	2.00[d]	620
Ternary compounds		
α-$AgGaO_2$	2.40[b]	517
α-FeOOH	2.20[b]	564
β-$AgGaO_2$	2.10[b]	590

(continued)

Table 1.1 (continued)

Semiconductor	Band gap (eV)	Absorption edge (nm)
$AgAlO_2$	3.20[b]	388
$AgCrO_2$	1.68[b]	738
Ag_2CrO_4	1.75[b]	709
Ag_2GeO_3	1.80[b]	689
$AgInO_2$	2.00[b]	620
$BaBiO_3$	2.05[b]	605
$BiFeO_3$	2.10[b]	590
Bi_2MoO_6	2.70[e]	459
$Bi_2Mo_2O_9$	3.10[e]	400
$Bi_2Mo_3O_{12}$	2.88[e]	431
BiOBr	2.75[b]	451
BiOCl	3.19[b]	389
Bi_3O_4Cl	2.80[b]	443
BiOF	3.40[b]	365
BiOI	1.76[b]	705
$BiVO_4$	2.34[h]	530
Bi_2WO_6	2.8[b]	459
$Ca_2Sb_2O_7$	4.02[b]	308
$CaIn_2O_4$	2.00[b]	580
$FeTiO_3$	2.80[c]	443
GaAsP	2.25[c**]	551
$In(OH)_3$	5.15[b]	241
InOOH	3.70[b]	335
$InVO_4$	2.00[b]	620
$KTaO_3$	3.40[a]	365
$NaBiO_3$	2.60[b]	477
$Pb_3Nb_2O_8$	2.72[b]	456
$Pb_3Nb_4O_{13}$	2.95[b]	420
$PbFe_{12}O_{19}$	2.30[c]	539
$Sr_2Sb_2O_7$	3.86[b]	321
$SrTiO_3$	3.20[c*]	388
$ZnAl_2O_4$	4.11[i]	301
Zn_2SnO_4	3.60[b]	344
$ZnBi_{12}O_{20}$	2.69[b]	461
$ZnIn_2S_4$	2.20[b]	564
Quaternary compounds		
Bi_4NbO_8Cl	2.38[b]	521
$LiBi_4Nb_3O_{14}$	3.00[b]	413

(continued)

Table 1.1 (continued)

Semiconductor	Band gap (eV)	Absorption edge (nm)
$LiBi_4Ta_3O_{14}$	3.50[b]	354
$Na_{0.5}Bi_{1.5}O_2Cl$	3.04[b]	408
$PbBi_2Nb_2O_9$	2.88[b]	431
$LaTaO_2N$	2.00[f]	620
$LaTiO_2N$	2.10[f]	590
$Y_2Ta_2O_5N_2$	2.20[f]	564

[a]Wang et al. 2013 (at pH 0)
[b]Di Paola et al. 2012
[c]Serpone and Pellizetti 1989, [c*]at pH 0, [c**]at pH 1
[d]Maeda 2011
[e]Shimodaira et al. 2006
[f]Kudo and Miseki 2009
[g]Luo et al. 2013
[h]Hernández-Alonso et al. 2009
[i]Sampath et al. 1999

band of these semiconductors is mainly formed by the overlapping of the oxygen *2p* orbitals, whereas the lower part of the conduction band is mainly constituted by the *3d* orbitals, with t_{2g} symmetry, of Ti^{4+} cations (Hernández-Alonso et al. 2009).

Anatase is the most active phase and is thermodynamically less stable than rutile, but its formation is kinetically favored at lower synthesis temperature (<600 °C). Indeed, the photocatalytic activities of various anatase and rutile samples overlap because other parameters influence these activities (Pichat 2003). Due to oxygen vacancies present in its structure, TiO_2 is an *n*-type semiconductor. Interband transitions of TiO_2 are indirect, but factors like the crystalline size or the presence of dopants can modify the type of transition. This characteristic directly affects the photonic efficiency because indirect semiconductors present reduced photon absorption and consequently require a higher mass of photocatalyst to obtain the same effect. In any case, it is worth emphasizing that as much as a 90 % of the electron–hole pairs recombine in less than 10 ns and consequently photogenerated carriers available for surface reactions are quite limited (Hernández-Alonso et al. 2009).

In general, TiO_2 has several advantages such as its high chemical stability, excellent functionality, nontoxicity, and relatively low price, but one of the most important advantages of TiO_2 among other semiconductors is that it allows the reduction and the oxidation of water for the process of water splitting. In this case anatase gives better results than rutile for hydrogen production in powder photocatalysis. One of the main reasons for the higher efficiency is probably the higher reduction potential of photogenerated electrons in the former than in the latter, i.e., the bottom of the conduction band of anatase is located 0.1 V more negative than that of rutile (Hashimoto et al. 2005). However, it has often been reported that the photocatalytic activities of anatase and rutile are greatly affected by the kind of substrates and rutile exhibits higher activities than anatase for some reactions (Park et al. 2013).

The wide band gap energies of both TiO_2 phases show that the major drawback of using TiO_2 as a photocatalyst is that it can only be excited by photons in the near-UV rather than the visible spectral region. The absorption properties are not well matched to the frequency spectrum of the solar light; therefore, TiO_2 photocatalysts can make use of less than 5 % of the solar energy that reaches the surface of the earth, and then a serious disadvantage is that only UV light can be used for its photocatalytic capabilities.

Despite the inherently poor performance of TiO_2 for solar-driven reactions, titanium dioxide-based photocatalysts are the materials of choice in most photocatalytic processes. Various strategies have been adopted for improving the photocatalytic efficiency of TiO_2: morphological modifications, such as increasing surface area and porosity, or chemical modifications, by incorporation of additional components in the TiO_2 structure. These strategies have been implemented since it is of great interest to find ways to extend the absorption wavelength range of TiO_2 into the visible region without decreasing the photocatalytic activity.

Visible light active (VLA) TiO_2 has been regarded as the second generation of TiO_2 photocatalysts and attracted much attention recently. The components, structures, and physiochemical properties of "doped" TiO_2 are always varied with source materials and synthesis methods. Then, it is more accurate in many cases using the nomenclature of "modification" instead of "doping" for new VLA TiO_2-based catalyst (Long and Cai 2011). Among the different ways for improving the photocatalytic activity of TiO_2, metal deposition, dye sensitization, coupled semiconductors, and nonmetal doping, among others, can be mentioned (Pelaez et al. 2012; Rehman et al. 2009). One of the most promising modifications of TiO_2 is the incorporation of nonmetals as an effective method to extend its absorption to visible range, and it will be discussed in the next section.

1.3.1 Visible Light Active TiO_2–NM (NM = Nonmetal)

At this point, it can be easily understood that the development of VLA TiO_2 materials should allow the use of solar or interior illumination energy for the activation of materials, reducing considerably the cost of the technology. There are different approaches for the production of VLA TiO_2 materials. Recently, Serpone et al. (2010) reviewed the general aspects and applications of second-generation VLA TiO_2 materials. Herein, we will focus on the latest reported work related to the production, physicochemical properties, and applications of TiO_2–NM materials.

To achieve efficient photocatalytic activity in the visible light spectral region, it is required to reduce the band gap of titania. Doping with nonmetals such as nitrogen, sulfur, carbon, iodine, phosphorous, fluorine, and boron has been explored in order to enhance the photocatalytic activity of TiO_2 under visible light irradiation. When employing dopants to change the absorption properties of titanium dioxide, there are at least two factors to be considered: First, it is desirable to maintain the integrity of the crystal structure of the pristine titanium dioxide material. Second, it is important to ensure that doping will produce favorable changes in the electronic structure of the doped titanium dioxide material.

The crystal structure of a semiconducting material can be rationalized to a first approximation by considering the relative sizes and relative numbers of the ions present. The substitution of Ti^{4+} in TiO_2 with other transition metal that possess similar ionic radii and positive charge is easier than to replace O^{2-} anion with other anions due to differences in charge states and ionic radii. However, in the case of nanostructured materials, anion doping can be possible since the inherent lattice strain in these materials provides an opportunity to dope TiO_2 to a larger extent (Chen et al. 2005). Also, experimental and theoretical studies show that anion doping offer advantages over cation metal doping, mainly because of the photo-leaching of metal ions into solution during the photocatalytic reaction (Asahi et al. 2001; Khan et al. 2002; Di Valentin et al. 2005b).

In the past few years, it has been demonstrated that nonmetal doping is effective for decreasing band gap and increasing the photocatalytic activity of TiO_2 materials (Serpone et al. 2010; Henderson 2011). Theoretical studies, based on density functional theory (DFT), indicate that the band gap of doped TiO_2 can be narrowed mainly due to: (a) modification of the valence band (VB) edge of TiO_2 (pushing it upward) due to the interactions between the dopant states and the O2*p* states in TiO_2 and (b) introduction of isolated dopant states within the band gap of TiO_2. The substitution of oxygen atoms in TiO_2 with other nonmetal elements is supposed to affect only the VB which is mainly made up of the O2*p* states. However, also the conduction band can be altered by nonmetal doping.

The substitution of oxygen atoms with lighter elements such as nitrogen, carbon, or boron has two main effects: The VB is depopulated by one, two, or three electrons, respectively, but also the new states are expected to be progressively less bound given the smaller nuclear effective charge. Thus, the smaller the atomic number, the higher the energy of the 2*p* states. On the contrary, in the case of the heavier and more electronegative fluorine dopant, the F2*p* states lay 1.19 eV below the bottom of the O2*p* VB: Since fluorine is monovalent, the extra electron formally belonging to the O atom which has been replaced must be transferred to the empty titanium states (Fig. 1.5) (Di Valentin and Pacchioni 2013). Experimental evidence regarding the optical properties in the visible spectral region for TiO_2 after doping mainly relies on diffuse reflectance spectroscopy studies. Experimental information on the nature of the doping bonding has been obtained from XPS and EPR spectroscopies. In the following paragraphs we will describe the theoretical principles that govern anion doping and enumerate some successful examples that have appeared recently in the literature regarding this topic.

1.3.1.1 Nitrogen-Doped TiO_2

The production of visible light active photocatalysts doped with nonmetals started to be explored using nitrogen as dopant. This element can be considered as "suitable

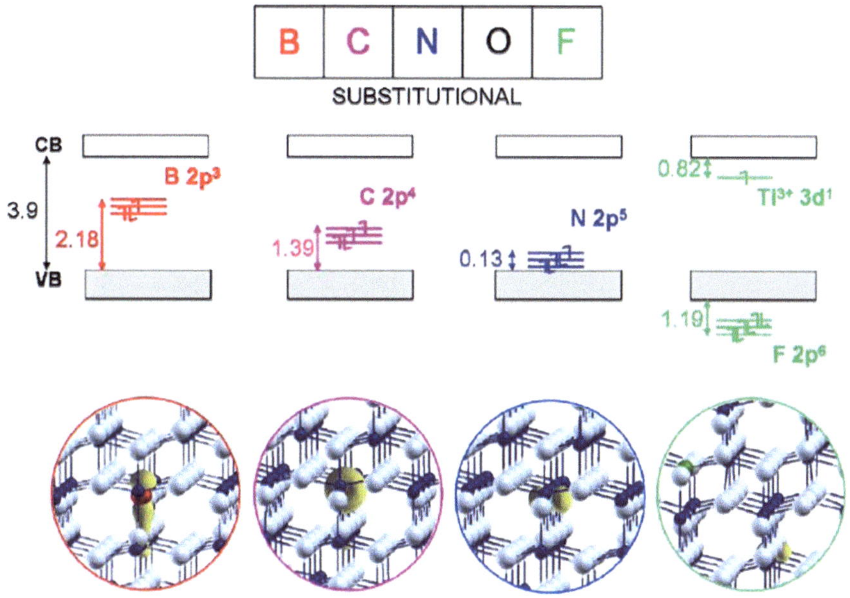

Fig. 1.5 Introduction of new electronic states within the anatase TiO_2 band gap due to nonmetal substitutional doping. With permission from Di Valentin and Pacchioni (2013)

dopant" due its ionic radii and band structure similar to oxygen. As the electronegativity of nitrogen is lower than that of oxygen, its 2*p* states generate shallow traps that contribute to the narrowing of the band gap of nitrogen-doped titanium dioxide materials ($TiO_{2-x}N_x$). From the experimental point of view, several low-cost nitrogen precursors can be used in order to prepare $TiO_{2-x}N_x$ materials; examples include ammonium hydroxide, amines, nitrogen gas, and urea, among others. Also, numerous synthetic techniques have been optimized for the controlled production of the abovementioned materials.

Asahi et al. (2001) pioneered theoretical and experimental studies regarding $TiO_{2-x}N_x$ materials. It was demonstrated that nitrogen doping produces states in the band gap of TiO_2 that overlap sufficiently with the band states of TiO_2, favoring the transfer of e^-/h^+ species to reactive sites at the TiO_2 surface within their lifetime. Theoretical studies regarding the substitutional doping of C, N, F, P, or S for O in the anatase TiO_2 crystal, using the full-potential linearized augmented plane wave (FLAPW) formalism in the framework of the local density approximation (LDA), show that N-doping is the most effective because its *p* states contribute to the band gap narrowing by mixing with O2*p* states (Fig. 1.6). The optimum doping level encountered by these researchers (which depends on the preparation process employed) was found at ~0.25 atomic%. Later, by means of more accurate hybrid functionals, spin-polarized calculations, and larger supercell models, it was demonstrated that the effect of N-doping proposed by Asahi was not correct since it was encountered that the N2*p* impurity states are almost fully localized on the N atom

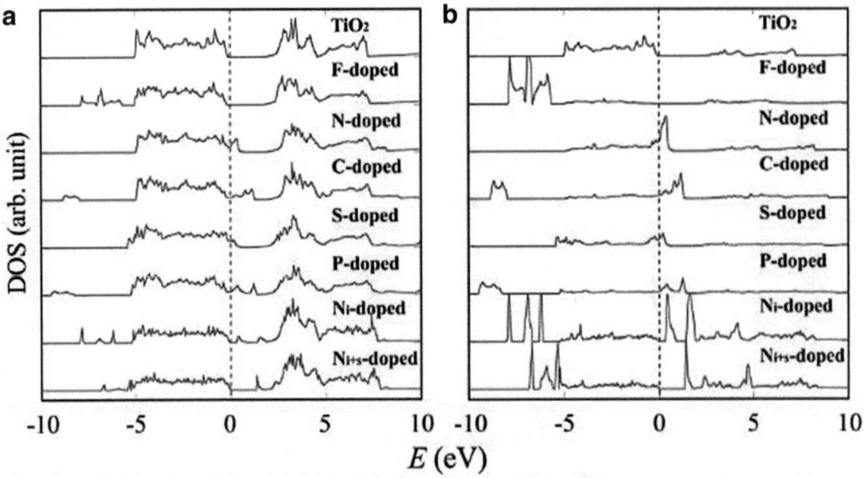

Fig. 1.6 Diagrams of calculated densities of states (DOSs) caused by the substitutional doping of F, N, C, S, or P for O in the anatase TiO_2 crystal. With permission of Asahi et al. (2001)

(Di Valentin et al. 2004). EPR analysis of nitrogen-doped TiO_2 show that the unpaired electron is highly localized on the nitrogen dopant, a strong proof that impurity states are not mixed with the O2*p* VB (Di Valentin et al. 2005a).

Nowadays, scientists devoted to the development of $TiO_{2-x}N_x$ materials have gained insight regarding the nature of the photoactive species in the nitrogen-doped titanium dioxide materials by means of X-ray photoelectron spectroscopy (XPS). This technique provides information about the electronic states and electronic structure in solids, liquids, and gases. In the last 15 years, XPS has made fundamental contribution to the understanding of a large variety of phenomena and properties of pure and doped titanium dioxide materials. By means of the use of this technique, the N1*s* binding energies for N-doped TiO_2 can be obtained and its chemical bonding elucidated. The XPS analysis of N-doped TiO_2 shows more than one peak for the N1*s* region. Several authors have reported a broad peak from 397 to 403 eV, which after curve fitting analysis deconvolutes into two peaks at 396 and 400 eV. The peak at 396 eV can be assigned to the nitridic nitrogen in the O–Ti–N linkage (Cong et al. 2007a; Mitoraj and Kisch 2008; Wu et al. 2005). Signal at 400 eV result from N–N, N–O, or N–C groups (Mitoraj and Kisch 2008; Nolan et al. 2012) and chemisorbed N_2 (Saha and Tompkins 1992; Yang et al. 2006). In general, as the amount of nitrogen doping incorporated into the lattice of the material increases, the position of the $Ti2p_{3/2}$ line shifts gradually toward lower binding energy values. Also, the intensity of the 396 eV component (representing bonded atomistic β-N) of the N1*s* line increases with the increase of nitrogen incorporated into the TiO_2 lattice (Sauthier et al. 2010). Figure 1.7 shows the XPS analyses of undoped TiO_2 and doped $TiO_{2-x}N_x$ films. The spectrum of undoped material (Fig. 1.7a) consists of two peaks for the Ti2*p* region ($Ti2p_{3/2}$, $Ti2p_{1/2}$). The resonance at 458.6 eV corresponds to the Ti–O bonding. For the N1*s*

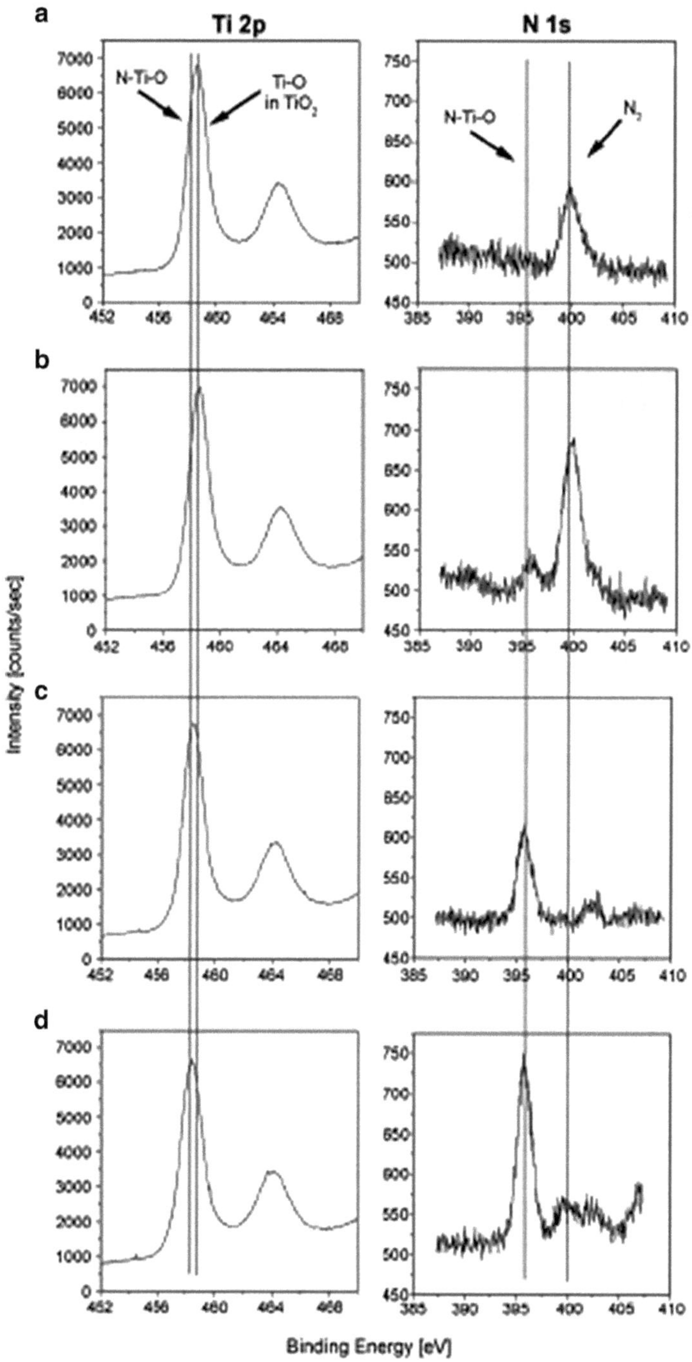

Fig. 1.7 Schematic representation of the XPS spectra of the Ti2*p* (doublet) and N1*s* peaks recorded at the top surface of the reference: (**a**) pure TiO_2 thin film deposited at 10 Pa, (**b**) nitrogen-doped TiO_2 film deposited at 10 Pa total pressure and nitrogen/oxygen partial pressure ratios of 6:4, (**c**) nitrogen-doped TiO_2 film deposited at 10 Pa total pressure and nitrogen/oxygen partial pressure ratios of 7:3, and (**d**) nitrogen-doped TiO_2 film deposited at 10 Pa total pressure and nitrogen/oxygen partial pressure ratios of 8:2. With permission of Sauthier et al. (2010)

region, only one peak is encountered and corresponds to chemisorbed N_2 in the surface of the film. As the amount of nitrogen doping increases (Fig. 1.7b–d), the Ti–O resonance (Ti2*p* region) shifts to lower binding energy values, and a second resonance appears in the N1*s* spectra, demonstrating the incorporation of the nitrogen anion into the TiO_2 lattice.

A fact that has not been clearly elucidated is regarding the physicochemical fundamental principles that justify the observed optical redshift of $TiO_{2-x}N_x$ materials. Lately, Serpone (2006) reported a detailed study on the causes that lead the absorption properties of titanium dioxide to be shifted to the visible region upon anion (particularly, nitrogen) doping. Contrary to the generally made assumption that the nitrogen species are responsible of VLA photocatalyst, it was proposed that the nitrogen precursor during the modification procedure induces formation of oxygen vacancies and color centers, which themselves are responsible for the visible light and enhanced photoactivity. Also, N-containing surface complexes have been reported to possibly be at the origin of the visible light activity of the materials (Livraghi et al. 2006). In the same direction, more recently, Di Valentin and Pacchioni (2013) have reported that substitutional nitrogen doping in anatase TiO_2 introduces defect states localized at the impurity site which may induce absorption in the visible region. On the other hand, interstitial doping of nitrogen in anatase TiO_2 can also be encountered. Nitrogen is found to bind only in a 1:1 (N:O) mole ratio within the TiO_2 lattice. The π NO states lie below the bottom of the VB, while the π^* NO states lie inside the band gap (Fig. 1.8). These shallow states can contribute to the visible light absorption of $TiO_{2-x}N_x$ materials.

Besides visible light activity, enhanced photoactivity has been reported for $TiO_{2-x}N_x$ materials. However, a solid and robust explanation for the enhancement of photoactivity is not yet formulated, and a debate is open on the chemical nature of the doping centers, on their role on the band structure modifications of the solid, and, consequently, on the mechanisms of photoactivation (Livraghi et al. 2008). Several nitrogen-containing species can be encountered in N-doped TiO_2 materials. The nature of these species depends upon on the synthesis procedure and can be broadly divided as diamagnetic (NH_4^+, CN^-) and paramagnetic (NO, $N_b^{\bullet}$ or NO^{-2}; b = bulk) species. Among these nitrogen-containing compounds, $N_b^{\bullet}$ seems to be responsible for visible light absorption and enhanced photoactivity due to the promotion of electrons from the band gap localized states to the conduction band or to surface adsorbed species (Livraghi et al. 2008).

The number of reports regarding $TiO_{2-x}N_x$ nanostructured materials (powders, thin films, composites) is showing an exponential increase in the past 10 years, and lately nitrogen co-doped (e.g., $Ti_{1-x}Ce_xO_{2-y}N_y$, $TiO_{2-x-y}N_xS_y$, N-*x*Fe-TiO_2, $TiO_{2-x-y}N_xF_y$) materials have been synthesized and evaluated (Yu et al. 2010; Rengifo-Herrera et al. 2008; Cong et al. 2007b; Hao and Zhang 2009; Livraghi et al. 2009). From literature review, it is known that the three most important methods for the synthesis of nitrogen-doped titanium dioxide are (a) sputtering and implantation techniques, (b) high-temperature sintering of titanium dioxide under nitrogen-containing atmospheres generated by low boiling point nitrogen compounds (urea, ammonia), and (c) sol–gel methods (Chen and Mao 2007).

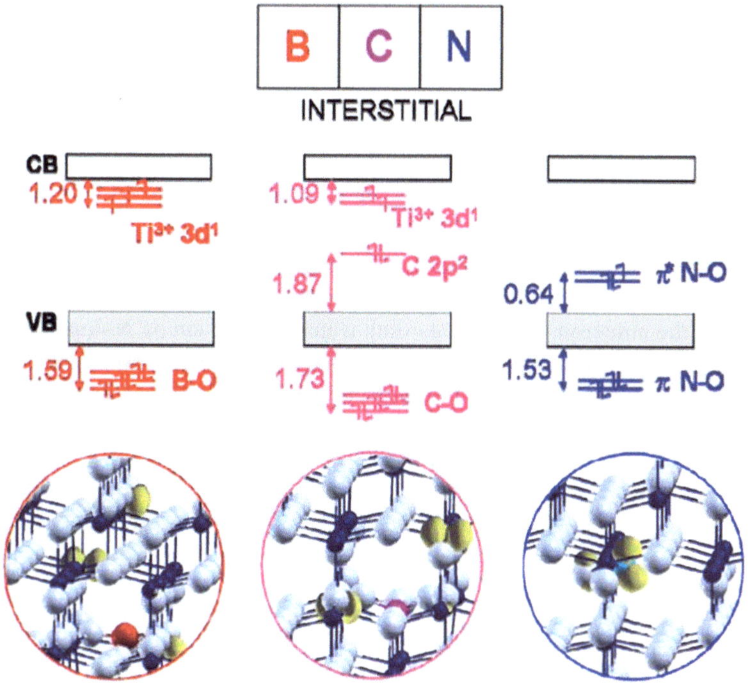

Fig. 1.8 Schematic representation of interstitial nonmetal doping in anatase TiO_2. Nitrogen binds to only one lattice oxygen forming NO species that bind to the Ti lattice through its π bonding system. The π NO states lie below the bottom of the VB, while the π^* NO states lie inside the TiO_2 band gap. With permission of Di Valentin and Pacchioni (2013)

The chemical nature of the produced material depends upon the preparation method and nitrogen precursor. It has been found that nitrogen dopant can be present in different oxidation states or chemical forms (i.e., N^{-3}, NO_x, NH_x).

Wang et al. (2009) reported on the preparation and photocatalytic activity evaluation of N-doped TiO_2 nanobelts. The material was prepared via hydrothermal processing and subsequent heat treatment in NH_3. Visible light absorption arises from the N2*p* levels near the VB and from the color centers induced by the oxygen vacancies and the Ti^{3+} species. The *3d* states of Ti^{3+} acted as the electron–hole recombination centers, leading to the *reduction* of photocatalytic activity.

The use of SiO_2 as support for the preparation of heterogeneous N-doped TiO_2 materials has been reported. For instance, N-doped SiO_2/TiO_2 core/shell nanoparticles were fabricated using SiO_2 template sol–gel method (Kim et al. 2010). The materials were fully characterized and the photocatalytic activity evaluated. Enhanced photoactivity under visible light was encountered due to high nitrogen doping level and surface area. Hydrogen can be produced by water splitting under visible light using N-doped mesoporous TiO_2 catalyst (Liu and Syu 2012). The efficiencies of different TiO_2 materials were evaluated for H_2 production: P25, mesoporous TiO_2 (MT), and nitrogen-doped TiO_2, with variable amounts of nitrogen doping (NMT-0.5, NMT-0.75, NMT-1.25, and NMT-1.5). The highest H_2 evolution rate (14.9 μmol g^{-1} h^{-1}) was achieved using NMT-1.25 catalyst. The enhanced photoactivity of this material relies on its excellent physicochemical properties (high surface area, small size, and crystalline phase).

Also, some examples of composite N-doped TiO_2 materials have appeared lately in the literature. A N-doped TiO_2 on multiwalled carbon nanotube (MWCNT) composite was prepared using a sol–gel method (Huang et al. 2010). It was observed that MWCNTs acted as support and as electron sink, retarding the recombination process and enhancing photoactivity. The optimal weight ratio of MWCNTs/N-doped TiO_2 was found to be 0.1 %.

Despite the numerous reports regarding water purification by N-doped TiO_2, the numbers of reports related to air decontamination are scarce. An annular reactor coated with N-doped TiO_2 material was constructed in order to degrade gaseous volatile organic compounds (VOCs) at indoor levels. Degradation of ethyl benzene and *o*-, *m*-, and *p*-xylenes was evaluated. High efficiencies were reported for the target compounds. The degradation of these compounds can be conducted under visible light and standard humidity conditions (50–60 %); thus, this technique should be easily implemented in order to improve the indoor air quality (Jo and Kim 2009).

In summary, numerous examples of N-doped TiO_2 materials (powders, films, composites) have appeared lately in the literature. In general, it has been demonstrated that nitrogen doping of TiO_2 results in enhanced photocatalytic activity under visible light. The improvement in photoactivity is associated to the increased surface area, the reduction in the recombination of the (e^-/h^+) pair, and the introduction of shallow states within the band gap of crystalline anatase or rutile TiO_2 compounds. It has also been reported that irradiating N-doped TiO_2 with UV light results in a higher quantum efficiency than radiation with visible light. This trend is plausible because visible light excitation only allows electronic transitions to the introduced shallow states (created by N-doping), whereas UV light affords excitation to both shallow states and the conduction band (Irie et al. 2003).

1.3.1.2 Carbon-Doped Titanium Dioxide Materials

In contrast to the numerous reports regarding $TiO_{2}\text{-}xN_x$ materials, few studies appear in the literature that referred to carbon-doped TiO_2 visible light active photocatalysts. This material was discovered serendipitously from the observation that coke containing TiO_2 powders presented visible light activity. First observations indicate that a titanium dioxide material prepared from $Ti(^iOPr)_4$ by an acid-catalyzed sol–gel process in the absence of any dopants was an active photocatalyst under visible light. Visible light activity could not be explained at first; chemical analysis of the material indicated the presence of carbonaceous species due to the incomplete degradation of titanium alkoxide precursors and alcohols employed in the preparation of TiO_2 materials. In order to demonstrate the incorporation of carbon-containing species into TiO_2 materials, commercial reference catalyst

Hombikat UV 100 was chemically modified. The commercial catalyst was immersed on different alcohols and calcinated at 250 °C. Chemical analysis indicated that carbon was incorporated in the material; after doping the commercial catalyst showed visible light activity; hence, it was concluded that carbon doping was responsible for the unusual photocatalytic activity of these materials. Further investigations demonstrated that the presence of carbon dopant into the TiO_2 produce a redshift to the forbidden band gap of the material. Different alkoxide precursors [$Ti(OEt)_4$, $Ti(O\text{-}iPr)_4$, $Ti(O\text{-}nPr)_4$, $Ti(O\text{-}nBu)_4$, $Ti(O\text{-}iBu)_4$, and $Ti(O\text{-}tBu)_4$] were employed for the preparation of TiO_2 materials by an acid (HCl)-catalyzed sol–gel process and heat treatment at low temperatures (250 °C × 3 h). The photocatalytic activity of these materials was evaluated and compared vs. commercially available titanium dioxide materials (P25 and Hombikat) for the degradation of 4-chlorophenol (4CP). The prepared powders were more active than commercial materials, and the best results were obtained when $Ti(O\text{-}iBu)_4$ was used as precursor (Lettmann et al. 2001).

Subsequently, Khan et al. (2002) reported on the preparation of a $TiO_{1.85}C_{0.15}$ material that showed efficient photoactivity for water splitting. The material was prepared by controlled combustion of Ti metal in a natural gas flame at 850 °C. The insertion of carbon into the TiO_2 lattice was evaluated by XPS analysis. The produced material shows visible light activity, rutile crystal structure, and a photoconversion efficiency for water splitting of 8.35 %.

After experimental evidence showing that carbon doping renders improved photocatalytic activity of TiO_2 under visible light activation, lately, theoretical studies based on DFT calculations have been used to investigate the modifications of the electronic band structure of TiO_2 due to carbon doping (Di Valentin et al. 2005b). In contrast to nitrogen doping, in which due to physicochemical similarities among N and O, nitrogen species are generally found in substitutional position in TiO_2, introduction of carbon into TiO_2 can be done in several ways depending upon the synthesis conditions (the concentration of doping C atoms, the oxygen pressure, and temperature). There are three possible ways to include carbon atoms into the crystalline TiO_2 lattice: (a) by replacing lattice oxygen with a carbon, (b) by replacing a Ti atom with a C atom, and (c) by introducing a C atom at an interstitial position. The relative stability of the carbon doping species varies as a function of the oxygen chemical potential (μ_O), which is a parameter that characterizes the oxygen environment during synthesis. Figure 1.9 illustrates the energies of formation of all the carbon-doped and nitrogen-doped models. For carbon doping, the energy diagram clearly shows that under oxygen-poor conditions, substitutional (to oxygen) carbon and oxygen vacancies are favored, whereas under oxygen-rich conditions, interstitial and substitutional (to Ti) C atoms are preferred. In the case of nitrogen doping, substitutional (to O) doping is favored under oxygen-poor conditions. Interstitial doping is independent of chemical oxygen potential.

Besides the consideration of energy consumption (or release), carbon doping involves distortion of the TiO_2 lattice and modification of the band gap. As illustrated in Fig. 1.9, carbon substitutional doping introduces energy states

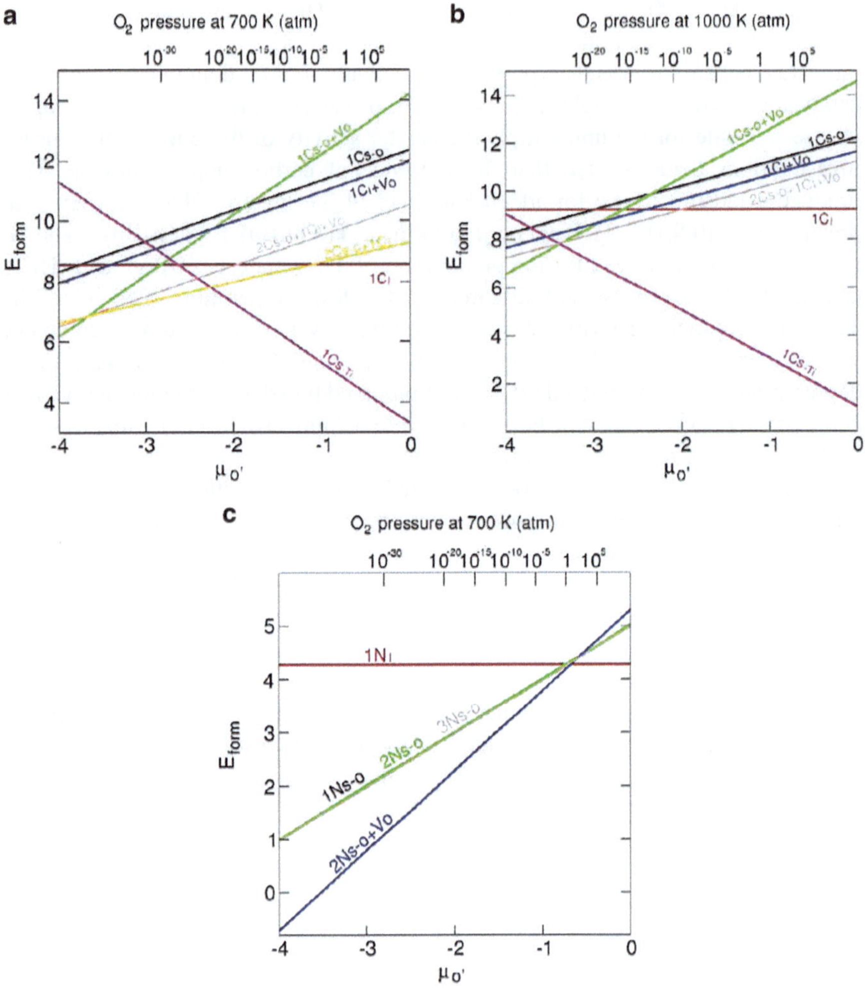

Fig. 1.9 Diagrams of energies of formation (E_{form}, in eV) as a function of the oxygen chemical potential (μ_O) or as a function of the oxygen pressure at fixed temperature ($T = 700$ K) (*top x*-axis), for (**a**) different carbon species in anatase, (**b**) different carbon species in rutile, and (**c**) different nitrogen species in anatase. With permission from Di Valentin et al. (2005b). Copyright 2005 American Chemical Society

(C2*p* states) within 1.39 eV above the VB of TiO_2 (anatase). The distortion of the lattice results from the differences on the anionic radius of carbon with respect to that of O. Axial and equatorial bonds (Ti–C) are elongated: 2.275 Å (axial) and 2.031 Å (equatorial) vs. 2.014 Å (axial) and 1.942 Å (equatorial) in pure anatase. Equatorial bond angles, Ti–C–Ti, are also elongated: 161° vs. 153° in pure anatase (Di Valentin and Pacchioni 2013). Addition of carbon atoms at interstitial positions can be achieved easily due to the small size of the neutral atom. Once introduced, the formation of three C–O bonds that involve the carbon dopant and three lattice oxygen follows; also, carbon donates two electrons to the host crystal, giving rise to the formation of two Ti^{3+} ions and a carbon ion that holds a lone pair with two electrons. In this configuration, carbon is formally oxidized and forms strong covalent bonds with the oxygen atoms; the three C–O bonding states lie below the bottom of the VB, the C lone pair is in the middle of the band gap, and the two Ti^{3+} states are about 1 eV below the conduction band (CB) of TiO_2, as illustrated in Fig. 1.8 (Di Valentin et al. 2005b; Di Valentin and Pacchioni 2013).

A green route for the synthesis of C-doped TiO_2 materials was recently reported (Dong et al. 2011). Contrary to the numerous reports regarding the use of organometallic titanium precursors, Dong and co-workers implemented a low-cost, nontoxic, and easy to scale up hydrothermal synthetic route based on $Ti(SO_4)_2$ and sucrose as molecular precursors to titanium and carbon, respectively. Post-thermal treatment (100, 200, and 300 °C) was found to be effective to promote visible light activity of C-doped TiO_2 materials. The incorporation of carbon into the TiO_2 lattice was confirmed by XPS analysis. Morphological structure determination and size distribution were conducted by TEM microscopy. The materials showed good photocatalytic activity for the degradation of toluene in air.

Recently, examples of N,C-co-doped titanium dioxide photocatalyst have appeared in literature (Dolat et al. 2012). A hydrothermal method was employed in the preparation of N,C-co-doped TiO_2. Different alcohols and ammonia were evaluated as carbon and nitrogen precursors, respectively. The physicochemical properties and photocatalytic activity of the materials were evaluated. Phenol was used as model contaminant, and the photoactivity of the materials was evaluated at a constant volumetric rate of photon absorption (VRPA) in an annular reactor under UV light. The authors propose the implementation of the VRPA method in order to evaluate different catalyst activity at an approximately equal rate of photon absorption.

Nanoporous crystalline materials of carbon- and sulfur-doped TiO_2 were prepared at room temperature using a sonochemical method (Lee et al. 2013). Anionic sulfur doping was demonstrated by XPS. Environmental applications (dye and bacteria degradation) of the materials were evaluated under visible light activation. Doped materials showed dye degradation efficiencies up to 99 % after 70 min of treatment. Also, degradation of *E. coli* was achieved within 180 min of photocatalytic treatment using C–TiO_2 and S–TiO_2 as catalysts. Carbon- and sulfur-co-doped TiO_2 materials were prepared by hydrolysis of tetrabutyl titanate in the presence of thiourea and urea (Sun et al. 2006). To investigate the states of doping ions, XPS analysis was performed. Elemental carbon, carbonate, S^{4+} ions (replacing Ti^{4+}), and sulfide ion were encountered. The co-doped material showed photocatalytic activity under visible light for the degradation of 4-chlorophenol.

Supported C-doped TiO_2 materials have been synthesized by oxidation of titanium carbide (TiC)-coated multiwalled carbon nanotubes (MWCNTs) (Cong et al. 2011). The MWCNTs were used as a reaction template and carbon source, and titanium powder as the titanium source. XPS analysis of the prepared material indicates that the chemical environment of Ti in TiO_2-coated MWCNTs is different

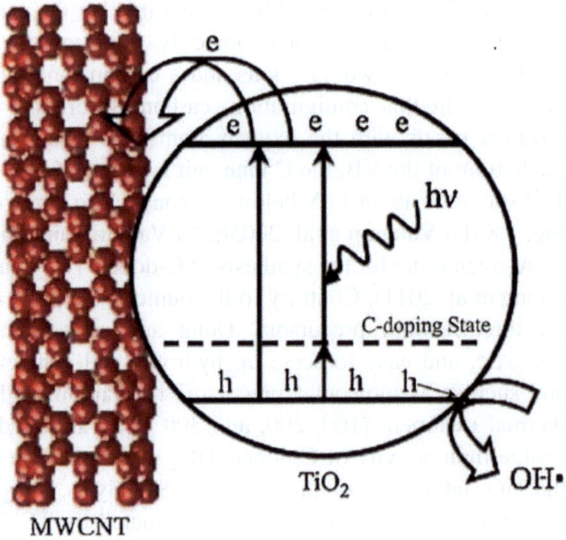

Fig. 1.10 Schematic representation of the mechanistic model for the enhanced photoactivity of MWCNTs–TiO_2 material. With permission from Cong et al. (2011)

from that of pure anatase, which is probably due to the strong interaction between TiO_2 nanoparticles and the MWCNTs and the formation of a Ti–O–C bond. The photocatalytic activity of MWCNTs–TiO_2 material was evaluated under visible light for the degradation of methylene blue, and a model for the enhanced photoactivity of the material was proposed (Fig. 1.10). It was concluded that MWCNTs display several functions in the hybrid material: act as support, absorbent, photogenerated transfer station, and carbon doping source to narrow the band gap of TiO_2.

Microwave (MW) plasma technique was evaluated as an alternative to magnetron sputtering deposition method for the fabrication of TiO_2 and carbon-doped TiO_2 films (Dang et al. 2012). MW plasma-produced TiO_2 films exhibited better physicochemical properties and photoactivity in comparison to the ones produced by magnetron sputtering technique; thus, MW plasma deposition can be envisioned as a potential processing technology for the fabrication of photoactive TiO_2 coatings. Carbon-doped TiO_2 nanopowders were prepared using an oxidative annealing procedure and TiC as precursor (McEvoy et al. 2013). XPS analysis showed that carbon was present in the form of carbonate species. The material showed good bactericidal activity against *E. coli* k12 (80 % inactivation in 30 min).

In summary, carbon doping of TiO_2 has been reported to improve photocatalytic activity and visible light response. Also, some studies have shown that carbon doping can be more effective than nitrogen doping (Khan et al. 2002; Di Valentin et al. 2005b; Dong et al. 2009; Di Valentin and Pacchioni 2013). It has been postulated that the improved photoactivity of carbon doped TiO_2 materials is a result of the combination of several modifications induced by carbon doping, such as: formation of color centers (including Ti^{3+} centers and defects associated with oxygen vacancies that give rise to color centers displaying visible light absorption bands), the introduction of several localized occupied states in the gap, and the higher surface area of the doped materials (Teh and Mohamed 2011). Synthesis of carbon-doped and carbon-co-doped (N, S, F) materials has been conducted via many routes, including simple mixing of a carbon precursor with TiO_2 (Lettmann et al. 2001; Lee et al. 2013), sol–gel synthesis (Park et al. 2009; Liu et al. 2012), direct oxidation of Ti metal in a burner flame (Khan et al. 2002), hydrothermal synthesis (Dolat et al. 2012), and deposition techniques such as MW plasma oxidation, reactive magnetron sputtering, physical vapor deposition (PVD), chemical vapor deposition (CVD) (Dang et al. 2012; Hsu et al. 2007), and electrophoretic deposition (Macak et al. 2007), among others.

1.3.1.3 Fluorine-Doped Titanium Dioxide Materials

In the previous paragraphs, we have been discussing the effects of nitrogen and carbon doping into the lattice of crystalline TiO_2 materials. Nitrogen and carbon are situated to the left of oxygen in the periodic table; thus the introduction of these impurities generates one or two holes (one for nitrogen and two for oxygen) in the VB of the material, and also due to the smaller nuclear effective charge of C and N compared to O, the new states are expected to be less bound. Thus, the position of the impurity states lies 0.13 eV (in the case of N) and 1.39 eV (in the case of carbon) above the top of the valence band maximum of the material (Di Valentin and Pacchioni 2013). As opposed to B,C or N electronic configuration with respect to oxygen, fluorine is situated to the right of oxygen in the periodic table and contains one extra electron and the nuclear effective charge is bigger than the one in oxygen; hence, the position of the impurity F2*p* states lies 1.19 eV below the bottom of the O2*p* VB. Fluorine easily enters in the TiO_2 lattice in substitutional positions taking the place of oxygen atoms; the extra electron associated to the heteroatom is transferred to the empty *3d* titanium states, generating reduced Ti^{3+} centers (shallow Ti^{3+} states, situated 0.8 eV below the conduction band of TiO_2). UV-Vis and EPR spectroscopic analyses indicate that no absorption in the visible is observed for F-doped TiO_2 (Czoska et al. 2008; Livraghi et al. 2008; Di Valentin and Pacchioni 2013).

Experimental evidence of fluorine doping indicates that F–TiO_2 present two different chemical species of fluorine: (a) fluoride ions at the solid surface (detected by XPS at ~684.5 eV) where they substitute OH^- hydroxyl groups yielding a terminal Ti–F bond without the generation of Ti^{3+} reduced centers and (b) fluoride ions that substitute O^{2-} ions in the solid lattice (detected by XPS at ~688.5 eV) yielding a bridging Ti–F–Ti bond and generating bulk Ti^{3+} species in an octahedral environment that introduce localized states just below the conduction band (Czoska et al. 2008). In regard to the lattice distortion upon fluorine doping, it has been observed that the crystallinity of the samples is generally improved after F^- doping. The fact that the ionic radius of F^- ion (1.33 nm) is close to that of the O^{-2} ion (1.40 nm) explains that fluoride ion can fill oxygen vacancies without decreasing the anatase crystallinity (Hattori et al. 1999).

Early in the 1990s the first reports regarding F–TiO_2 appeared in the literature (Wang and Mallouk 1990). Later, Hattori and co-workers (1999) reported on the preparation of F–TiO_2 films. The films were evaluated for the photocatalytic degradation of acetaldehyde, and an enhanced photoactivity in comparison to analogous TiO_2 films was observed. The improved activity of doped material can be justified as follows: (a) diffusion of Na^+ ions into the TiO_2 film from the soda-lime glass substrate that is almost completely blocked by the F:SiO_2 underlayer, (b) the improved crystallinity of doped material, and (c) decreasing oxygen vacancies via the action of HF from the gel–TiO_2 film upon heating.

More recently, increased adsorption capacity of F–TiO_2 due to the larger number of acid sites in the doped material was reported (Li et al. 2005; Huang et al. 2007). The higher electronegativity of fluorine makes the neighboring titanium atom more "positively charged" acting as Lewis acidic sites and improving the interaction with polar organic compounds. The effect of F/Ti atomic ratio on the crystallinity of the material and photodegradation rate of formaldehyde was evaluated (Huang et al. 2007). It was observed that a F:Ti = 0.03 was optimal for degradation, whereas the crystallinity of the materials was increased linearly with dopant concentration.

As previously mentioned, despite the improved activity of F–TiO_2, no visible light absorption is encountered for the material. In order to overcome this problem, fluorine-co-doped TiO_2 has been investigated. N–F-co-doped system has been widely investigated (Di Valentin et al. 2008; Pelaez et al. 2010; Wu et al. 2010; Zhao et al. 2013). It has been proven that N–F co-doping reduces the energy cost of doping and also the amount of defects in the lattice, as a consequence of the charge compensation between the nitrogen (p-dopant) and the fluorine (n-dopant) impurities. Air purification (removal of NO_2) was successfully conducted under both visible (room lighting) and UVA lighting using thin films of F,S–TiO_2 catalyst (O'Keeffe et al. 2013). Umadevi and co-workers (2013) investigated the antimicrobial activity of F–TiO_2 against *S. aureus* and *P. aeruginosa*. The material was found to be a good antibacterial agent.

Co-doping with metals has also been reported (Lin et al. 2012; Yang et al. 2012). Ag/F–TiO_2 powders were prepared using a two-step procedure: sol–gel and photoreduction (Lin et al. 2012). The co-doped material showed good activity for the degradation of phenol under visible light. Silver acts as electron sink reducing the recombination of the e^-/h^+ pair, thus improving the performance of the catalyst.

In summary, nanopowders (Yu et al. 2002; Huang et al. 2007; Czoska et al. 2008), thin films (Hattori et al. 1999; Pelaez et al. 2010; O'Keeffe et al. 2013), nanotubes (Yu et al. 2008), and mesoporous (Wu et al. 2010) materials of fluorine-doped and fluorine-co-doped TiO_2 have been fabricated using different synthetic approaches. In general, the doped materials show a photocatalytic activity that exceeds that of commercial (Degussa P25) or undoped TiO_2 prepared under similar synthetic conditions. This enhanced photoactivity is ascribed to the crystalline nature of the material, increased number of active sites, and diminished e^-/h^+ recombination reaction, among others.

1.3.1.4 Boron-Doped Titanium Dioxide Materials

In the last decade, boron-doped TiO_2 has been investigated. Pioneering work by Zhao and collaborators (2004) regarding $Ni_2O_3/TiO_{2-x}B_x$ materials demonstrated that incorporation of boron into TiO_2 lattice can extend the spectral response to the visible region. Later, using a sol–gel method, B–TiO_2 (boron-doped TiO_2) nanopowders were prepared and fully characterized. Different atomic ratios of B to Ti (from 1 to 20 %) were evaluated in order to optimize the activity of B–TiO_2 materials. The highest photoactivity was achieved for the 5 % boron-doped TiO_2 (Chen et al. 2006). Grabowska et al. (2009) evaluated the influence of the preparation method (sol–gel vs. grinding procedure) and B doping content on the physicochemical properties and photocatalytic activity of the prepared materials. It was found that the best catalyst under visible light was B–TiO_2 (2 wt % of boron) prepared by grinding ST-01 with boric acid triethyl ester followed by calcinations at 400 °C.

By comparing the ionic radius of B^{3+} (0.023 nm) to that of Ti^{4+} (0.064 nm), it is easily inferred that it is difficult to replace B^{3+} for Ti^{4+} into the TiO_2 lattice; thus, it is estimated that most boron ions are doped in the interstitial of TiO_2 matrix. Although boron doping has been less explored, theoretical and experimental studies indicate similarities to nitrogen, where different chemical species (NO_x or NH_x in the case of nitrogen; B_2O_3, [BO_3], and [BO_4] for boron) can be introduced as interstitial doping. It has been reported that segregated (in the surface or within the lattice) B_2O_3 agglomerates confer visible light activity but reduced photocatalytic capacity. On the other hand, the formation of [BO_3] or [BO_4] species can render both visible light response and enhanced activity. The formation of these interstitial impurities can be explained as follows: Boron interacts with titanium ions as reducing agent, transferring its three valence electrons. The energy cost for the oxidation of B is compensated by the formation of strong bonds between the B atom and the neighboring oxygens. As a result of this interaction, the oxygen atoms relax significantly from their lattice positions toward the interstitial B impurity (Finazzi et al. 2009). Substitutional boron doping (TiO_2-xBx) can also be encountered. As previously discussed for substitutional nitrogen doping where it was explained that the mode of N incorporation depends on the prevailing oxygen activity, a similar effect may operate in the case of boron; thus, a reduced amount of oxygen will favor the incorporation of boron as substitutional impurity (In et al. 2007).

In order to elucidate the chemical nature of boron within the TiO_2 lattice and understand its role on the photocatalytic activity of the material, In and co-workers (2007) implemented a method to synthesize B–TiO_2 that allows good control of the level of B doping. B–TiO_2 and B,N–TiO_2 were synthesized under same conditions in order to compare visible light absorption and photocatalytic activity. Visible light absorption was demonstrated by means of UV-Vis diffuse reflectance spectroscopy. Figure 1.11 illustrates the reflectance spectra of different materials. It is shown that an increase of the boron content give rise to a bigger absorption in the

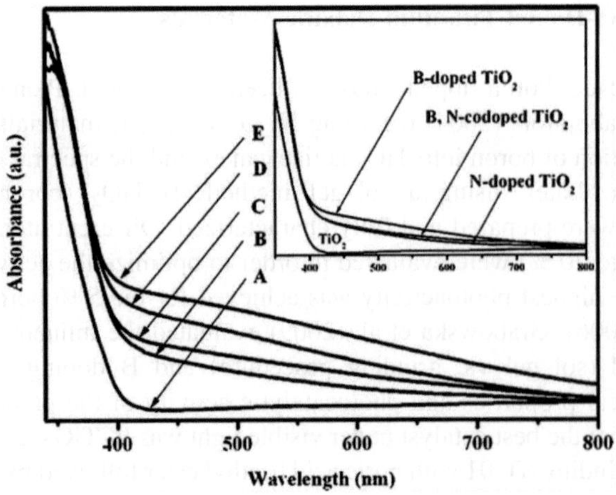

Fig. 1.11 UV-Vis diffuse reflectance spectra of bare TiO_2 (anatase), B–TiO_2 (at different dopant levels), and B,N–TiO_2 powders. (a) TiO_2; (b) B–TiO_2 (1.13 atom%); (c) B–TiO_2 (3.23 atom%); (d) B–TiO_2 (5.82 atom%) (e) B–TiO_2 (11.9 atom%). *Inset* shows comparison between B-doped (3.23 atom%), N-doped (3.23 atom%), and B, N-co-doped (1.5 atom% of both N and B) TiO_2 catalysts. With permission from In et al. (2007)

visible region. The insertion of boron into the TiO_2 lattice, replacing oxygen at an optimum level of 1.13 % was also postulated.

Co-doping of TiO_2 with boron and metallic ions (V^{3+}, Bi^{3+}, Y^{3+}) has been explored (Bettinelli et al. 2007; Bagwasi et al. 2013; Shi et al. 2010). For instance, Bagwasi and co-workers reported on the synthesis, characterization, and applications of Bi–B–TiO_2 nanoparticles (Bagwasi et al. 2013). B and Bi^{3+} single doped materials were also prepared using the same modified sol–gel method. Co-doped catalyst showed better performance for the degradation of acid orange and 2,4-dichlorophenol. The promoting effects of the dopants on the photocatalytic activity of TiO_2 were studied. The doping of Bi^{3+} results in reduced recombination of e^-/h^+ pairs whereas B doping renders enhanced visible light absorption.

Lately, reports on the synthesis, characterization, and application of NM–B–TiO_2-co-doped materials (NM = nonmetal) have been published. N,B–TiO_2 co-doping has been explored by several groups (Zhang et al. 2014; In et al. 2007; Gopal et al. 2008; Sun and Searles 2013). Although still controversial, the enhanced visible light photoactivity of optimized co-doped materials can be rationalized as follows: Boron and nitrogen coexist as interstitial and substitutional dopants in a 1:3 (B:N) atomic ratio (boron donates its valence electrons to nitrogen) in order to compensate the net charges of dopants. Since theoretically as experimentally it is impossible to keep the 1:3 atomic ratio of dopants, the co-doped material will present impurity bands (introduced by boron doping) that play a key role in the strong absorption of visible light over the full visible spectrum (Sun and Searles 2013). The photoactivity of the material is enhanced due to the following: (a) the synergistic effect of the interstitial B and N species and (b) the nonmetal doping that can introduce residual charge, which may result in an increase in the number of surface O–H groups. These hydroxyl groups react with the photogenerated (e^-/h^+) species, producing highly active oxidants such as $^{\bullet}OH$ an O^{2-}. (c) In addition, the surface boron species may also introduce Bronsted and Lewis acid centers on the surface of co-doped materials. All these factors suit B,N–TiO_2 materials as promising efficient photocatalysts under a wide range of light sources (Feng et al. 2011).

Concluding Remarks

It was discussed that in heterogeneous photocatalysis, the understanding of a semiconductor features is of great interest. In this chapter some fundamental principles regarding their band structure and the role of dopants were introduced. It has been also illustrated that numerous visible light active TiO_2 photocatalysts have been developed recently. These materials present different chemical compositions and nanostructures which determine their physicochemical properties and photoactivity. Some examples of photocatalytic materials include metal oxides (ternary or quaternary), oxynitrides, nonmetal co-doped TiO_2, etc. Numerous approaches have been explored in order to fabricate VLA TiO_2 materials; in this chapter, nonmetal doping was emphasized. Numerous studies have been conducted in order to offer a clear mechanism regarding the activity of NM-doped materials. Still a general consensus regarding the photoactivity of a particular material (i.e., N-doped TiO_2) cannot be formulated since dopant ions can be incorporated as substitutional or interstitial states into the TiO_2 lattice, rendering different mode of activity. It was also revealed that these materials can be used to degrade recalcitrant (and/or emergent) pollutants and pathogen microorganisms. Scaling up photocatalytic processes involving the use of novel materials will contribute for the construction of sustainable systems for solving numerous environmental and energy problems.

References

Ahmed S, Rasul MG, Brown R, Hashib MA (2011) Influence of parameters on the heterogeneous photocatalytic degradation of pesticides and phenolic contaminants in wastewater: a short review. J Environ Manage 92:311–330

Anchieta CG, Sallet D, Foletto EL, da Silva SS, Chiavone-Filho O, do Nascimento CAO (2014) Synthesis of ternary zinc spinel oxides and their application in the photodegradation of organic pollutant. Ceram Int 40:4173–4178

Asahi R, Morikawa T, Ohwaki T, Aoki K, Taga Y (2001) Visible-light photo-catalysis in nitrogen-doped titanium oxides. Science 293:269–271

Aslam M, Ismail IM, Chandrasekaran S, Hameed A (2014) Morphology controlled bulk synthesis of disc-shaped WO_3 powder and evaluation of its photocatalytic activity for the degradation of phenols. J Hazard Mater 276C:120–128

Asthana A, Momeni K, Prasad A, Yap YK, Yassar RS (2011) On the correlation of crystal defects and band gap properties of ZnO nanobelts. Appl Phys A 105:909–914

Atkins PW, Overton TL, Rourke JP, Weller MT, Armstrong FA (eds) (2010) Inorganic chemistry, 5th edn. Oxford University Press, New York

Bagwasi S, Tian B, Zhang J, Nasir M (2013) Synthesis, characterization and application of bismuth and boron co-doped TiO_2: a visible light active photocatalyst. Chem Eng J 217:108–118

Bahnemann D (2004) Photocatalytic water treatment: solar energy applications. Sol Energy 77:445–459

Bai W, Cai L, Wu C, Xiao X, Fan X, Chen K, Lin J (2014) Alcohothermal synthesis of flower-like ZnS nano-microstructures with high visible light photocatalytic activity. Mater Lett 124:177–180

Bakardjieva S, Stengl V, Subrt J, Houskova V, Kalenda P (2007) Photocatalytic efficiency of iron oxides: degradation of 4-chlorophenol. J Phys Chem Solids 68:721–724

Bandara J, Mielczarski JA, Lopez A, Kiwi J (2001) Sensitized degradation of chlorophenols on iron oxides induced by visible light: comparison with titanium oxide. Appl Catal B Environ 34:321–333

Bandara J, Klehm U, Kiwi J (2007) Raschig rings-Fe_2O_3 composite photocatalyst activate in the degradation of 4-chlorophenol and Orange II under daylight irradiation. Appl Catal B Environ 76:73–81

Bettinelli M, Dallacasa V, Falcomer D, Fornasiero P (2007) Photocatalytic activity of TiO_2 doped with boron and vanadium. J Hazard Mater 146:529–534

Chakrabarti S, Dutta B (2004) Photocatalytic degradation of model textile dyes in wastewater using ZnO as semiconductor catalyst. J Hazard Mater 112:269–278

Chen X, Mao SS (2007) Titanium dioxide nanomaterials: synthesis, properties, modifications and applications. Chem Rev 107(7):2891–2959

Chen X, Lou YB, Samia ACS, Burda C, Gole JL (2005) Formation of oxynitride as the photocatalytic enhancing site in nitrogen-doped titania nanocatalysts: comparison to a commercial nanopowder. Adv Funct Mater 15(1):41–49

Chen D, Yang D, Wang Q, Jiang Z (2006) Effects of boron doping on photo-catalytic activity and microstructure of titanium dioxide nanoparticles. Ind Eng Chem Res 45:4110–4116

Cong Y, Zhang J, Chen F, Anpo M (2007a) Synthesis and characterization of nitrogen-doped TiO_2 nanophotocatalyst with high visible light activity. J Phys Chem C 111(19):6976–6982

Cong Y, Zhang J, Chen F, Anpo M, He D (2007b) Preparation, photocatalytic activity, and mechanism of nano-TiO_2 co-doped with nitrogen and iron (III). J Phys Chem C 111 (28):10618–10623

Cong Y, Li X, Qin Y, Dong Z, Yuan G, Cui Z, Lai X (2011) Carbon-doped TiO_2 coating on multiwalled carbon nanotubes with higher visible light photocatalytic activity. Appl Catal B Environ 107:128–134

Czoska AM, Livraghi S, Chiesa M, Giamello E, Agnoli S, Granozzi G, Finazzi E, Di Valentin C, Pacchioni G (2008) The nature of defects in fluorine-doped TiO_2. J Phys Chem C 112 (240):8951–8956

Dang BH, Rahman M, MacElroy D, Dowling DP (2012) Evaluation of microwave plasma oxidation treatments for the fabrication of photoactive un-doped and carbon-doped TiO_2 coatings. Surf Coat Technol 206:4113–4118

Delgado-Balderas R, Hinojosa-Reyes L, Guzmán-Mar JL, Garza-González MT, López-Chuken UJ, Hernández-Ramírez A (2012) Photocatalytic reduction of Cr(VI) from agricultural soil column leachates using zinc oxide under UV light irradiation. Environ Technol 33:2673–2680

Di Paola A, García-López E, Marcì G, Palmisano L (2012) A survey of photocatalytic materials for environmental remediation. J Hazard Mater 211–212:3–29

Di Valentin C, Pacchioni G (2013) Trends in non-metal doping of anatase TiO_2: B, C, N and F. Catal Today 206:12–18
Di Valentin C, Pacchioni G, Selloni A (2004) Origin of the different photoactivity of N-doped anatase and rutile TiO_2. Phys Rev B 70(8):085116
Di Valentin C, Pacchioni G, Selloni A, Livraghi S, Giamello E (2005a) Characterization of paramagnetic species in N-doped TiO_2 powders by EPR spectroscopy and DFT calculations. J Phys Chem B 109(23):11414–11419
Di Valentin C, Pacchioni G, Selloni A (2005b) Theory of carbon doping of titanium dioxide. Chem Mater 17:6656–6665
Di Valentin C, Finazzi E, Pacchioni G, Selloni A, Livraghi S, Czoska AM, Paganini MC, Giamello E (2008) Density functional theory and electron paramagnetic resonance study on the effect of N-F codoping of TiO_2. Chem Mater 20(11):3706–3714
Diebold U (2003) The surface science of titanium dioxide. Surf Sci Rep 48:53–229
Doerfler W, Hauffe K (1964) Heterogeneous photocatalysis II. The mechanism of the carbon monoxide oxidation at dark and illuminated zinc oxide surfaces. J Catal 3:171–178
Dolat D, Quici N, Kusiak-Nejman E, Morawski AW, Li Puma G (2012) One-step, hydrothermal synthesis of nitrogen, carbon co-doped titanium dioxide (N, C-TiO2) photocatalysts. Effect of alcohol degree and chain length as carbon dopant precursors on photocatalytic activity and catalyst deactivation. Appl Catal B Environ 115–116:81–89
Dong F, Wang H, Wu Z (2009) One-step "green" synthetic approach for mesoporous C-doped titanium dioxide with efficient visible light photocatalytic activity. J Phys Chem 113 (38):16717–16723
Dong F, Guo S, Wang H, Li X, Wu Z (2011) Enhancement of the visible light photocatalytic activity of C-doped TiO_2 nanomaterials prepared by a green synthetic approach. J Phys Chem C 115(27):13285–13292
Farbod M, Jafarpoor E (2012) Fabrication of different ZnO nanostructures and investigation of morphology dependence of their photocatalytic properties. Mater Lett 85:47–49
Feng N, Zheng A, Wang Q, Ren P, Gao X, Liu S-B, Shen Z, Chen T, Deng F (2011) Boron environments in B-doped and (B, N)-codoped TiO_2 photocatalysts: a combined solid-state NMR and theoretical calculation study. J Phys Chem C 115:2709–2719
Finazzi E, Di Valentin C, Pacchioni G (2009) Boron-doped anatase TiO_2: pure and hybrid DFT calculations. J Phys Chem C 113:220–228
Friedmann D, Mendive C, Bahnemann D (2010) TiO_2 for water treatment: parameters affecting the kinetics and mechanisms of photocatalysis. Appl Catal B Environ 99:398–406
Fujishima A, Honda K (1972) Electrochemical photolysis of water at a semiconductor electrode. Nature 238:37–38
Fujishima A, Zhang X, Tryk D (2008) TiO_2 photocatalysis and related surface phenomena. Surf Sci Rep 63:515–582
Giribabu K, Suresh R, Manigandan R, Vijayaraj A, Prabu R, Narayanan V (2012) Cadmium sulphide nanorods: synthesis, characterization and their photocatalytic activity. Bull Korean Chem Soc 33:2910–2916
Gopal NO, Lo H, Ke S (2008) Chemical state and environment of boron dopant in B, N-codoped anatase TiO2 nanoparticles: an avenue for probing dia-magnetic dopants in TiO2 by electron paramagnetic resonance spectroscopy. J Am Chem Soc 130:2760–2761
Grabowska E, Zaleska A, Sobczak JW, Gazda M, Hupka J (2009) Boron-doped TiO_2: characteristics and photoactivity under visible light. Procedia Chem 1:1553–1559
Hao H, Zhang J (2009) The study of iron (III) and nitrogen co-doped mesoporous TiO_2 photocatalysts: synthesis, characterization and activity. Microporous Mesoporous Mater 121 (1–3):52–57
Hashimoto K, Irie H, Fujishima A (2005) TiO_2 photocatalysis: a historical overview and future prospects. Jpn J Appl Phys 44:8269–8285

Hattori A, Shimoda K, Tada H, Ito S (1999) Photoreactivity of sol-gel TiO_2 films formed on soda-lime glass substrates: effect of SiO_2 underlayer containing fluorine. Langmuir 15(16):5422–5425

He GH, Liang CJ, Ou YD, Liu DN, Fang YP, Xu YH (2013) Preparation of novel Sb_2O_3/WO_3 photocatalysts and their activities un-der visible light irradiation. Mater Res Bull 48:2244–2249

Henderson MA (2011) A surface science perspective on TiO_2 photocatalysis. Surf Sci Rep 66:185–297

Hernández-Alonso MD, Fresno F, Suárez S, Coronado JM (2009) Development of alternative photocatalysts to TiO_2: challenges and opportunities. Energy Environ Sci 2:1231–1257

Herrmann JM (2005) Heterogeneous photocatalysis: state of the art and present applications. Top Catal 34:49–65

Herrmann JM (2010) Photocatalysis fundamentals revisited to avoid several misconceptions. Appl Catal B Environ 99:461–468

Hong SJ, Jun H, Borse PH, Lee JS (2009) Size effects of WO_3 nanocrystals for photooxidation of water in particulate suspension and photoelectrochemical film systems. Int J Hydrogen Energy 34:3234–3242

Hsu SW, Yang TS, Chen TK, Wang MS (2007) Ion assisted electron beam evaporation of carbon-doped titanium oxide films as visible-light photocatalyst. Thin Solid Films 515:3521–3526

Huang D, Liao S, Quan S, Liu L, He Z, Wan J, Zhou W (2007) Preparation of anatase F doped TiO_2 sol and its performance for photodegradation of formaldehyde. J Mater Sci 42(19):8193–8202

Huang BS, Chang FY, Wey MY (2010) An efficient composite growing N-doped TiO_2 on multi-walled carbon nanotubes through sol–gel process. J Nanopart Res 12(7):2503–2510

In S, Orlov A, Berg R, García F, Pedrosa-Jimenez S, Tikhov M, Wright D, Lambert R (2007) Effective visible light-activated B-doped and B, N-codoped TiO2 photocatalysts. J Am Chem Soc 129:13790–13791

Irie H, Watanabe Y, Hashimoto K (2003) Nitrogen-concentration dependence on photocatalytic activity of $TiO_{2-x}N_x$ powders. J Phys Chem B 107:5483–5486

Ismail AA, Bahnemann DW (2014) Photochemical splitting of water for hydrogen production by photocatalysis: a review. Sol Energy Mater Sol Cells 128:85–101

Jacobsson TJ, Edvinsson T (2012) Photoelectrochemical determination of the absolute band edge positions as a function of particle size for ZnO quantum dots. J Phys Chem C 116:15692–15701

Jo W, Kim J (2009) Application of visible-light photocatalysis with nitrogen-doped or unmodified titanium dioxide for control of indoor-level volatile organic compounds. J Hazard Mater 164 (1):360–366

Kansal SK, Ali AH, Kapoor S, Bahnemann DW (2011) Synthesis of flower like zinc oxide nanostructure and its application as a photocatalyst. Sep Purif Technol 80:125–130

Khan SUM, Al-Shahry M, Ingler WB Jr (2002) Efficient photochemical water splitting by a chemically modified n-TiO_2. Science 297(27):2243–2245

Kim J, Lee CW, Choi W (2010) Platinized WO_3 as an environmental photocatalyst that generates OH radicals under visible light. Environ Sci Technol 44:6849–6854

Kondarides DI (2010) Photocatalysis. In: Centi G (ed) Catalysis. Encyclopedia of life support systems (EOLSS), developed under the auspices of the UNESCO. EOLSS Publishers, Paris

Kudo A, Miseki Y (2009) Heterogeneous photocatalyst materials for water splitting. Chem Soc Rev 38:253–278

Lee HU, Lee SC, Choi SH, Son B, Lee SJ, Kim HJ, Lee J (2013) Highly visible-light active nanoporous TiO_2 photocatalysts for efficient solar photocatalytic applications. Appl Catal B Environ 129:106–113

Lettmann C, Hildenbrand K, Kisch H, Macyk W, Maier WF (2001) Visible light photodegradation of 4-chlorophenol with a coke-containing titanium dioxide photocatalyst. Appl Catal B Environ 32(4):215–227

Lezner M, Grabowska E, Zaleska A (2012) Preparation and photocatalytic activity of iron-modified titanium dioxide photocatalyst. Physicochem Probl Miner Process 48(1):193–200
Li D, Haneda H, Hishita S, Ohashi N, Nitin K, Labhsetwar NK (2005) Fluorine-doped TiO_2 powders prepared by spray pyrolysis and their improved photocatalytic activity for decomposition of gas-phase acetaldehyde. J Fluor Chem 126(1):69–77
Li X, Zhu Z, Zhao Q, Wang L (2011) Photocatalytic degradation of gaseous toluene over $ZnAl_2O_4$ prepared by different methods: a comparative study. J Hazard Mater 186:2089–2096
Lin X, Rong F, Fu D, Yuan C (2012) Enhanced photocatalytic activity of fluorine doped TiO_2 by loaded with Ag for degradation of organic pollutants. Powder Technol 219:173–178
Litter MI (1999) Heterogeneous photocatalysis. Transition metal ions in photocatalytic systems. Appl Catal B Environ 23:89–114
Liu SH, Syu HR (2012) One step fabrication of N-doped mesoporous TiO_2 nanoparticles by self-assembly for photocatalytic water splitting under visible light. Appl Energy 100:148–154
Liu Z, Qi Y, Lu C (2010) High efficient ultraviolet photocatalytic activity of $BiFeO_3$ nanoparticles synthesized by a chemical coprecipitation process. J Mater Sci Mater Electron 21:380–384
Liu PS, Xia FJ, Chen YM, Cui G (2012) An anatase film with improved structure of titanium dioxide modified by carbon black. Mater Lett 72:5–8
Livraghi S, Paganini MC, Giamello E, Selloni A, Di Valentin C, Pacchioni G (2006) Origin of photoactivity of nitrogen-doped titanium dioxide under visible light. J Am Chem Soc 128:15666–15671
Livraghi S, Chierotti MR, Giamello E, Magnacca G, Paganini MC, Cappelletti G, Bianchi CL (2008) Nitrogen-doped titanium dioxide active in photocatalytic reactions with visible light: a multi-technique characterization of differently prepared materials. J Phys Chem C 112 (44):17244–17252
Livraghi S, Elghniji K, Czoska AM, Paganini MC, Giamello E, Ksibi M (2009) Nitrogen-doped and nitrogen–fluorine-codoped titanium dioxide. Nature and concentration of the photoactive species and their role in determining the photocatalytic activity under visible light. J Photochem Photobiol A Chem 205(2–3):93–97
Long M, Cai W (2011) Visible light responsive TiO_2 modification with nonmetal elements. Front Chem China 6:190–199
Luo Y, Chen H, Li X, Gong Z, Wang X, Peng X, He M, Sheng Z (2013) Wet chemical synthesis of Bi_2S_3 nanorods for efficient photocatalysis. Mater Lett 105:12–15
Macak JM, Gong BG, Hueppe M, Schmuki P (2007) Filling of TiO_2 nanotubes by self-doping and electrodeposition. Adv Mater 19:3027–3031
Maeda K (2011) Photocatalytic water splitting using semiconductor particles: history and recent developments. J Photochem Photobiol C 12:237–268
Malato S, Fernández-Ibáñez P, Maldonado MI, Blanco J, Gernjak W (2009) Decontamination and disinfection of water by solar photocatalysis: recent overview and trends. Catal Today 147:1–59
McEvoy JG, Cui W, Zhang Z (2013) Degradative and disinfective properties of carbon-doped anatase-rutile TiO_2 mixtures under visible light irradiation. Catal Today 207:191–199
Mi Q, Ping Y, Li Y, Cao B, Brunschwig BS, Khalifah PG, Galli GA, Gray HB, Lewis NS (2012) Thermally stable N_2-intercalated WO_3 photoanodes for water oxidation. J Am Chem Soc 134:18318–18324
Mitoraj D, Kisch H (2008) The nature of nitrogen-modified titanium dioxide photocatalysts active in visible light. Angew Chem Int Ed 47(51):9975–9978
Morkoç H, Özgür Ü (2009) Zinc oxide. Fundamentals, materials and device technology. Wiley VCH Verlag GmbH & Co. KGaA, Weinheim
Nolan NT, Synnott DW, Seery MK, Hinder SJ, Van Wassenhoven A, Pillai SC (2012) Effect of N-doping on the photocatalytic activity of sol-gel TiO_2. J Hazard Mater 211–212:88–94
O'Keeffe C, Gannon P, Gilson P, Kafizas A, Parkin IP, Binions R (2013) Air purification by heterogeneous photocatalytic oxidation with multi-doped thin film titanium dioxide. Thin Solid Films 537:131–136

Ohtani B (2008) Preparing Articles on Photocatalysis-Beyond the Illusions, Misconceptions, and Speculation. Chem Lett 37:216–229

Park Y, Kim W, Park H, Tachikawa T, Majima T, Choi W (2009) Carbon-doped TiO_2 photocatalyst synthesized without using an external carbon precursor and the visible light activity. Appl Catal B Environ 91:355–361

Park H, Park Y, Kim W, Choi W (2013) Surface modification of TiO_2 photocatalyst for environmental applications. J Photochem Photobiol C 15:1–20

Pelaez M, Falaras P, Likodimos V, Kontos AG, De la Cruz AA, O'Shea K, Di-onysiou DD (2010) Synthesis, structural characterization and evaluation of sol–gel-based NF-TiO_2 films with visible light-photoactivation for the removal of microcystin-LR. Appl Catal B Environ 99:378–387

Pelaez M, Nolan NT, Pillai SC, Seery MK, Falaras P, Kontos AG, Dunlop PSM, Hamilton JWJ, Byrne JA, O'Shea K, Entezari MH, Dionysiou DD (2012) A review on the visible light active titanium dioxide photocatalysts for environmental applications. Appl Catal B Environ 125:331–349

Pichat P (2003) Photocatalytic degradation of pollutants in water and air: basic concepts and applications. In: Tarr MA (ed) Chemical degradation methods for wastes and pollutants: environmental and industrial applications. Marcel Dekker, New York, pp 77–119

Pichat P (2013) Photocatalysis and water purification: from fundamentals to recent applications. Wiley-VCH, Weinheim

Rehman S, Ullah R, Butt AM, Gohar ND (2009) Strategies of making TiO_2 and ZnO visible light active. J Hazard Mater 170:560–569

Rengifo-Herrera JA, Mielczarski J, Castillo NC, Kiwi J, Pulgarin C (2008) *Escherichia coli* inactivation by N, S co-doped commercial TiO_2 powders under UV and visible light. Appl Catal B Environ 84(3–4):448–456

Rivera-Reyna N, Hinojosa-Reyes L, Guzmán-Mar JL, Cai Y, O'Shea K, Hernández-Ramírez A (2013) Photocatalytical removal of inorganic and organic arsenic species from aqueous solution using zinc oxide semiconductor. Photochem Photobiol Sci 12:653–659

Saha NC, Tompkins HG (1992) Titanium nitride oxidation chemistry: an X-ray photoelectron spectroscopy study. J Appl Phys 72(7):3072–3079

Salvador P (2007) On the nature of photogenerated radical species active in the oxidative degradation of dissolved pollutants with TiO_2 aqueous suspensions: a revision in the light of the electronic structure of adsorbed water. J Phys Chem C 111:17038–17043

Sampath SK, Kanhere DG, Pandey R (1999) Electronic structure of spinel oxides: zinc aluminate and zinc gallate. J Phys Condens Matter 11:3635–3644

Sauthier G, Ferrer FJ, Figueras A, György E (2010) Growth and characterization of nitrogen-doped TiO_2 thin films prepared by reactive pulsed laser deposition. Thin Solid Films 519 (4):1464–1469

Sayama K, Hayashi H, Arai T, Yanagida M, Gunji T, Sugihara H (2010) Highly active WO_3 semiconductor photocatalyst prepared from amorphous peroxo-tungstic acid for the degradation of various organic compounds. Appl Catal B Environ 94:150–157

Schiavello M (1987) Photocatalysis and environment: trends and applications, NATO ASI series C, vol 238. Kluwer Academic Publishers, London

Serpone N (2006) Is the band gap of pristine TiO_2 narrowed by anion- and cation-doping of titanium dioxide in second generation photocatalysts? J Phys Chem B 110(48):24287–24293

Serpone N, Pellizetti E (1989) Photocatalysis: fundamentals and applications. Wiley, New York

Serpone N, Emeline AV, Kuznetsov VN, Ryabchuk VK (2010) Second generation visible-light active photocatalysts: preparation, optical properties, and consequences of dopants on the band gap energy of TiO_2. In: Anpo M, Kamat PV (eds) Environmentally benign photocatalysts. Springer, New York, pp 35–111

Serway RA, Jewett JW (2014) Physics for scientists and engineers with modern physics, 9th edn. Cengage Learning, New York

Sharma P, Kumar R, Chauhan S, Singh D, Chauhan MS (2014) Facile growth and characterization of α-Fe_2O_3 nanoparticles for photocatalytic degradation of methyl orange. J Nanosci Nanotechnol 14:6153–6157
Shi Z, Liu F, Yao S (2010) Preparation and photocatalytic activity of B, Y co-doped nanosized TiO_2 catalyst. J Rare Earths 28(5):737–741
Shimodaira Y, Kato H, Kobayashi H, Kudo A (2006) Photophysical properties and photocatalytic activities of bismuth molybdates under visible light irradiation. J Phys Chem B 110:17790–17797
Soga T (ed) (2006) Nanostructured materials for solar energy conversion. Elsevier, Amsterdam
Sreethawong T, Ngamsinlapasathian S, Yoshikawa S (2013) Synthesis of crystalline mesoporous-assembled ZrO_2 nanoparticles via a facile surfactant-aided sol-gel process and their photocatalytic dye degradation activity. Chem Eng J 228:256–262
Sun C, Searles DJ (2013) Origin of the visible light absorption of boron/nitrogen co-doped anatase TiO_2. J Phys Chem C 117:26454–26459
Sun H, Bai Y, Cheng Y, Jin W, Xu N (2006) Preparation and characterization of visible-light-driven carbon-sulfur-codoped TiO_2 photocatalysts. Ind Eng Chem Res 45(14):4971–4976
Teh CM, Mohamed AR (2011) Roles of titanium dioxide and ion-doped titanium dioxide photocatalytic degradation of organic pollutants (phenolic compounds and dyes) in aqueous solutions: a review. J Alloys Compd 509:1648–1650
Tobaldi DM, Pullar RC, Škapin AS, Seabra MP, Labrincha JA (2014) Visible light activated photocatalytic behaviour of rare earth modified commercial TiO_2. Mater Res Bull 50:183–190
Tong H, Ouyang S, Bi Y, Umezawa N, Oshikiri M, Ye J (2012) Nano-photocatalytic materials: possibilities and challenges. Adv Mater 24:229–251
Topkaya E, Konyar M, Yatmaz C, Öztürk K (2014) Pure ZnO and composite ZnO/TiO_2 catalyst plates: a comparative study for the degradation of azo dye, pesticide and antibiotic in aqueous solutions. J Colloid Interface Sci 430:6–11
Umadevi M, Sangari M, Parimaladevi R, Sivanantham A, Mayandi J (2013) Enhanced photocatalytic, antimicrobial activity and photovoltaic characteristics of fluorine doped TiO_2 synthesized under ultrasound irradiation. J Fluor Chem 156:209–213
Wang H, Lewis JP (2006) Second-generation photocatalytic materials: anion-doped TiO_2. J Phys Condens Matter 18:421–434
Wang CM, Mallouk TE (1990) Wide-range tuning of the titanium dioxide flat-band potential by adsorption of fluoride and hydrofluoric acid. J Phys Chem 94(10):4276–4280
Wang J, Tafen N, Lewis JP, Hong Z, Manivannan A, Zhi M, Li M, Wu N (2009) Origin of photocatalytic activity of nitrogen-doped TiO_2 nanobelts. J Am Chem Soc 131:12290–12297
Wang Y, Wang Q, Zhan X, Wang F, Safdar M, He J (2013) Visible light driven type II heterostructures and their enhanced photocatalysis properties: a review. Nanoscale 5:8326–8339
Widiyandari H, Purwanto A, Balgis R, Ogi T, Okuyama K (2012) CuO/WO_3 and Pt/WO_3 nanocatalysts for efficient pollutant degradation using visible light irradiation. Chem Eng J 180:323–329
Wu PG, Ma CH, Shang JK (2005) Effects of nitrogen doping on optical properties of TiO_2 thin films. Appl Phys A 81:1411–1417
Wu Y, Xing M, Tian B, Zhang J, Chen F (2010) Preparation of nitrogen and fluorine co-doped mesoporous TiO_2 microsphere and photodegradation of acid orange 7 under visible light. Chem Eng J 162:710–717
Xie J, Wu Q, Zhao D (2012) Electrospinning synthesis of $ZnFe_2O_4/Fe_3O_4/Ag$ nanoparticle-loaded mesoporous carbon fibers with magnetic and photocatalytic properties. Carbon 50:800–807
Yang K, Dai Y, Huang B, Han S (2006) Theoretical study of N-doped TiO_2 rutile crystals. J Phys Chem B 110(47):24011–24014
Yang L, Liu P, Li X, Li S (2012) The photocatalytic activities of neodymium and fluorine doped TiO_2 nanoparticles. Ceram Int 38(6):4791–4796

Yu JC, Yu J, Ho W, Jiang Z, Zhang L (2002) Effects of F-doping on the photocatalytic activity and microstructures of nanocrystalline TiO_2 powders. Chem Mater 14:3808–3816

Yu Y, Wu HH, Zhu BL, Wang SR, Huang WP, Wu SH, Zhang SM (2008) Preparation, characterization and photocatalytic activities of F-doped TiO_2 nanotubes. Catal Lett 121:165–171

Yu T, Tan X, Zhao L, Yin Y, Chen P, Wei J (2010) Characterization, activity and kinetics of a visible light driven photocatalyst: cerium and nitrogen co-doped TiO_2 nanoparticles. Chem Eng J 157(1):86–92

Yu Y, Ding Y, Zuo S, Liu J (2011) Photocatalytic activity of nanosized cadmium sulfides synthesized by complex compound thermolysis. Int J Photoenergy 2011:1–5

Zhang Z, Hossain MF, Takahashi T (2010a) Fabrication of shape-controlled α-Fe_2O_3 nanostructures by sonoelectrochemical anodization for visible light photocatalytic application. Mater Lett 64:435–438

Zhang Z, Hossain MF, Takahashi T (2010b) Self-assembled hematite (α-Fe_2O_3) nanotube arrays for photoelectrocatalytic degradation of azo dye under simulated solar light irradiation. Appl Catal B 95:423–429

Zhang K, Wang X, He T, Guo X, Feng Y (2014) Preparation and photocatalytic activity of B–N co-doped mesoporous TiO_2. Powder Technol 253:608–613

Zhao W, Ma W, Chen C, Zhao J, Shuai Z (2004) Efficient degradation of toxic organic pollutants with Ni_2O_3/TiO_2-xBx under visible irradiation. J Am Chem Soc 126:4782–4783

Zhao C, Pelaez M, Duan X, Deng H, O'Shea K, Fatta-Kassinos D, Dionysiou DD (2013) Role of pH on photolytic and photocatalytic degradation of antibiotic oxytetracycline in aqueous solution under visible/solar light: kinetics and mechanism studies. Appl Catal B Environ 134–135:83–92

Chapter 2
New Visible-Light Active Semiconductors

Roberto Candal and Azael Martínez-de la Cruz

Abstract So far, the anatase TiO_2 polymorph has been the most studied semiconductor photocatalyst due to its high activity under UV irradiation, high stability against photocorrosion process, and low cost. Nevertheless, from the whole solar energy spectrum that radiates the earth, UV irradiation only represents 4 %. In the same way, other semiconductors such as ZnO, Fe_2O_3, CdS, ZnS, Nb_2O_5, Ta_2O_5, and $BiTaO_4$ have been reported with excellent performance as photocatalysts, among others. In particular, oxides with perovskite structure formed by TaO_6 or NbO_6 octahedra layers have shown the capability to develop an important catalytic activity have reported photocatalytic activity in tantalates and niobates of the type $NaMO_3$ (M = Ta and Nb) for the stoichiometric decomposition of water. In the same way, photocatalytic activity better than TiO_2 has been observed on laminar oxides such as $BaLi_2Ti_6O_{14}$, $MTaO_3$ (M = Li, Na, K), and SrM_2O_7 (M = Nb, Ta) for the degradation of organic pollutants. However, most of these have the same drawbacks that TiO_2 has in relation with their limited range of light absorption and an inefficient charge separation which leads to a high recombination process with the concomitant diminishing of their photocatalytic activity.

For this reason, different alternatives have been proposed to gather the solar energy and then develop large-scale technological applications. Numerous efforts have been made for the development of new visible-light-induced photocatalysts, and some oxides have shown visible-light-driven catalytic activity, such as $In_{1-x}Ni_xTaO_4$, $CaIn_2O_4$, $InVO_4$, $BiVO_4$, and Bi_2MoO_6.

In this chapter, some alternatives to the use of TiO_2 as photocatalyst will be discussed. In Sect. 2.2, the use of semiconductors doped with a transition metal as photocatalysts will be reviewed, in particular the effect of the introduction of metal in the physicochemical properties of semiconductor and in its improvement as

R. Candal
Instituto de Química Física de Materiales, Ambiente y Energía Ciudad Universitaria, Pabellón II, 1428-CABA

Instituto de Investigación e Ingeniería Ambiental Universidad Nacional de San Martín, Campus Miguelete, B1650HMP General San Martín, Buenos Aires, Argentina

A. Martínez-de la Cruz (✉)
Facultad de Ingeniería Mecánica y Eléctrica, Universidad Autónoma de Nuevo León, Av. Universidad s/n, Ciudad Universitaria, San Nicolás de los Garza, N.L. MEXICO, CP 66451
e-mail: azael.martinezdl@uanl.edu.mx

A. Hernández-Ramírez, I. Medina-Ramírez (eds.), *Photocatalytic Semiconductors*,
DOI 10.1007/978-3-319-10999-2_2

photocatalyst. Other interesting possibility in the development of materials with high photocatalytic activity is the doping with nonmetal elements as is discussed in Sect. 2.3. The sensitization of a semiconductor photocatalyst by the action of chemical species allows taking advantage of the free visible-solar energy. This point is included in Sect. 2.4 for the different types of sensitizers.

The association of two or more nanocrystalline semiconductors allows the design of semiconductors heterostructures that are potentially useful in photocatalysis, water splitting, microelectronics, and others. The possibilities of this type of semiconductor coupled photocatalyst are showed in Sect. 2.5. Finally, in Sect. 2.6, the formation of nanostructured semiconductors with photocatalytic activity due to the control of some of their physicochemical properties such as particle size and composition is described.

2.1 Ion-Doped Semiconductors (M^+ = Transition Metal Ion)

The use of semiconductors doped with a transition metal as photocatalysts has been extended in order to improve their physical properties, in particular to induce a narrow energy band gap in its electronic structure. This situation can lead to produce materials with better photocatalytic activity than the well-known TiO_2, in particular to take advantage of the energy that comes from the visible region of the solar spectrum. Some binary and ternary oxides have been proposed in its pristine form, in particular metal transition oxides, as alternative to TiO_2 in the visible region. Nevertheless, the poor photo efficiency in the process of separation of charges under light excitation has limited its future application. To solve this problem, the doped of metal transition oxides with a great variety of transition metal ions represents an interesting expectative, in particular to enhance the photocatalytic properties of the oxide. Additionally, the synthesis of ion-doped semiconductor oxides is easy accomplished and inexpensive and produces large amount of material. The use of wet chemical methods for this purpose, such as coprecipitation, allows the synthesis of materials with specific textural and morphologic properties which can be associated with the photocatalytic activity developed.

The doped of semiconductor oxides for photocatalytic applications included binary oxides such as ZnO, WO_3, Bi_2O_3, and Ta_2O_5 (Bamwenda and Arakawa 2001; Cheng et al. 2007; Hong et al. 2009; Tang et al. 2003) and ternary oxides of the type $BiVO_4$, $InVO_4$, $InTaO_4$, and Bi_2MoO_6 (Kudo et al. 1999; Zhang et al. 2006; Marschall and Wang 2014, Kato et al. 2004), among others. A great variety of synthesis route (see Chap. 3) are used for the synthesis of doped semiconductors which included mainly coprecipitation (Liu et al. 2005), sol–gel (Jaimy et al. 2011), hydrothermal (Jiang et al. 2014), solvothermal (Sun et al. 2013), Pechini method (Wang and Cao 2011), combustion (Ahmad et al. 2013), microwave (Li et al. 2014a), ultrasound (Omidi et al. 2013), and in the form of thin films by sol–gel (Wang and Cao 2011) deposited by dip coating (Zou et al. 2011).

The binary oxide ZnO with wurtzite structure is one of the most studied photocatalyst after TiO_2 due to its high electrochemical stability and nontoxicity. Several works revealed this oxide as a promising candidate to solve environmental pollution problems related with water purification. Yayapao et al. (2013a) reduced the energy band gap of the oxide by doping the crystal structure with 3 % of Ce by a sonochemical method. The improvement in the activity of Ce-doped ZnO to the photocatalytic degradation of methylene blue in solution was attributed to the presence of a higher concentration of oxygen defects on the surfaces of crystalline-doped material. By the same synthesis route, Yayapao et al. (2013b) prepared Nd-doped ZnO nanoneedles finding the best results for a concentration of 1 % Nd with a performance 2.5 times higher than the undoped ZnO. The origin of this enhancement was the presence of Nd^{3+} ions, which acted as electron scavengers and suppressed the electron–hole recombination.

The sonochemical method was also successfully used for the synthesis of three-dimensional flowerlike Ho-doped ZnO microstructure with improved photocatalytic activity in the degradation of methylene blue (Phuruangrat et al. 2014). The optimum concentration value of Ho^{3+} was established at 3 %, and as was previously described for other lanthanides, the presence of Ho^{3+} prevents the hole-electron recombination due to the ion that acts as an efficient scavenger.

The use of rare earths to doping ZnO also includes europium ions. Europium-doped ZnO hierarchical micro-/nanospheres were synthesized via simple coprecipitation. Eu-doped ZnO was more effective for the degradation of phenol under sunlight irradiation than pure ZnO. This situation was explained on the basis of a better charge separation efficiency and due to hydroxyl radical generation ability as was evidenced by photoluminescence.

Similar successful results were obtained with La^{3+} and Sm^{3+}-doped ZnO materials in the degradation reaction of 4-nitrophenol under UV irradiation (Khatamian et al. 2012).

Although the unfavorable position of E_{CB} is relative to the redox potential of the H_2/H^+, tungsten oxide (WO_3) is a semiconductor oxide of great interest. WO_3 is a material with high stability in aqueous solutions under acidic conditions, it does not undergo photocorrosion process, and it exhibits polymorphism. Indeed, WO_3 has been shown good photocatalytic activity under visible light for degradation of some organic compounds (Takeuchi et al. 2011). Among the polymorphs of WO_3, the phases that have attracted the most attention are the hexagonal (h-WO_3) and monoclinic phase (m-WO_3). The monoclinic phase is an attractive candidate for photocatalytic applications under solar irradiation because it exhibits the proper band-gap energy for the absorption of visible light (Bamwenda and Arakawa 2001; Cheng et al. 2007; Hong et al. 2009).

Different metallic ions such as Mg^{2+}, Al^{3+}, In^{3+}, Fe^{3+}, and Zr^{4+}, among others, have been successfully used to dope the monoclinic crystalline structure of WO_3. The results showed an important change in their photocatalytic properties with respect to pristine oxide, showing the metallic-doped WO_3 materials better activity than pure WO_3 (Tang et al. 2003). In the same way, the effect of doping with different transition metals (Fe, Co, Ni, Cu, and Zn), at different concentrations, was previously reported for the development of doped WO_3 photocatalysts for splitting of water into hydrogen and oxygen under UV laser irradiation (Hameed et al. 2004). On the other hand, Radecka et al. (2005) have reported an important enhancement in the photoelectrochemical properties of WO_3 when it is doped with Ti ion. This situation was attributed to the fact that Ti^{4+} ions can act as acceptor-type centers in crystalline structure of WO_3 to induce an improvement of light absorption spectra.

Following this line of search, Cheng et al. (2007) have prepared WO_3 and Zn-doped WO_3 thin films on indium-tin oxide glass by a dip coating. Different experimental conditions in the preparation of samples were analyzed, such as Zn (II) concentration, annealing temperature, and number of layers, in order to study the response of the system in the generated photocurrent (Cheng et al. 2007). The photocatalytic activity of the Zn-doped WO_3 was evaluated by using it in a doped photoanode in terms of decay rate of nitrite ions (NO_2^-) under visible light. The best results were obtained with a Zn concentration of 2 %, where an enhancement of both photocurrent and photoactivity of WO_3 was achieved. The authors proposed a photochemical oxidation mechanism where the light absorption is more efficient in Zn-doped WO_3 which allows the production of more photocarriers and consequently the formation of highly active species such as $^{\bullet}OH$ radicals.

The doped of crystalline structure of WO_3 has been extended to rare earth elements. Liu et al. (2007) reported the preparation and photocatalytic activity of dysprosium-doped WO_3 nanoparticles via precipitation. Rhodamine B (rhB) was used as model compound for degradation reaction. They found an important correlation among the photoactivity, photostability, and surface acidity with the adsorbed amount of rhB on the photocatalyst. The mechanism proposed to explain the improvement in the activity of Dy-doped material suggests that doped Dy^{3+} can give an electron to O_2 adsorbed on the surface of doped material to form a superoxide radical ($^{\bullet}O_2^-$) and Dy^{4+}. The Dy^{4+} generated is able to receive a photogenerated electron in the conduction band of WO_3 to form Dy^{3+} inhibiting its recombination with photogenerated holes. This point is reinforced with the fact that Dy-doped WO_3 can enhance the reactivity of rhB on the photocatalyst surface and, therefore, improve its photoactivity.

In the same way, Liu et al. (2005) reported high photoactivity and photostability of Tb-doped WO_3 nanoparticles with respect to pristine WO_3 oxide. WO_3 also was doped with Eu^{3+} with good results from the point of view of photocatalytic applications. Following the Pechini method, Wang and Cao (2011) prepared Eu-doped WO_3 nanoparticles and evaluated its performance as photocatalyst by the degradation reaction of rhB under sunlight-type radiation. Among different physical properties, the authors attributed the improvement of the photocatalytic activity of Eu-doped WO_3 to the distortion of the crystalline structure of the oxide due to the presence of Eu^{3+} ions. The presence of Eu^{3+} ions in the crystalline structure of WO_3 allows an energy band gap of the material ($E_g = 2.65$ eV) suitable to absorb in the visible region of solar spectrum.

The doped of WO_3 with metallic ions also has produced photocatalyst with high performance in the photocatalytic oxidation of water to produce H_2 under UV and visible-light radiation. For this purpose, Zou et al. (2011) reported the microwave synthesis of mesoporous Bi-doped WO_3 materials in presence of cetyltrimethyl ammonium bromine (CTAB) as structure-directing agent. The amount of Bi_2O_3 loading was optimized in a 7.0 wt % for the production of H_2. The incorporation of Bi^{3+} ions seems to avoid the electron–hole pair recombination and can create a visible absorption center and an active surface site.

Another promising example of binary oxide with potential application in photocatalysis and photoelectrocatalysis under visible-light illumination is Bi_2O_3, which can be combined with Nb and doped with different transition metals (Li et al. 2012; Ropero-Vega et al. 2010).

2.2 Nonmetal-Doped Materials

Several semiconductors (SC) with potential applications in photocatalysis, such as ZnO, Ta_2O_5, and Nb_2O_5 (and derived compounds such as niobates and tantalates), are transparent to visible light being activated only with UV (A or B) light. In the case of other colored SC, such as WO_3, Bi_2WO_6, and $BiVO_4$, it is desirable to increase the range of light useful for its activation as photocatalyst. For all these SC, as in the case of TiO_2 discussed in Chap. 1, *nonmetal doping* represents an interesting alternative to improve their photocatalytic performance under visible-light illumination.

The origin of visible-light absorption in these systems can be a consequence of several phenomena. When the energy of the *2p* electronic states of the nonmetal atom and oxygen is similar (as in the case of N in the doping of TiO_2), their mixing possibly leads to the formation of new valence band (vb) with an upward shifted edge that diminishes the band gap energy. However, this is not the only possibility. The nonmetal dopant can also form localized states above the vb, which play an important role in electron transfer to reducible compounds. Other possible explanations are related with oxygen vacancies, color centers, sensitization with nonmetal polymers, and other as shown in Fig. 2.1, adapted from the recently published

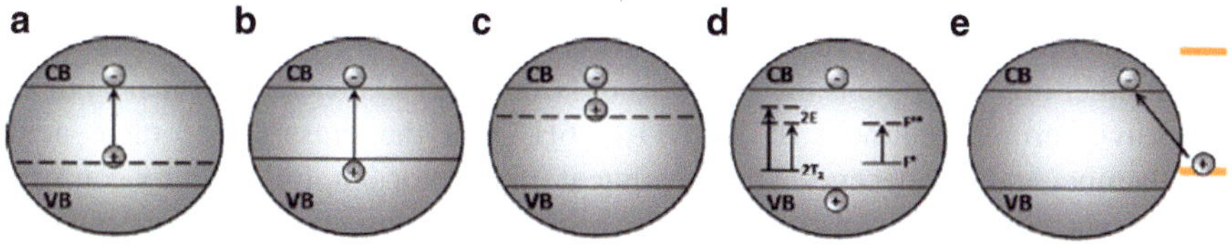

Fig. 2.1 Scheme describing different doping alternatives for SC. (**a**) Localized states above vb. (**b**) Nonmetal-doped SC with narrowed E_g. (**c**) Localized states below CB. (**d**) Color centers formed in the E_g. (**e**) Surface modification with N-containing compounds (From Marshall and Wang, 2014)

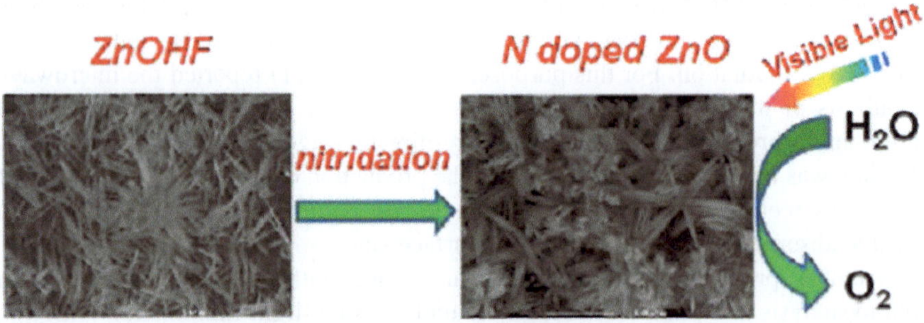

Fig. 2.2 Nitridation of ZnOHF nanobundles led to N-ZnO with similar morphology. This new material displays photocatalytic activity for water splitting under visible light (From Zong et al. 2013. Copyright 2013 American Chemical Society)

review of Marschall and Wang (2014). These authors concluded that homogeneous doping is necessary to shift the absorption edge to visible-light wavelength with band to band excitation, being the amount of dopant related with the threshold of the extended absorption. Homogeneous doping is mostly depending on particle size, crystalline structure, and doping procedure. The type of doping also may depend on the synthesis procedure. In the case of N-TiO_2, Di Valentin et al. (2005), based in DFT theoretical calculations, suggested the existence of two types of N doping: interstitial and substitutional. The first case is favored when calcinations are done under air atmosphere, while the second in the absence of air. Experimental information about the localization of the nonmetal dopant can be obtained by XPS analysis of doped samples prepared under different conditions, as in the work reported by Pulgarin and coworkers (Rengifo-Herrera et al. 2008).

One of the most studied SC other than TiO_2 is ZnO. In spite of its low stability under regular environmental conditions, ZnO is a very attractive material due to its potential applications in disinfection and because of the diversity of shapes in which it can be obtained. An interesting example is represented by N-ZnO nanobundles with visible-light photocatalytic activity, prepared by thermal treatment of ZnOHF nanobundles under NH_3, see Fig. 2.2 (Zong et al. 2013). X-ray photoelectron spectroscopy (XPS) analysis indicated that N was bound to Zn as nitride (Zn-N) and oxynitride (O-Zn-N). The band gap of ZnO was reduced from 2.20 to 1.95 eV as a consequence of substitutional nitrogen doping and the contribution of nitrogen to the top of the vb of ZnO. Water oxidation was performed on this photocatalyst under visible-light ($\lambda > 420$ nm) illumination.

Another nice example is the synthesis of N-doped ZnO mesoporous nanospheres by solvothermal treatment of $Zn(NO_3)_2 \cdot 6H_2O$ in the presence of oleic acid, oleylamine, and octadecene. After heating at 260 °C for 2 h, followed by centrifugation, washing of the solid, and firing at 400 °C, mesoporous ZnO spheres with diameters in the range 100–300 nm were obtained. XPS characterization indicates that N is involved in O-Zn-N bonds. This material displayed enhanced photocatalytic activity for rhodamine degradation under UVA illumination when

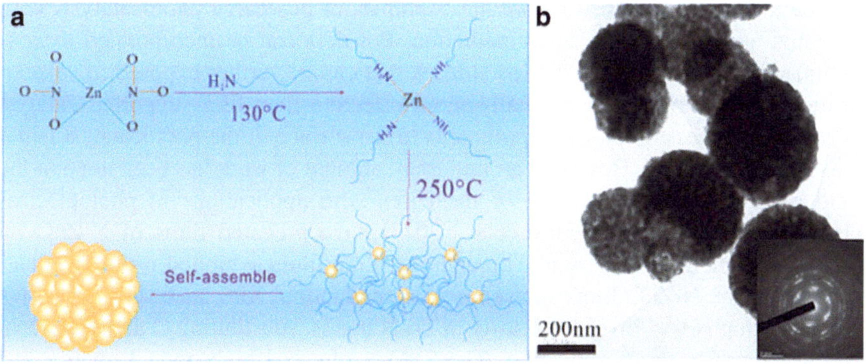

Fig. 2.3 Scheme of the synthesis process (**a**) and TEM image of the N-ZnO mesoporous particles (**b**) (From Zhang et al. 2013)

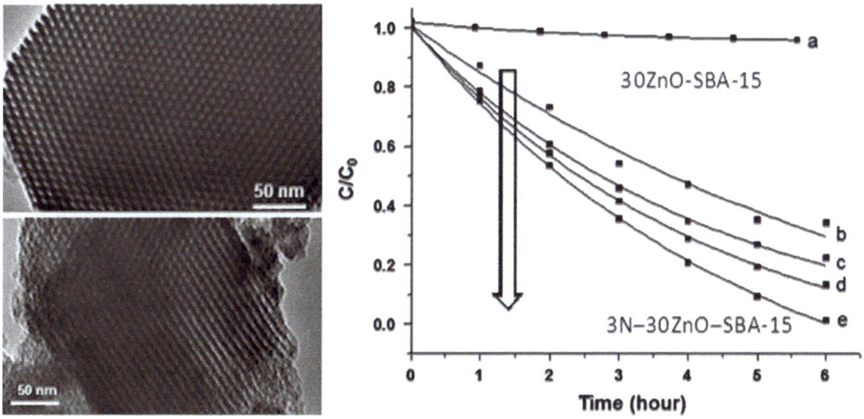

Fig. 2.4 TEM images of SBA-15. (**a**) 3N–30ZnO–SBA-15. (**b**) Photocatalytic degradation of MB under visible-light illumination, using photocatalysts with different level of N doping. The *arrow* indicates the increment in N doping (From Vo et al. 2014)

compared with pure ZnO. Figure 2.3 shows a scheme of the synthetic procedure and TEM images of the obtained particles.

N-ZnO was successfully supported on organized mesoporous silica, in an attempt to develop a photocatalyst with high surface area (Vo et al. 2014). Mesoporous Santa Barbara Amorphous-15 (SBA-15) silica was prepared in the laboratory following well-established procedures and firing at 500°C to remove organics. ZnO was incorporated to SBA-15 by impregnation with $Zn(NO_3)_2$, followed by firing at 550 °C, doping with urea as N source, drying at 100 °C, and annealed at 500 °C. The level of doping was controlled through the repetition of the step involving urea incorporation. The material displayed photocatalytic activity for the discoloration of methylene blue (MB) under visible light. Figure 2.4 shows TEM images of SBA-15 and N-ZnO-SBA-15 and degradation curves of MB.

The materials described before are examples of powdered photocatalysts with potential application in water remediation. Powders can be incorporated to contaminated water producing diluted slurries that can be further exposed to solar or artificial light. Powders are usually more effective than film supported catalyst because the accessibility of the pollutants to the surface of the photocatalyst is not mass transfer limited. However, an important amount of work has been done in the field of N-ZnO films. In most of the cases with applications in photoelectrochemistry, sensors, etc., Wang et al. (2014) have prepared films of N-ZnO by reactive magnetron sputtering. The undoped ZnO films exhibited n-type conduction, while the N-ZnO films showed p-type conduction. XPS analysis of the films indicated that N was involved mostly in Zn-N bonds, substituting O atoms to form NO acceptors in the N-ZnO film. The ZnO:N film has high optical quality and displays the stronger near band edge (NBE) emission in the temperature-dependent photoluminescence spectrum, and the acceptor energy level was estimated to be located 110 meV above the valence band. N-ZnO films were also prepared by other high vacuum techniques, as high vacuum plasma-assisted chemical vapor deposition (HVP-CVD; see, e.g., Barnes et al. 2005) and pulsed-filtered cathodic vacuum arc deposition (PFCVAD; see e.g., Tuzemen et al. 2014). N_2O or N_2 was used as doping agent, respectively. In both cases, as doping level increased, the semiconductor changed from n- to p-type. The stability of the p-type doping depends on the film synthesis. Typically, with aging, the hole conduction decreased and films reverted to n-type conductivity (usually accompanied by an increase in the lattice constant). The films prepared by PFCVAD preserved the p-type doping for at least 12 month.

Tuomisto et al. (2013) prepared N-ZnO films by chemical vapor deposition (CVD) using metallic zinc, NO_2, and NH_3 as precursors. The introduction of N impurities into ZnO films led to the formation of stable vacancy clusters and negative-type defects. N-ZnO and ZnO films prepared by pulsed laser deposition (PLD) were successfully used as platform for the development of biosensors for the determination of uric acid (Jindal et al. 2013). The introduction of defect states upon N doping produced a film with excellent charge transfer characteristics. Besides, doping increased the surface-to-volume ratio providing higher sensitivity to uric acid in comparison with pure ZnO. The exact location of introduced N is unknown, but should be at either substitutional or interstitial sites in the host ZnO lattice. At N low concentrations, due to low formation energy, it is expected to substitute at the O sites improving sensitivity. Sol–gel process, in combination with spin coating technique, was used for the synthesis of Al and N co-doped ZnO. Zinc acetate dihydrate, aluminum nitrate nonahydrate, and ammonium acetate were used as a source for Zn, Al, and N, respectively, in an isopropanol solution with diethanolamine as stabilizer. The spin-coated glass substrates were annealed at 450 °C for 1 h. The optical band gap was reduced from 3.17 to 3.10 due to nitrogen doping. The Al, N co-doped films showed p-type conductivity and high photoelectrical response with a fast recovery of 12 s (Saravanakumar et al. 2013).

Sulfur doping is another approach used to modify the band gap of ZnO. It was reported that sulfur incorporation increases the energy of the vb, while the energy of

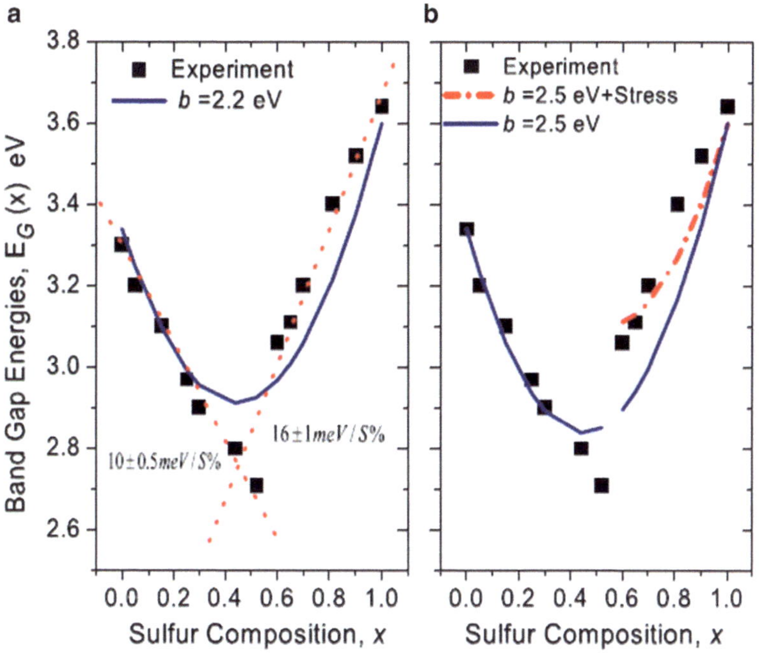

Fig. 2.5 Band gap dependence on the sulfur composition of the $ZnO_{1-x}S_x$ system due to anion substitution with (**a**) $b = 2.2$ eV, (**b**) $b = 2.5$ eV plus stress (b = bowing parameter) (From Thankalekshmi and Rastogi 2012)

the cv remains almost constant (Persson et al. 2006). Electronic states in the band gap are also created. These phenomena lead to the diminution of the band gap energy of ZnO, shifting the absorption of light to the red. Incorporation of sulfur is limited by its low solubility in ZnO and its ZnS phase segregation use to take place at high temperature. To avoid this problem, low temperatures and homogeneously mixed precursors should be used. Films of sulfur-doped ZnO (S-ZnO) were successfully prepared by techniques as pulse laser deposition (Yoo et al. 2002; Xu et al. 2011), reactive sputtering (Meyer et al. 2004), and chemical spray pyrolysis (Thankalekshmi and Rastogi 2012). In the last paper, simple chemicals as $ZnCl_2$ and thiourea dissolved in isopropanol/water 70/30 mixture were used as precursors. When the films deposition was done at 300 °C, the decomposition of zinc(bis) thiourea $Zn[(CS(NH_2)_2]_2Cl_2)$ complex occurred concurrently with the decomposition of Zn(OH)Cl (produced by thermolysis of $ZnCl_2.(H_2O)_4$ on the hot substrate) in the presence of ZnO. Under this condition, the incorporation of S at the O sites is highly probable, and alloyed $ZnO_{1-x}S_x$ films were formed without ZnO and ZnS phase separation. XPS analysis confirmed the formation of the mixed film by S^{2-} substitution at the O^{2-} lattice sites. The band gap decreased as the content of S increased from $0 \leq x \leq 0.5$ and increased for higher contents of sulfur (Fig. 2.5). Meyer et al. (2004) reported similar behavior for S-ZnO films prepared by reactive

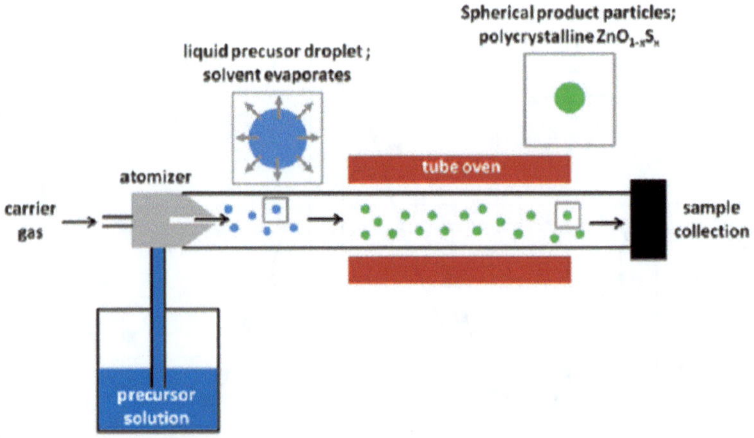

Fig. 2.6 Scheme of the device designed by Leher et al. for the synthesis of ZnO1-xSx spherical particles. The precursor solutions were homogeneous mixtures $[MeZn\text{-}S_{iso}Pr]_8$ and $[MeZnOEtOMe]_4$, in the appropriate proportions, diluted in toluene (From Lehr et al. 2012. Copyright 2012 American Chemical Society)

sputtering. The changes in band gap were approximately fitted by the expression: $E_G(x) = x{\cdot}E_G(\text{ZnS}) + (1-x){\cdot}E_G(\text{ZnO})\text{-b}{\cdot}(1-x){\cdot}x + \Delta E_G(\text{stress})$, where $\Delta E_G(\text{stress})$ is the stress-induced shift in the optical transition energy (Thankalekshmi and Rastogi 2012).

S-ZnO films are suitable for applications in electronics, spintronic, photonics, sensors, photoelectrochemistry, etc. But for photocatalytic applications, powders are more versatile than films. Efforts were realized to design synthesis protocols that can produce S-ZnO with different levels of doping in a reproducible way. The strategy followed in these cases was to use homogeneously mixed precursors and relatively low temperatures. Lehr et al. (2012) designed a synthesis that uses organometallic precursors and low temperatures. $[MeZn\text{-}S_{iso}Pr]_8$ was dissolved in $[MeZnOEtOMe]_4$ to prepare solutions in the range $0.5 - 25$ mol % followed by thermolysis at 350 °C. XRD analysis demonstrated that the materials were ZnO doped with different amounts of sulfur. The spectroscopic analysis by diffuse reflectance showed that the band gap shift to the red as the amount of sulfur increased. The band edge was blurred significantly indicating the presence of defect states. A method to prepare spherical particles with diameters in the range 100–200 nm was also developed (see Fig. 2.6). Patil et al. (2010) designed a method based in the mechanochemical synthesis of bisthiourea zinc oxalate (BTZO) powders, followed by thermal treatment at different temperatures. At lower temperatures, the obtained product was ZnS, but as the temperature rose, S-ZnO was produced by controlled oxidation of ZnS. The optimal temperature was 600 °C. The possibility to control the incorporation of sulfur was not reported. The powder displayed enhanced photocatalytic activity for degradation of resorcinol under sunlight when compared with pure ZnO (Patil et al. 2010).

Few other nonmetals were used to dope ZnO. For example, Zheng et al. (2011) reported the synthesis of I-ZnO by a solvothermal method, using zinc salts and iodic acid in polyol medium as precursors. The presence of I in the ZnO network was determined by XRD and EDX analysis. Incident photon to current conversion efficiency (IPCE) analysis of films prepared with ZnO:I showed a red shift, facilitating the utilization of sunlight in the full spectrum range (Zheng et al. 2011).

Tantalum oxide, Ta_2O_5, has also been reported as a semiconductor with applications in photocatalysis. The band gap is in the range 3.9–4.0 eV; thus UV light irradiation is needed for photocatalytic activity. Ta_2O_5-1 % RuO_2 displayed good activity for water splitting under UV irradiation (Sayama and Arakawa 1994). Several attempts were realized in order to shift the absorption of light to the visible part of the solar spectrum. Murase et al. (2004) reported the doping of Ta_2O_5 by thermal treatment under NH_3(g) atmosphere. The incorporation of N was controlled by the heating temperature: $Ta_2O_{5-x}N_x$ with $x=0.1$, 0.17, 0, 24, and 0.35 were obtained at 600, 620, 650, and 700 °C, respectively. The samples were active for the degradation of 2-propanol in air after long exposure times under visible light (400–530 nm). Very interestingly, by annealing Ta_2O_5 at 850 °C, only TaON was obtained, which absorbed visible light but could not decompose 2-propanol. The synthesis of N-doped Ta_2O_5 by a sol–gel method was recently reported (Ullah et al. 2013). N doping was also produced by annealing under NH_3 atmosphere. The visible and solar light photocatalytic activity of the N-doped material prepared by sol–gel was better than another prepared from commercial Ta_2O_5. The authors reported the conversion of 70 % of toluene under simulated solar light and 30 % of conversion under visible light ($\lambda > 400$ nm). Tantalates and N-doped tantalates were also tested as photocatalyst. $NaTaO_{3-x}N_x$ showed a broad adsorption at 550 nm and was able to decompose gaseous formaldehyde under visible-light illumination (Fu et al. 2008a). It was reported that a high number of lattice defects, obtained with a higher doping, decrease the catalytic activity due to the formation of recombination centers (Fu et al. 2008b; Zhu et al. 2008). Other tantalates, as $InTaO_4$ with perovskite structure and $Sr_2Ta_2O_7$, $Ba_5Ta_4O_{15}$, $CsTaWO_6$ with laminar structure, were doped with nitrogen. In the last cases, the laminar structure helped to reach high levels of doping (Marschall and Wang 2014).

Nb_2O_5 displayed photocatalytic activity for organics oxidation under UV illumination, but not for water splitting. Nitrogen-doped Nb_2O_5 was reported to be able to decompose formaldehyde under visible-light illumination (Murase et al. 2004). N-Nb_2O_5 was compared with N-Ta_2O_5 prepared under identical conditions, for toluene photocatalytic decomposition (Ullah et al. 2013). N-Nb_2O_5 displayed much lower activity than N-Ta_2O_5 due to low amount of defects and inefficient doping. Niobic acid, HNb_3O_8, displays a layered structure which facilitates incorporation of N. N-doped niobic acid was prepared by heating HNb_3O_8 with urea at 400 °C. The obtained material also presented a layered structure; light absorption shifted to the red and displayed enhanced photocatalytic activity for rhodamine B degradation when compared with pure niobic acid (Li et al. 2008).

As was mentioned in Sect. 2.2, WO_3 is a well-known semiconductor with photocatalytic activity. The band gap of WO_3 is in the range 2.4–2.8 eV, so it displays photoactivity with UV–Vis light. Although it is stable toward water oxidation, because the minimum of its conductive band is 0.5 V vs. NHE (more positive than the standard hydrogen potential), water splitting is not possible on WO_3. Interstitial nitrogen doping was found to change the CB position. N_2 placed in the hollow position of WO_3 crystal structure can shift the CB maximum toward more negative values increasing the band gap which, after further interstitial doping, decreases again. The preparation of N_2-WO_3 by thermolysis of ammonium metatungstate under oxygen stream was recently reported (Mi et al. 2012). The optical absorption of N_2-WO_3 was shifted to the red with respect to pure WO_3. Thermal decomposition of ammonium metatungstate can also produce (N^{3-})-doped WO_3 (Takeuchi et al. 2011). This material, after decorated with Pt nanoparticles, exhibited good photocatalytic activity for methanol degradation in gas phase under visible light ($\lambda > 450$ nm).

Tungstates, as Bi_2WO_6, were also doped with nonmetals to improve their photocatalytic activity for organic degradation. Doping with F- modified the surface of the catalyst, leading the reaction mechanism to the *N*-demethylation route (Fu et al. 2008a).

Bismuth vanadates, although they absorb visible light, were also doped with nonmetals to improve photocatalytic activity and to modify reaction mechanism. Such was the case for S-$BiVO_4$ (Zhao et al. 2013) and F-$BiVO_4$ (Liu et al. 2012), respectively. In both cases, the synthesis was carried out by hydrothermal process.

2.3 Dye-Sensitized Semiconductor

The sensitization of a semiconductor has been used for different technological purposes which includes the development of new solar cells (San Esteban and Enriquez 2013), production of hydrogen by water splitting (Li et al. 2014b), and for the degradation of organic dyes coming from industrial effluents (Chatterjee and Dasgupta 2005). In 1991, Grätzel and coworkers developed an efficient dye-sensitized solar cell based in a ruthenium complex adsorbed on nanocrystalline TiO_2 films (O'Regan and Grätzel 1991). This fact opened a new generation of solar cells with higher photon-electron efficiency with a considerable diminishing in the costs of production. In the same way, in recent years, the evolution of hydrogen by water splitting has attracted attention due to the importance of using this gas as energy vector in the future. In this field, the activity of photocatalyst oxides has been considerably improved by its sensitization in order to extend the range in the solar spectrum where it absorbs. Finally, the use of organic-dye/semiconductor-material systems in the process of wastewater treatment via a photosensitization mechanism is a promising field to solve environmental problems.

A great variety of chemical substances have been proposed as effective photosensitizers of a semiconductor. Among these, the most relevant photosensitizers belong to the families of inorganic sensitizers, coordination metal complexes, and organic dyes with molecular complex structures (Pei and Luan 2012). In the first case, the use of inorganic sensitizers involves semiconductors with narrow band gaps suitable for absorption in the visible region. Among the main physical properties of these compounds are their high stability to photocorrosion process and their absorption in a wide wavelength region. Inorganic sensitizers with considerable performance include metal ions of transition series such as V, Cr, Mn, Fe, and Ni (Chatterjee and Dasgupta 2005) as well as metallic alloys of the type Pt-Au and Au-Ag (Sreethawong and Yoshikawa 2012; Chen et al. 2013). The coordination metal complexes are a type of sensitizers widely used due to their high efficiency to extend the region of absorption of TiO_2; particularly, the ruthenium and osmium complexes have been intensely used for this purpose (Pei and Luan 2012).

An important disadvantage of the sensitization of semiconductor with an organic dye is the gradual decomposition of organic molecule by photocatalytic degradation, due to the natural tendency of organic dyes to undergo redox reactions. In order to solve this problem, the incorporation of electron donors to the reaction medium can be an effective solution, such is the case of EDTA (ethylenediaminetetraacetic acid) (Abe et al. 2004). In the past, numerous organic dyes have been tested with excellent performance to sensitize a semiconductor, in particular TiO_2, such as methyl orange (Liu et al. 2013), methylene blue (Maia et al. 2012), eosin Y (Chen et al. 2013), reactive red 198 (Kaur and Singh 2007), riboflavin (Chu et al. 2007), rose bengal (Whitehead and Hedges 2005), cyanine (Guo et al. 2005), cresyl violet (Dou et al. 2011), hemicyanine (Chen et al. 2005), and merocyanine (Abe et al. 2004). Moreover, some natural organic dyes have been proposed for several reasons. Among others, unlike artificial dyes, the natural ones are available, easy to prepare, low in cost, nontoxic, environmentally friendly, and fully biodegradable (Wongcharee et al. 2007). As an example of this type of compound, some authors have proposed the use of red Sicilian orange juice (*Citrus sinensis*) and the purple extract of eggplant peels (*Solanum melongena*, L.) as natural sensitizers of TiO_2 films for dye-sensitized solar cells (Calogero and Di Marco 2008). In the same way, carotene (Gómez-Ortíz et al. 2010), tannin (Espinosa et al. 2005), and chlorophyll (Kumara et al. 2006) have been utilized also as sensitizers in these devices. Zhou et al. (2011) have reported twenty natural dyes, extracted from natural materials such as flowers, leaves, fruits, traditional Chinese medicines, and beverages, as sensitizers to fabricate dye-sensitized solar cells (DSCs). As main conclusion from the use of natural dyes, they reported that the DSC sensitized by mangosteen pericarp extract offered the highest conversion efficiency of 1.17 % among the 20 extracts. Specifically, the voltage open circuit of mangosteen pericarp extract is comparable to that of the DSC sensitized by a Ru complex N719 (Zhou et al. 2011). An extended review about the use of natural dyes in DSCs has been published by Narayan (2012).

Although the use of organic dyes as sensitizers of semiconductors has the disadvantage of a gradual degradation of organic molecule, at the same time this fact can be used for the photocatalytic degradation of a great variety of organic pollutants from wastewater effluents. Taking advantage of catalytic activity of semiconductors with high energy band gap, i.e., above $E_g = 3$ eV, the photocatalytic degradation process of organic pollutants can be activated in the visible region of the solar spectrum for environmental applications. The undesirable degradation of the sensitizer in the course of redox reaction applied in devices to produce hydrogen or in a dye-sensitized solar cell can be exploited to remove colored organic pollutants from wastewater.

Nowadays, the industry of synthetic dyes has reached an important place in the development of the society because it supplies its products to a great variety of industries. Usually, these compounds are formed by complex organic molecules with one or more aromatic rings. Unfortunately, about 1–20 % of the total world production of dyes is lost during the dyeing process and is released into the textile effluents (Lopes et al. 2014). Dyehouse wastewater usually contains about 10–50 mg L^{-1} of dyes in solution (Xu et al. 2012). Such concentrations are high enough to induce a remarkable coloring of the receiving water bodies where they are discharged.

The degradation of an organic pollutant by the action of a photocatalyst can take place by two ways. The first of which is by a true photocatalytic process, where radiation on the photocatalyst promotes an electron from its valence band to the conduction band, and then the electron–hole pair is formed. The second possibility is through a photosensitization process, where the radiation excites an electron from the dye and, then, it is injected to the conduction band of the semiconductor oxide. In both processes, a series of consecutive reactions lead to the eventual mineralization of organic dyes to CO_2 and H_2O.

The photosensitization process by rhodamine B (rhB) is well known for different systems (Takizawa et al. 1978) and can be described in the case of a semiconductor oxide (SM) as follows:

$$\text{rhB (ads)} + h\nu\,(554\,\text{nm}) \rightarrow \text{rhB}^*\,\text{(ads) (1) excitation of dye} \tag{2.1}$$

$$\text{rhB}^*\,\text{(ads)} + \text{SM} \rightarrow \text{rhB}^{\bullet +}\text{(ads)} + \text{SM}\left(e^-_{cb}\right)\text{ (2) electron injection} \tag{2.2}$$

$$\text{SM}\left(e^-_{cb}\right) + O_2 \rightarrow \text{SM (2)} + O_2^{\bullet -}\text{(3) electron scavenging} \tag{2.3}$$

$$\text{rhB}^{\bullet +}\text{(ads)} + O_2^{\bullet -} \rightarrow\rightarrow\rightarrow CO_2 + H_2O\text{ (mineralization process)} \tag{2.4}$$

Of course, in order to reach a complete mineralization of the dye, it is necessary that a direct activation of SM by light irradiation occurs too.

$$\text{SM}\left(e^-_{cb}\right) + O_2^{\bullet -} + H^+ \rightarrow HO_2^- + \text{SM} \tag{2.5}$$

$$HO_2^- + H^+ \rightarrow H_2O_2 \tag{2.6}$$

$$H_2O_2 + e^- \rightarrow {}^{\bullet}OH + OH^- \tag{2.7}$$

$$\text{rhB}^{\bullet +}\text{(ads)} + {}^{\bullet}OH \rightarrow\rightarrow\rightarrow CO_2 + H_2O \tag{2.8}$$

2.4 Coupled Semiconductors

The association of two or more nanocrystalline semiconductors allows the design of semiconductors heterostructures that are potentially useful in photocatalysis, water splitting, microelectronics, and others (Kamat 1997; Beydoun et al. 1999). The advantage of using associations of semiconductors is twofold: (1) to extend the photoresponse by coupling a large band gap semiconductor with a short band gap semiconductor and (2) to retard the recombination of photogenerated charge carriers by injecting electrons into the lower lying conduction band of the large band-gap semiconductor.

There are two principal configurations: coupled and capped semiconductors heterostructures. The principle of charge separation in both structures is displayed in Fig. 2.7a, b.

While the mechanism of charge separation in coupled and capped semiconductors is similar, the interfacial charge transfer is notably different. In a coupled semiconductor system, the particles are in contact with each other and the holes and electrons are available for oxidation or reduction reactions on the surface of different particles. In capped semiconductors with a core shell structure, a charge rectification takes place, and only one of the charge carriers is accessible at the surface; the other charge carrier gets trapped inside the inner semiconductor (Fig. 2.7).

Coupled semiconductors can be prepared by combining colloid systems avoiding heterocoagulation. Nanostructure films of SnO_2/TiO_2 were prepared by deposition of SnO_2 water-based sol on conductive glass followed by deposition of a TiO_2 sol. The coupling of these two wide band gap semiconductors allowed a better charge separation. The CB of SnO_2 (0.0 V vs NHE at pH 7) is more positive than that of TiO_2 (−0.5 V vs NHE at pH 7) and can trap the photogenerated electron. On the other hand, the holes move to the opposite direction and accumulate on TiO_2 particles improving charge separation. This system displayed enhanced photocatalytic activity for acid orange 7 (Vinodgopal and Kamat 1995). In the same direction, the coupling of ZnO with SnO_2 resulted in an improvement of the photocatalytic efficiency of SnO_2 under UV illumination for the degradation of eosin Y (Tanasa et al. 2013).

By coupling small band gap semiconductors with large band gap semiconductors, it is possible to get hole-electron separation using longer wavelength. In the case of CdS/TiO_2, photocurrent generation was observed in the 400–600 nm range. The photogenerated electrons can be injected from CdS to TiO_2 particles, while the holes remain in the CdS particles (Kohtani et al. 1993). The synthesis by chemical route of self-assembled flowerlike CdS/ZnO nanocomposites was recently reported (Jana et al. 2014). Using this coupled system as photocatalyst, the photodegradation of rhodamine B in aqueous solution was enhanced, thanks to the efficiency of the coupled charge transfer mechanism. Figure 2.8 shows a scheme of the process and the synthesized particles. A ternary system containing Ag-ZnO-CdO was prepared by a coprecipitation method. The system was formed by hexagonal plate-like ZnO

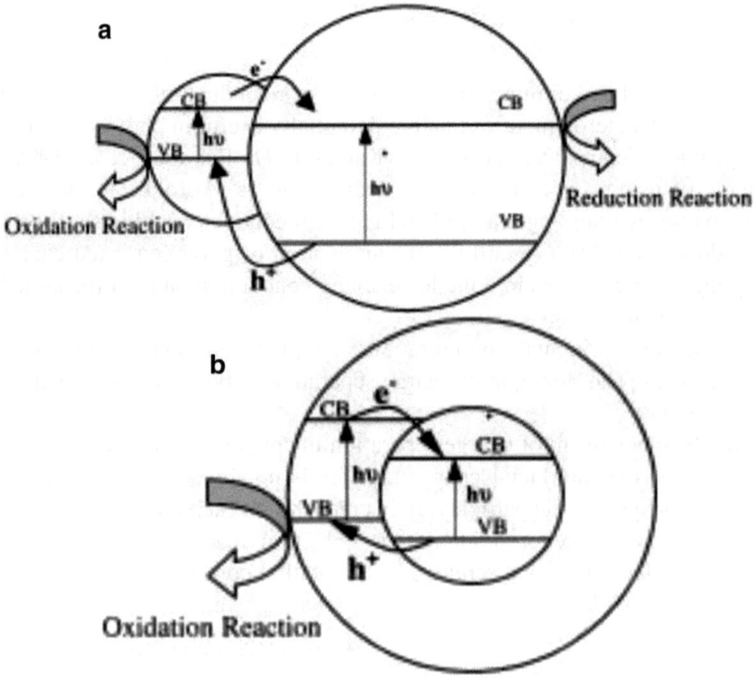

Fig. 2.7 Photoinduced charge separation in semiconductor structure: (**a**) coupled; (**b**) capped semiconductor nanocrystallites (From Beydoun et al. 1999)

structures, containing CdO and Ag clusters on its smooth surfaces. This system had an increased adsorption in the UV and visible region than ZnO prepared in similar way and displayed enhanced photocatalytic activity for the degradation of acid black 1 and acid violet 7 under sunlight (Balachandran et al. 2014).

Another ternary composite was successfully prepared by loading ZnO and CdS nanoparticles in graphene sheets via a one-step hydrothermal method. Photoelectrochemical experiments revealed that the ternary CdS/ZnO/graphene composite exhibits enhanced photocatalytic activity compared with the matrix binary composites and pure ZnO and CdS (Han et al. 2014). Mixed oxide coupled with oxide also shows some improved properties. ZnO-$ZnWO_4$ nanocomposites prepared by sol–gel method displayed higher photoactivity for the degradation of 4-nitrophenol than ZnO and $ZnWO_4$. The high efficiency of the mixed samples was likely due to an improved charge separation as consequence of the relative values of their conduction and valence bands, as shown in Fig. 2.9 (Hamrouni et al. 2014). The coupling of graphene with $BiFeO_3$ also displays photocatalytic enhanced properties under visible-light illumination. The coupling between $BiFeO_3$ nanoparticles and graphene was probably achieved by the formation of Fe-O-C bonds, as determined by XPS and Raman analysis (Li et al. 2014c).

The preparation of capped semiconductors involves the synthesis of the inner core semiconductor colloid of desire size, followed by the controlled deposition of a

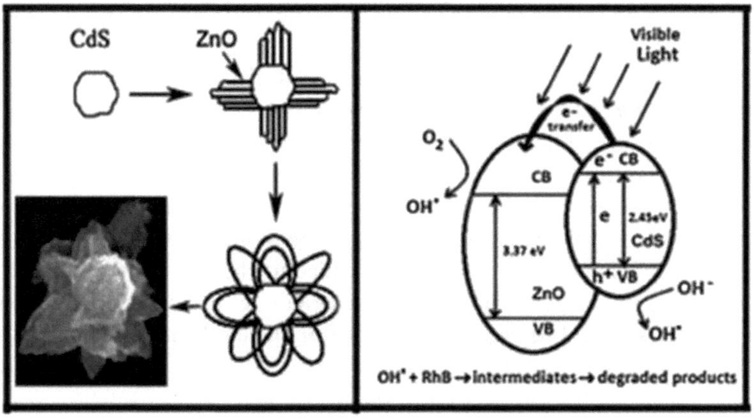

Fig. 2.8 Scheme showing the structure of the flower-shaped particles and the charge transfer process between the coupled semiconductors (From Jana et al. 2014)

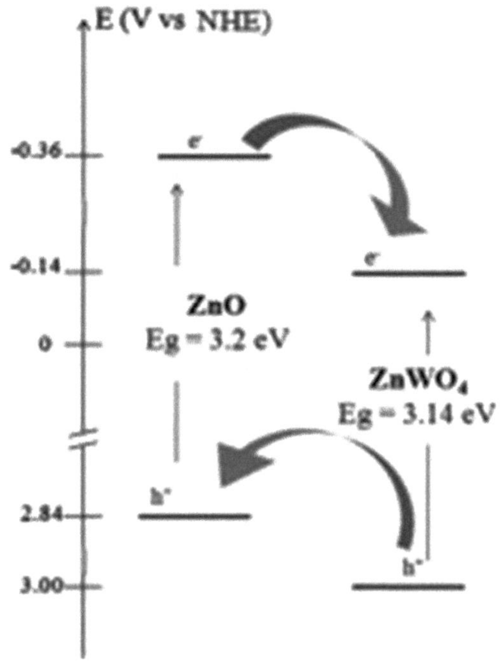

Fig. 2.9 Energy level diagram showing the direction of electron and hole transfer in $ZnO/ZnWO_4$ system (From Hamrouni et al. 2014)

second semiconductor layer. For example, for the synthesis of $TiO_2@SnO_2$, titanium isopropoxide was added dropwise to a SnO_2 water-based colloidal suspension. As the particles became coated with TiO_2, a precipitate containing the capped particles settled down. The thickness of the TiO_2 layer could be controlled through the total amount of added titanium isopropoxide. Due to the electronic band structure, upon UV excitation, the photogenerated electrons accumulate at an inner core of SnO_2, while holes accumulate at the TiO_2 surface. The holes can be rapidly scavenge away by easily oxidizable species like adsorbed I^- anions. The relative quantum efficiency for I^- oxidation at SnO_2@TiO_2 was two to three times higher than SiO_2@TiO_2 or TiO_2 alone (Kamat 1997).

Although the use of capped and coupled semiconductors present interesting and promising properties that may enhance the photocatalytic efficiency of these systems when compared with the parent semiconductors, it should be considered that the accumulation of holes or electrons may lead to the photocorrosion of one or both materials. Photocorrosion is a major concern when working with potentially toxic materials such as CdS and CdO, which can release soluble species to water. These materials should be carefully tested before using them as photocatalysts.

2.5 Nanostructured Semiconductors: Effect of Size and Composition

Since the discovery of heterogeneous photocatalysis (Honda and Fujishima 1972), numerous efforts have been carried out to understand the mechanism of reaction and how to select the main variables that affect its efficiency. Several authors have associated the efficiency of a semiconductor photocatalyst with electronic, structural, and morphological properties of the material such as the band-gap energy (E_g), crystalline structure, surface area, and morphology. The possibility of controlling these variables is very interesting from a technological point of view because the development of materials with high photocatalytic activity allows their application to environmental fields such as wastewater treatment (Malato et al. 2009).

In the search of semiconductor materials with photocatalytic activity under visible-light irradiation, important efforts have been carried out in the last decade. For example, the TiO_2 anatase polymorph has been doped with nitrogen in order to increase its absorption in the visible range (Yang and Gao 2004). This limitation has attracted the synthesis of new visible-light-driven photocatalysts in recent years. Some compounds such as Bi_2O_3, WO_3, Ag_3VO_4, Bi_2WO_6, and $BiVO_4$ have been recently studied because they have shown visible-light-driven catalytic activity (Eberl and Kisch 2008; Kominami et al. 2003; Konta et al. 2003; Shang et al. 2008; García-Pérez et al. 2012).

In this sense, the search of new photocatalysts with an important improvement in its photocatalytic activity has been focused in the synthesis of semiconductors for a great variety of chemical methods. For semiconductor oxides such as ZnO, $BiVO_4$, and $PbMoO_4$, several experimental methods were reported to prepare the materials with specific morphological and textural properties. The use of wet chemical methods, such as sol–gel, coprecipitation, hydrothermal, microwave, and sonochemical in the presence or absence of organic additives, has been reported as a successful way to develop nanostructured semiconductors. Frequently, the

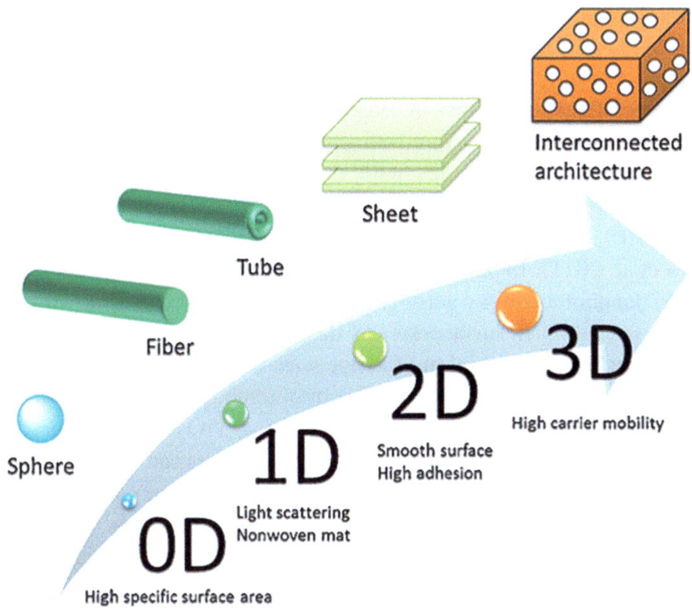

Fig. 2.10 Dimensionality of TiO_2 nanostructure related with the photocatalytic activity of the material. Taken from Nakata and Fujishima (2012)

solid-state reaction method employed to obtain the solid compounds leads to materials with poor photocatalytic activity due to the small surface area developed. The so-called soft chemistry methods have shown to be efficient to prepare better catalysts than those synthesized by classical methods.

The development of nanostructured semiconductors for catalytic applications provides materials with a high value in their surface/volume ratio, but maintaining the physical properties of the bulk material and at the same time including new physical properties due to the relation between surface and volume of the material (Anta 2012). In particular, due to small size of the particles, quantum effect affects directly some physical properties of material such as optical and electronic properties (Nozik et al. 2010). Such situation has opened numerous field of technological application of these materials, which include the development of new generations of solar cells, production of hydrogen, and elimination of pollutants from wastewaters.

From a point of view of photocatalytic applications, the small size of particles of nanostructured semiconductors plays an important role in the activity of photocatalysts. As is well known, the critical step in the heterogeneous photocatalysis is the undesirable recombination of pair electron–hole before they reach the catalyst surface. In this sense, the small size of prepared particles reduces the pathway from the site of generation of the pair electron–hole to the catalyst surface, as is shown in Fig. 2.10.

Among semiconductors with high activity, the anatase polymorph of TiO_2 is the compound most extensively synthesized in nanoscale size. In an interesting review, Nakata and Fujishima (2012) describe how the dimensionality of TiO_2 nanostructures affects its physical properties such as surface area, adsorption, reflectance, adhesion, and carrier transportation properties. They reviewed the formation of TiO_2 nanostructures with different shapes, which include spheres (Chen et al. 2011), nanorods (Yun et al. 2009), fibers (Cheng et al. 2010), tubes (Yu et al. 2010), sheets (Aoyama et al. 2012), and interconnected architectures (Hasegawa et al. 2011). In particular, the dimensionality of TiO_2 nanostructure can be related to the photocatalytic activity of the material, as can be shown in Fig. 2.10.

For example, a higher surface area is desirable in a heterogeneous catalyst, in order to get a higher concentration of active sites where a molecule of pollutant can be anchored to undergo its decomposition by interchange of carriers coming from excitation of semiconductor. To reach this condition, TiO_2 nanostructured with zero dimensionality due to the shape of sphere of its particles can be developed. On the other hand, the development of one-dimensional fiber or nanotube structures promotes a less recombination of pair hole-electron due to the short distance between the place of its generation and the surface of material where pollutant species react. The application of nanostructured TiO_2 for wastewater treatment usually includes the elimination of organic compounds with large and complex molecular structure. For this reason, steric effects in the adsorption of molecules of organic pollutants over the surface of photocatalyst should be taken into account. In these sense, two-dimensional nanosheets have smooth surfaces which leads to a high adhesion of organic molecule. Finally, the presence of channels interconnected in three-dimensional structures promotes a high mobility of carriers to reach the surface of photocatalyst and, then, induce oxidation or reduction reactions.

Zhang et al. (2014) have prepared nanostructured TiO_2 hollow fiber photocatalytic membranes via a facile spinning-sintering method. The photocatalytic activity, permeability, and separation efficiency of the membranes were evaluated using both acid orange 7 and raw sewage as pollutants. For a series of prepared TiO_2 membranes, the calcination temperature was the factor that ruled the photocatalytic activity. This fact had a direct relation with the physical properties of the samples, including the morphology and crystal phase.

Nano-sized TiO_2 for photocatalytic applications also have been synthesized by solvothermal methods under supercritical conditions (Wang et al. 2013). Nanostructured TiO_2 samples were prepared by using a mixture of tetrabutyl orthotitanate (TBOT) and ethanol in supercritical CO_2. The photocatalytic activity of the samples tested with methyl orange was strongly influenced by its corresponding morphology. In particular, the activity of the samples decreased with the increasing particle size.

The synthesis of nanostructures of TiO_2 has been successfully applied for photocatalytic coatings for corrosion protection and surface repair (He et al. 2013). These were developed by the incorporation of a corrosion inhibitor, 8-hydroxyquinoline (8-HQ), into TiO_2 nanoparticles.

The formation of nanostructures has been extended to materials with good performance to absorb in the region visible of solar spectrum. Min et al. have reported the synthesis of nanostructured ZnO/Bi_2WO_6 heterojunction for photocatalytic applications (Min et al. 2012). The formation of this type of heterojunction nanostructure promotes an efficient charge separation and consequently an important improvement of the photocatalytic activity of the material with respect to the pristine oxides.

In the same way, nanostructured WO_3/Bi_2WO_6 heterojunction was prepared by one-step hydrothermal method with high visible-light photocatalytic activity for degradation of rhodamine B (Gui et al. 2012).

Concluding Remarks

The number of pure and doped metal oxide semiconductors available for photocatalysis and other applications that use UV or visible light as a source of energy had notably increased during the last ten years. The force that drives this situation is the need to systematically count hydrogen production or pollution control based on the use of solar energy.

The most studied SC oxides with photocatalytic activity other than TiO_2 are ZnO, WO_3, Ta_2O_5, Nb_2O_5, Bi_2O_3, CdO, and SnO_2. Ternary systems such as bismutates, vanadates, molybdates, and niobates are being increasingly prepared and studied for photocatalytic applications in both pristine and doped form. The great development in SC metal oxide was helped by the advances in wet chemistry synthesis that allows the production of nanostructured powders and films in a relatively economic way. Not only composition can be controlled by an appropriate synthesis design but also the shape, size, and dimensionality of the particles that constitute the powders or the films. Nanostructured materials (powders or films) are essential to reach photoefficiencies higher enough to be applied in industrial processes. The dimensionality of the particles also has a crucial role in the mobility of charge carriers.

Different strategies were used to shift the absorption spectrum to longer wavelength with different results. In some cases, the doping with metal ion was very successful, but in others hole-electron recombination was increased, with a diminution in photocatalytic efficiency. Nonmetal doping increases the absorption of visible light, but the oxidation power of the photogenerated holes is lower than when the system is illuminated by UV light. Sensitization with dyes is successful in the use of visible light, but the dyes are finally degraded after long-term illumination, which suggest that is a good method to eliminate colored pollutants but not very useful for hydrogen production or energy conversion. Coupling of SC is a very interesting possibility, but care should be taken to avoid photocorrosion and dissolution of toxic metals into water. At present, there is not a unique strategy that works for all the systems.

(continued)

(continued)

Each case should be analyzed to find the best approach to improve the use of solar light for water or air remediation, hydrogen production, or energy conversion.

The general trend indicates an improvement in the development of SC materials with high photocatalytic efficiency under visible-light illumination and a better understanding of the physicochemical processes involved in photocatalysis. This knowledge will undoubtedly help to increase the photoefficiency of the new materials that will be designed in the near future.

References

Abe R, Sayama K, Arakawa H (2004) Dye-sensitized photocatalysts for efficient hydrogen production from aqueous I^- solution under visible light irradiation. J Photochem Photobiol A 166:115–122

Ahmad M, Ahmed E, Zhang Y, Khalid NR, Xu J, Ullah M, Hong Z (2013) Preparation of highly efficient Al-doped ZnO photocatalyst by combustion synthesis. Curr Appl Phys 13:697–704

Anta JA (2012) Electron transport in nanostructured metal-oxide semiconductors. Curr Opin Colloid Interface Sci 17:124–131

Aoyama Y, Oaki Y, Ise R, Imai H (2012) Mesocrystal nanosheet of rutile TiO_2 and its reaction selectivity as a photocatalyst. Cryst Eng Comm 14:1405–1411

Balachandran S, Praveen SG, Velmurugan R, Swaminathan M (2014) Facile fabrication of highly efficient, reusable heterostructured Ag–ZnO–CdO and its twin applications of dye degradation under natural sunlight and self-cleaning. RSC Adv 4:4353–4362

Bamwenda GR, Arakawa H (2001) The visible light induced photocatalytic activity of tungsten trioxide powders. Appl Catal A 210:181–191

Barnes TM, Olson K, Wolden CA (2005) On the formation and stability of p-type conductivity in nitrogen-doped zinc oxide. Appl Phys Lett 86:112112

Beydoun D, Amal R, Low G, McEvoy S (1999) Role of nanoparticles in photocatalysis. J Nanopart Res 1:439–458

Calogero G, Di Marco G (2008) Red Sicilian orange and purple eggplant fruits as natural sensitizers for dye-sensitized solar cells. Sol Energy Mater Sol Cells 92:1341–1346

Chatterjee D, Dasgupta S (2005) Visible light induced photocatalytic degradation of organic pollutants. J Photochem Photobiol C 6:186–205

Chen YS, Li C, Zeng ZH, Wang WB, Wang XS, Zhang BW (2005) Efficient electron injection due to a special adsorbing group's combination of carboxyl and hydroxyl: dye-sensitized solar cells based on new hemicyanine dyes. J Mater Chem 15:1654–1661

Chen JS, Chen C, Liu J, Xu R, Qiao SZ, Lou XW (2011) Ellipsoidal hollow nanostructures assembled from anatase TiO_2 nanosheets as a magnetically separable photocatalyst. Chem Commun 47:2631–2633

Chen L, Tran TTT, Huang C, Li J, Yuan L, Cai Q (2013) Synthesis and photocatalytic application of Au/Ag nanoparticle-sensitized ZnO films. Appl Surf Sci 273:82–88

Cheng XF, Leng WH, Liu DP, Zhang JQ, Cao CN (2007) Enhanced photoelectrocatalytic performance of Zn-doped WO_3 photocatalysts for nitrite ions degradation under visible light. Chemosphere 68:1976–1984

Cheng Y, Huang W, Zhang Y, Zhu L, Liu Y, Fan X, Cao X (2010) Preparation of TiO_2 hollow nanofibers by electrospinning combined with sol-gel process. Cryst Eng Comm 12:2256–2260

Chu W, Chan KH, Jafvert CT, Chan YS (2007) Removal of phenylurea herbicide monuron via riboflavin-mediated photosensitization. Chemosphere 69:177–183

Di Valentin C, Pacchioni G, Selloni A, Livraghi S, Giamello E (2005) Characterization of paramagnetic species in N-doped TiO_2 powders by EPR spectroscopy and DFT calculations. J Phys Chem B 109:11414–11419

Dou C, Wen P, Kong X, Nakanishi S, Feng Q (2011) The nonlinear refraction sign turned to reverse by intercalating cresyl violent dye into layered titanate nanosheets. Opt Commun 284:1067–1071

Eberl J, Kisch H (2008) Visible light photo-oxidations in the presence of α-Bi_2O_3. Photochem Photobiol Sci 7:1400–1406

Espinosa R, Zumeta I, Santana JL, Martínez-Luzardo F, González B, Docteur S, Vigil E (2005) Nanocrystalline TiO_2 photosensitized with natural polymers with enhanced efficiency from 400 to 600 nm. Sol Energ Mat Sol C 85:359–369

Fu H, Zhang S, Xu T, Zhu Y, Chen J (2008a) Photocatalytic degradation of RhB by fluorinated Bi_2WO_6 and distribution of the intermediate products. Environ Sci Technol 42:2085–2091

Fu H, Zhang S, Zhang L, Zhu Y (2008b) Visible-light-driven $NaTaO_{3-x}N_x$ catalyst prepared by a hydrothermal process. Mater Res Bull 43:864–872

García-Pérez UM, Sepúlveda-Guzmán S, Martínez-de la Cruz A (2012) Nanostructured $BiVO_4$ photocatalysts synthesized via a polymer-assisted coprecipitation method and their photocatalytic properties under visible-light irradiation. Solid State Sci 14:293–298

Gómez-Ortíz NM, Vázquez-Maldonado IA, Pérez-Espadas AR, Mena-Rejón GJ, Azamar-Barrios JA, Oskam G (2010) Dye-sensitized solar cells with natural dyes extracted from achiote seeds. Sol Energ Mat Sol C 94:40–44

Gui MS, Zhang WD, Chang YQ, Yu YX (2012) One-step hydrothermal preparation strategy for nanostructured WO_3/Bi_2WO_6 heterojunction with high visible light photocatalytic activity. Chem Eng J 197:283–288

Guo M, Diao P, Ren YJ, Meng F, Tian H, Cai SM (2005) Photoelectrochemical studies of nanocrystalline TiO_2 co-sensitized by novel cyanine dyes. Sol Energy Mater Sol Cells 88:23–35

Hameed A, Gondal MA, Yamani ZH (2004) Effect of transition metal doping on photocatalytic activity of WO_3 for water splitting under laser illumination: role of 3d-orbitals. Catal Commun 5:715–719

Hamrouni A, Moussa N, Di Paola A, Parrino F, Houas A, Palmisano L (2014) Characterization and photoactivity of coupled ZnO–$ZnWO_4$ catalysts prepared by a sol-gel method. Appl Catal B Environ 154–155:379–385

Han W, Ren L, Qi X, Liu Y, Wei X, Huang Z, Zhong J (2014) Synthesis of CdS/ZnO/graphene composite with high-efficiency photoelectrochemical activities under solar radiation. Appl Surf Sci 299:12–18

Hasegawa G, Morisato K, Kanamori K, Nakanishi K (2011) New hierarchically porous titania monoliths for chromatographic separation media. J Sep Sci 34:3004–3010

He X, Chiu C, Esmacher MJ, Liang H (2013) Nanostructured photocatalytic coatings for corrosion protection and surface repair. Surf Coat Technol 237:320–327

Honda K, Fujishima A (1972) Electrochemical photolysis of water at a semiconductor electrode. Nature 238:37–38

Hong SJ, Jun H, Borse PH, Lee JS (2009) Size effects of WO_3 nanocrystals for photooxidation of water in particulate suspension and photoelectrochemical film systems. Int J Hydrogen Energy 34:3234–3242

Jaimy KB, Ghosh S, Sankar S, Warrier KGK (2011) An aqueous sol-gel synthesis of chromium (III) doped mesoporous titanium dioxide for visible light photocatalysis. Mater Res Bull 46:914–921

Jana TK, Pal A, Chatterjee K (2014) Self assembled flower like CdS–ZnO nanocomposite and its photo catalytic activity. J Alloys Compd 583:510–515

Jiang H, Wang Q, Zang S, Li J, Wang X (2014) Hydrothermal synthesis of high-efficiency Pr, N, P-tridoped TiO_2 from $TiCl_4$ hydrolysis and mechanism of its enhanced photoactivity. J Alloys Compd 600:34–42

Jindal K, Tomar M, Gupta V (2013) Nitrogen-doped zinc oxide thin films biosensor for determination of uric acid. Analyst 138:4353–4362

Kamat P (1997) Native and surface modified semiconductor nanoclusters, Progress in inorganic chemistry. In: Karlin D (ed) Progress in Inorganic Chemistry: Molecular Level Artificial Photosynthetic Materials, vol 44. John Wiley and Son, Hoboken, NJ, pp 273–343. ISBN ISBN 0-471-12535-0

Kato H, Hori M, Konta R, Shimodaira Y, Kudo A (2004) Construction of Z-scheme type heterogeneous photocatalysis systems for water splitting into H_2 and O_2 under visible light irradiation. Chem Lett 33:1348–1349

Kaur S, Singh V (2007) Visible light induced sonophotocatalytic degradation of Reactive Red dye 198 using dye sensitized TiO_2. Ultrason Sonochem 14:531–537

Khatamian M, Khandar AA, Divband B, Haghighi M, Ebrahimiasl S (2012) Heterogeneous photocatalytic degradation of 4-nitrophenol in aqueous suspension by Ln (La^{3+}, Nd^{3+} or Sm^{3+}) doped ZnO nanoparticles. J Mol Catal A Chem 365:120–127

Kohtani S, Kudo A, Sakata T (1993) Spectral sensitization of a TiO_2 semiconductor electrode by CdS microcrystals and its photoelectrochemical properties. Chem Phys Lett 206:166–170

Kominami H, Kato J, Murakami S, Ishii Y, Kohno M, Yabutani K, Yamamoto T, Kera Y, Inoue M, Inui T, Ohtani B (2003) Solvothermal syntheses of semiconductor photocatalysts of ultra-high activities. Catal Today 84:181–189

Konta R, Kato H, Kobayashi H, Kudo A (2003) Photophysical properties and photocatalytic activities under visible light irradiation of silver vanadates. Phys Chem Chem Phys 5:3061–3065

Kudo A, Omori K, Kato H (1999) A novel aqueous process for preparation of crystal form-controlled and highly crystalline $BiVO_4$ powder from layered vanadates at room temperature and its photocatalytic and photophysical properties. J Am Chem Soc 121:11459–11467

Kumara GRA, Kaneko S, Okuya M, Onwona-Agyeman B, Konno A, Tennakone K (2006) Shiso leaf pigments for dye-sensitized solid-state solar cell. Sol Energy Mat Sol C 90:1220–1226

Lehr D, Luka M, Wagner MR, Bügler M, Hoffmann A, Polarz S (2012) Band gap engineering of zinc oxide colloids via lattice substitution with sulfur leading to materials with advanced properties for optical applications like full inorganic UV protection. Chem Mater 24:1771–1778

Li X, Kikugawa N, Ye J (2008) Nitrogen-doped lamellar niobic acid with visible light responsive photocatalytic activity. Adv Mater 20:3816–3819

Li K, Yang C, Xu Y, Ying D, Wanga Y, Jia J (2012) Effect of inorganic anions on Rhodamine B removal under visible light irradiation using Bi_2O_3/Ti rotating disk reactor. Chem Eng J 211–212:208–221

Li D, Huang JF, Cao LY, LI JY, OuYang HB, Yao CY (2014a) Microwave hydrothermal synthesis of Sr^{2+} doped ZnO crystallites with enhanced photocatalytic properties. Ceram Int 40:2647–2653

Li T, Chen Y, Fu WF (2014b) Photocatalytic H_2 production from water based on platinum (II) Schiff base sensitizers and a molecular cobalt catalyst. Catal Commun 45:91–94

Li Z, Shen Y, Guan Y, Hu Y, Lin Y, Nan CW (2014c) Bandgap engineering and enhanced interface coupling of graphene–$BiFeO_3$ nanocomposites as efficient photocatalysts under visible light. J Mater Chem A 2:1967–1973

Liu H, Peng T, Xiao J, Zhao D, Peng Z (2005) Preparation and photocatalytic activity of nanoscale Tb^{3+}-doped WO_3. J Wuhan Univ (Mater Sci Ed) 4:397–401

Liu H, Peng T, Ke D, Peng Z, Yan C (2007) Preparation and photocatalytic activity of dysprosium doped tungsten trioxide nanoparticles. Mater Chem Phys 104:377–383

Liu S, Yin K, Ren W, Cheng B, Yu J (2012) Tandem photocatalytic oxidation of Rhodamine B over surface fluorinated bismuth vanadate crystals. J Mater Chem 22:17759–17767

Liu L, Lv J, Xu G, Wang Y, Xie K, Chen Z, Wu Y (2013) Uniformly dispersed CdS nanoparticles sensitized TiO_2 nanotube arrays with enhanced visible-light photocatalytic activity and stability. J Solid State Chem 208:27–34

Lopes OF, Paris EC, Ribeiro C (2014) Synthesis of Nb_2O_5 nanoparticles through the oxidant peroxide method applied to organic pollutant photodegradation: A mechanistic study. Appl Catal B Environ 144:800–808

Maia DLS, Pepe I, Ferreira da Silva A, Silva LA (2012) Visible-light-driven photocatalytic hydrogen production over dye-sensitized β-$BiTaO_4$. J Photochem Photobiol A:Chem 243:61–64

Malato S, Fernández-Ibánez P, Maldonado MI, Blanco J, Gernjak W (2009) Decontamination and disinfection of water by solar photocatalysis: Recent overview and trends. Catal Today 147:1–59

Marschall R, Wang L (2014) Non-metal doping of transition metal oxides for visible light photocatalysis. Catal Today 225:111–135

Meyer BK, Polity A, Farangis B, He Y, Hasselkamp D, Krämer T, Wang C (2004) Structural properties and bandgap bowing of ZnO_{1-x} S_x thin films deposited by reactive sputtering. Appl Phys Lett 85:4929–4931

Mi Q, Ping Y, Li Y, Cao B, Brunschwig BS, Khalifah PG, Galli GA, Gray HB, Lewis NS (2012) Thermally stable N_2-intercalated WO_3 photoanodes for water oxidation. J Am Chem Soc 134:18318–18324

Min Y, Zhang K, Chen Y, Zhang YG, Zhao W (2012) Synthesis of nanostructured ZnO/Bi_2WO_6 heterojunction for photocatalysis application. Sep Purif Technol 92:115–120

Murase T, Irie H, Hashimoto K (2004) Visible light sensitive photocatalysts, nitrogen-doped Ta_2O_5 powders. J Phys Chem B 108:15803–15807

Nakata K, Fujishima A (2012) TiO_2 photocatalysis: design and applications. J Photochem Photobiol C 13:169–189

Narayan MR (2012) Review: Dye sensitized solar cells based on natural photosensitizers. Renewable Sustainable Energy Rev 16:208–215

Nozik AJ, Beard MC, Luther JM, Law M, Ellingson RJ, Johnson JC (2010) Semiconductor quantum dots and quantum dot arrays and applications of multiple exciton generation to third-generation photovoltaic solar cells. Chem Rev 110:6873–6890

O'Regan B, Grätzel M (1991) A low-cost, high-efficiency solar cell based on dye-sensitized colloidal TiO_2 films. Nature 353:737–740

Omidi A, Habibi-Yangjeh A, Pirhashemi M (2013) Application of ultrasonic irradiation method for preparation of ZnO nanostructures doped with Sb^{+3} ions as a highly efficient photocatalyst. Appl Surf Sci 276:468–475

Patil A, Patil K, Pardeshi S (2010) Ecofriendly synthesis and solar photocatalytic activity of S-doped ZnO. J Hazard Mater 183:315–323

Pei D, Luan J (2012) Development of visible light-responsive sensitized photocatalysts. Int J Photoenergy ID 262831:13

Persson C, Platzer-Bjorkman C, Malmstrom J, Torndahl T, Edoff M (2006) Strong valence band offset bowing of $ZnO_{1-x}S_x$ enhances p-type nitrogen doping of ZnO like alloys. Phys Rev Lett 97:146403

Phuruangrat A, Yayapao O, Thongtem T, Thongtem S (2014) Preparation, characterization and photocatalytic properties of Ho doped ZnO nanostructures synthesized by sonochemical method. Superlattices Microstruct 67:118–126

Radecka M, Sobas P, Wierzbicka M, Rekas M (2005) Photoelectrochemical properties of undoped and Ti-doped WO_3. Physica B 364:85–92

Rengifo-Herrera JA, Mielczarski E, Mielczarski J, Castillo NC, Kiwi J, Pulgarin C (2008) Escherichia coli inactivation by N, S co-doped commercial TiO_2 powders under UV and visible light. Appl Catal B Environ 84:448–456

Ropero-Vega JL, Rosas-Barrera KL, Pedraza-Avella JA, Laverde-Cataño DA, Pedraza-Rosas JE, Niño-Gómez ME (2010) Photophysical and photocatalytic properties of Bi_2MNbO_7 (M = Al, In, Ga, Fe) thin films prepared by dip-coating. Mater Sci Eng B 174:196–199

San Esteban ACM, Enriquez EP (2013) Graphene-anthocyanin mixture as photosensitizer for dye-sensitized solar cell. Sol Energy 98:392–399

Saravanakumar B, Mohan R, Thiyagarajan K, Kim SJ (2013) Investigation of UV photoresponse property of Al, N co-doped ZnO film. J Alloys Compd 580:538–543

Sayama K, Arakawa H (1994) Effect of Na_2CO_3 addition on photocatalytic decomposition of liquid water over various semiconductor catalysis. J Photochem Photobiol A:Chem 77:243–247

Shang M, Wang W, Sun S, Zhou L, Zhang L (2008) Bi_2WO_6 nanocrystals with high photocatalytic activities under visible light. J Phys Chem C 112:10407–10411

Sreethawong T, Yoshikawa S (2012) Impact of photochemically deposited monometallic Pt and bimetallic Pt-Au nanoparticles on photocatalytic dye-sensitized H_2 production activity of mesoporous-assembled TiO_2-SiO_2 mixed oxide nanocrystal. Chem Eng J 197:272–282

Sun T, Liu E, Fan J, Hu X, Wu F, Hou W, Yang Y, Kang L (2013) High photocatalytic activity of hydrogen production from water over Fe doped and Ag deposited anatase TiO_2 catalyst synthesized by solvothermal method. Chem Eng J 228:896–906

Takeuchi M, Shimizu Y, Yamagawa H, Nakamuro T, Anpo M (2011) Preparation of the visible light responsive N^{3-}-doped WO_3 photocatalyst by a thermal decomposition of ammonium paratungstate. Appl Catal B:Environ 110:1–5

Takizawa T, Watanabe T, Honda K (1978) Photocatalysis through excitation of adsorbates. 2. A comparative study of Rhodamine B and methylene blue on cadmium sulfide. J Phys Chem 82:1391–1396

Tanasa DE, Piuleac CG, Curteanu S, Popovici E (2013) Photodegradation process of Eosin Y using ZnO/SnO_2 nanocomposites as photocatalysts: experimental study and neural network modeling. J Mater Sci 48:8029–8040

Tang JW, Wang DF, Zou ZG, Ye JH (2003) Modification of photophysical properties of WO_3 by doping different metals. Mater Sci Forum 423–424:163–166

Thankalekshmi RR, Rastogi AC (2012) Structure and optical band gap of $ZnO_{12x}S_x$ thin films synthesized by chemical spray pyrolysis for application in solar cells. J Appl Phys 112: 063708-063708-10

Tuomisto F, Rauch C, Wagner MR, Hoffmann A, Eisermann S, Meyer BK, Kilanski L, Tarun MC, McCluskey MD (2013) Nitrogen and vacancy clusters in ZnO. J Mater Res 28:1977–1983

Tuzemen ES, Kara K, Elagoz S, Takci DK, Altuntas I, Esen R (2014) Structural and electrical properties of nitrogen-doped ZnO thin films. Appl Surf Sci http://dx.doi.org 10.1016/j.apsusc.2014.02.118

Ullah R, Sun H, Ang HM, Tade MO, Wang S (2013) Comparative investigation of photocatalytic degradation of toluene on nitrogen doped Ta_2O_5 and Nb_2O_5 nanoparticles. Ind Eng Chem Res 52:3320–3328

Vinodgopal K, Kamat P (1995) Enhanced rates of photocatalytic degradation of an azo dye using SnO_2/TiO_2 coupled semiconductor thin films. Environ Sci Technol 29:841–845

Vo V, Thi TPT, Kim HY, Kim SJ (2014) Facile post-synthesis and photocatalytic activity of N-doped ZnO–SBA-15. J Phys Chem Solids 75:403–409

Wang C, Cao L (2011) Preparation, spectral characteristics and photocatalytic activity of Eu^{3+}-doped WO_3 nanoparticles. J Rare Earths 29:727–731

Wang M, Chen C, Zhao B, Zeng Q, He D (2013) Solvothermal synthesis of nanostructured TiO_2 photocatalyst in supercritical CO_2 fluids. Mater Lett 109:104–107

Wang Z, Wang ZW, Yue Y, Cao Y (2014) Preparation and properties of nitrogen doped p-type zinc oxide films by reactive magnetron sputtering. Vacuum 101:313–316

Whitehead K, Hedges JI (2005) Photodegradation and photosensitization of mycosporine-like amino acids. J Photochem Photobiol B 80:115–121

Wongcharee K, Meeyoo V, Chavadej S (2007) Dye-sensitized solar cell using natural dyes extracted from rosella and blue pea flowers. Sol Energy Mater Sol Cells 91:566–571
Xu H, Zhou YN, Lu F, Fu ZW (2011) Electrochemistry of $ZnO_{1-x}S_x$ thin film with lithium. J Electrochem Soc 158:A285–A290
Xu L, Shi W, Guan J (2012) Preparation of crystallized mesoporous CdS/Ta_2O_5 composite assisted by silica reinforcement for visible light photocatalytic hydrogen evolution. Catal Commun 25:54–58
Yang S, Gao L (2004) New method to prepare nitrogen-doped titanium dioxide and its photocatalytic activities irradiated by visible light. J Am Ceram Soc 87:1803–1805
Yayapao O, Thongtem S, Phuruangrat A, Thongtem T (2013a) Sonochemical synthesis, photocatalysis and photonic properties of 3 % Ce-doped ZnO nanoneedles. Ceram Int 39: S563–S568
Yayapao O, Thongtem T, Phuruangrat A, Thongtem S (2013b) Ultrasonic-assisted synthesis of Nd-doped ZnO for photocatalysis. Mater Lett 90:83–86
Yoo YZ, Jin ZW, Chikyow T, Fukumura T, Kawasaki M, Koinuma H (2002) S doping in ZnO film by supplying ZnS species with pulsed laser deposition methods. Appl Phys Lett 81:3798–3790
Yu J, Dai G, Cheng B (2010) Effect of crystallization methods on morphology and photocatalytic activity of anodized TiO_2 nanotube array films. J Phys Chem C 114:19378–19385
Yun HJ, Lee H, Joo JB, Kim W, Yi J (2009) Influence of aspect ratio of TiO_2 nanorods on the photocatalytic decomposition of formic acid. J Phys Chem C 113:3050–3055
Zhang L, Fu H, Zhang C, Zhu Y (2006) Synthesis, characterization, and photocatalytic properties of $InVO_4$ nanoparticles. J Solid State Chem 179:804–811
Zhang D, Gong J, Ma JJ, Han G, Tong Z (2013) A facile method for synthesis of N-doped ZnO mesoporous nanospheres and enhanced photocatalytic activity. Dalton Trans 42:16556–16561
Zhang X, Wang DK, Schmeda López DR, Diniz da Costa JC (2014) Fabrication of nanostructured TiO_2 hollow fiber photocatalytic membrane and application for wastewater treatment. Chem Eng J 236:314–322
Zhao Z, Dai H, Deng J, Liu Y, Au CT (2013) Enhanced visible-light photocatalytic activities of porous olive-shaped sulfur-doped $BiVO_4$-supported cobalt oxides. Solid State Sci 18:98–104
Zheng YZ, Tao X, Hou Q, Wang DT, Zhou WL, Chen JF (2011) Iodine-doped ZnO nanocrystalline aggregates for Improved dye-sensitized solar cells. Chem Mater 23:3–5
Zhou H, Wu L, Gao Y, Ma T (2011) Dye-sensitized solar cells using 20 natural dyes as sensitizers. J Photochem Photobiol A:Chem 219:188–194
Zhu S, Fu H, Zhang S, Zhang L, Zhu Y (2008) Two-step synthesis of a novel visible-light-driven $K_2Ta_2O_{6-x}N_x$ catalyst for the pollutant decomposition. J Photochem Photobiol A 193:33–41
Zong X, Sun C, Yu H, Chen ZG, Xing Z, Ye D, Lu GQ, Li X, Wang L (2013) Activation of photocatalytic water oxidation on N-doped ZnO bundle-like nanoparticles under visible light. J Phys Chem C 117:4937–4942
Zou L, Liu Q, Xu Q, Mi G (2011) Microwave synthesis of mesoporous WO_3 doping with bismuth and photocatalytic oxidation of water to H_2. Korean J Chem Eng 28:1299–1303

Chapter 3
Synthesis Methods for Photocatalytic Materials

Iliana Medina-Ramírez, Aracely Hernández-Ramírez, and M. Lourdes Maya-Treviño

Abstract As previously discussed, there exists interest for the large-scale production of nanostructured materials of bare or doped semiconductor photocatalysts. Numerous reports have been presented regarding the synthesis, characterization, and photocatalytic activity evaluation of undoped and doped (metal or nonmetal) titanium dioxide, for example. It has been demonstrated that the physicochemical properties and photocatalytic activity of semiconductors are mainly determined by the preparation technique used in its production. Photocatalytic semiconductors can be prepared in the form of powders, fibers, and films by different synthetic methods including sol-gel process, hydrothermal and solvothermal techniques, direct oxidation reactions, sonochemical method, microwave method, chemical vapor deposition method, and electrodeposition method, among others. In this section will be briefly reviewed the principles that govern some of the most used synthetic methodologies employed to obtain photocatalytic materials, most of them related to bare and modified TiO_2. We will also illustrate each synthetic approach with relevant examples of materials produced in the last years.

3.1 Sol-Gel Process

The sol-gel process has become the most widely used method for the synthesis of semiconductor photocatalysts. It is a powerful method for tailoring metal oxides to fit particular applications since a great number of parameters can be regulated: e.g., the nature of the precursors, pH, temperature and time of reaction, reagent concentrations, catalyst nature and concentration, aging temperature and time, addition of organic additives, and the amount of water added. The main advantage of the sol-gel method is the homogeneous mixing at the molecular level of

I. Medina-Ramírez (✉)
Departamento de Química, Universidad Autónoma de Aguascalientes, Av. Universidad 940, Ciudad Universitaria, 20131 Aguascalientes, Ags., Mexico
e-mail: iemedina@correo.uaa.mx

A. Hernández-Ramírez • M.L. Maya-Treviño
Universidad Autónoma de Nuevo León, Facultad de Ciencias Químicas, Cd. Universitaria, San Nicolás de los Garza, N.L., México

A. Hernández-Ramírez, I. Medina-Ramírez (eds.), *Photocatalytic Semiconductors*,
DOI 10.1007/978-3-319-10999-2_3

metal ion which enhances the formation of polycrystalline particles with special properties.

Another great advantage of the sol-gel method is that during some stage of the process, it is possible to incorporate different types of dopants. The incorporation of an active dopant in the sol during the gelation stage allows doping elements to have a direct interaction with support in such a way that the photocatalytic properties of the material can be enhanced (Zaleska 2008).

The sol-gel method can be defined as the conversion of a precursor solution into an inorganic solid via inorganic polymerization reactions induced by water; it uses as a precursor an aqueous or alcoholic mixture of metal-organic (alkoxides) or inorganic salts (chloride, nitrate, sulfate, acetate, etc.). The sol-gel process consists firstly of the preparation of a homogeneous solution followed by the next steps: (a) conversion of the homogeneous solution into a sol by treatment with a suitable reagent (generally water with or without any acid/base), (b) aging, (c) shaping, and (d) thermal treatment/sintering (Niederberger and Pinna 2009). The precursors are hydrolyzed and condensed to form inorganic polymers composed of M-O-M bonds (Brinker and Scherer 1990); further condensation results in a gel. The gels can be dried under hypercritical conditions and aerogels are produced. When the gel is dried under ambient conditions, a xerogel is obtained. The gel is then thermally treated to yield the desired material, and many forms such as monoliths, films, fibers, and monosized powders can be shaped.

In various studies, TiO_2 has been usually synthesized by sol-gel method using as precursor titanium (IV) n-butoxide (Su et al. 2004; Choi et al. 2011; Liu et al 2011; Han et al. 2012); depending on the synthesis conditions and temperature of calcination, anatase or rutile phase can be obtained. The effect of the nature of the oxide precursors has been demonstrated in a research work when the preparation of TiO_2 was carried out by the sol-gel method from two different alkoxide precursors: titanium (IV) n-butoxide and titanium (IV) isopropoxide (Saragiotto Colpini et al. 2008). The catalysts were used in the photocatalytic degradation of textile dye C.I. reactive orange 122 irradiating with a 300-W tungsten halogen lamp. A reduction of 65.8 % in TOC was observed with TiO_2 obtained from n-butoxide precursor, whereas with TiO_2 from isopropoxide, the maximum mineralization achieved was 27.7 % in 60 min. The authors attributed the improved activity to the high dispersion of the particles obtained when titanium (IV) n-butoxide was used as a precursor.

As already mentioned in the previous chapters, changes in the band structure can be achieved by doping TiO_2 or ZnO using different synthesis techniques. However, the sol-gel method has the advantage that does not require complicated instruments and provides simple and easy means for preparing doped nanosized particles. In this context, tungsten, silver, and tungsten/silver co-doped TiO_2 nanopowders were prepared by sol-gel method (Tobaldi et al. 2013). During the synthesis, titanium (IV) isopropoxide was hydrolyzed using a water/HNO_3 solution. In order to prepare W-TiO_2, Ag-TiO_2, and Ag-W-TiO_2 samples, silver nitrate or tungstic acid was added to the produced TiO_2 sol. The as-synthesized gels were dried at 120 °C, and afterward, the dried gels were thermally treated at 450 and 600 °C in a static air flow. The simultaneous presence of rutile, anatase, and brookite was detected in the gels thermally treated at 450 °C. At 600 °C, the only crystalline phases were anatase and rutile. The photocatalytic activity of the samples was evaluated under UVA light and visible light irradiation on the degradation of methylene blue solution. Undoped titania and the doped and co-doped powders, calcined at 450 °C, showed an appreciable photocatalytic activity under visible light irradiation, better than Aeroxide® TiO_2 P25 confirmed by TOC analysis.

In another work, gold particles were incorporated into $Ti_{1-x}Zn_xO_2$ films by sol-gel method (Linnik et al 2013). Titanium isopropoxide was added to a mixture of ethanol, acetylacetone, and nitric acid. Pluronic123 in ethanol was added to the mixture in order to produce a supramolecular structure of inorganic framework surrounded by organic template. Aqueous solution of zinc (II) acetate was added to titanium sol to be deposited on glass substrates or silica wafers by dip-coating technique. Preparation of the gold particles in the film structures was carried out by incorporation of tetrachloroauric acid (3 mol %) to the sol of titanium and zinc species. The films were prepared at pH 4 and the thermal treatment was carried out at different temperatures (200 and 500 °C).

The photocatalytic activity of $Au/Ti_{1-x}Zn_xO_2$ films was investigated using as test reaction the degradation of tetracycline hydrochloride under UV light. The results indicated that synthesis conditions affect the sizes and morphology of the obtained particles; hence, mesoporous surface structure was produced during mild thermal conditions which improve the photocatalytic degradation of tetracycline antibiotic.

On the other hand, many coupled semiconductor oxides have been also prepared by means of the sol-gel process including Fe_2O_3-ZnO (Hernández et al. 2007; Maya-Treviño et al. 2014), Cr_2O_3-TiO_2 (Jung et al. 2011), TiO_2-NiO (Sreethawong et al. 2012), Fe_2O_3-TiO_2 (Palanisamy et al. 2013), WO_3-TiO_2 (Ramos-Delgado et al. 2013), and ZrO_2-TiO_2 (Gionco et al. 2013), among others. In each case, precursors of both metal oxides were hydrolyzed together under stirring, allowing the simultaneous formation of the coupled oxides. The synthesis of mixed metal oxides by sol-gel method generally affects the surface area with respect to pure metal oxide, which is an important parameter involved in the photocatalytic activity of a semiconductor material.

3.2 Hydrothermal Method

The term hydrothermal usually refers to any heterogeneous reaction in the presence of aqueous solvents or mineralizers under high pressure and temperature conditions (Byrappa and Yoshimura 2001). This synthetic method is normally conducted in steel pressure vessels called autoclaves with or without Teflon liners under controlled temperature and/or pressure with the reaction in aqueous solutions. The temperature can be elevated above the boiling point of water, reaching the pressure of vapor saturation. The temperature and the amount of solution added to the autoclave largely determine the internal pressure produced (Chen and Mao 2007). When selecting a suitable autoclave, the most important parameters are the experimental temperature and pressure conditions and the corrosion resistance in that pressure-temperature range in a given solvent or hydrothermal fluid. If the reaction is taking place directly in the vessel, the corrosion resistance is a prime factor in the choice of autoclave material. The most successful materials which are corrosion resistant are high-strength alloys, such as 300 series (austenitic) stainless steel, iron, nickel, cobalt-based superalloys, and titanium and its alloys (Byrappa and Yoshimura 2008).

Many researchers have been using the hydrothermal method to prepare nanoparticles with photocatalytic properties (Livraghi et al. 2013; Tang et al. 2013; Yang et al. 2009). For example, TiO_2 nanoparticles can be obtained by hydrothermal treatment of peptized precipitates of a titanium precursor with water. The precipitates were prepared using titanium butoxide as precursor in the presence of tetraalkylammonium hydroxide (peptizer). Under the same concentration of peptizer, the particle size decreased with increasing alkyl chain length. The peptizers and their concentrations influenced the morphology of the particles (Chen and Mao 2007). Furthermore, TiO_2 nanotubes have received increasing attention, and an extensive review on TiO_2-based nanotubes synthesized via hydrothermal technique has been recently published by Liu et al. (2014). In this paper, formation mechanism, structure modification, and photocatalytic applications of TiO_2 nanotubes are discussed. The factors during the hydrothermal reaction that influence the formation of nanotubes (TiO_2 precursor, hydrothermal temperature, and duration), auxiliary methods (ultrasonication and microwave assistance), and posttreatment (acid washing and calcination) on the formation of titanate nanotubes are also reviewed (Liu et al. 2014).

On the other hand, as previously discussed in Chap. 1, doping TiO_2 with nonmetals has been investigated in order to enhance the photocatalytic activity of TiO_2 under visible light irradiation; in this context, hydrothermal method allows the preparation of NM-doped titania at low temperature; such is the case, for example, of boron-doped TiO_2 by using sodium borohydride as boron source. Compared to sol-gel method, hydrothermal B-doped sample exhibits larger surface area, enhanced adsorption ability, and better photocatalytic activity for the degradation of reactive brilliant red and 4-chlorophenol under visible light irradiation (Xu et al. 2009).

Recently, Guo et al. (2013) prepared phosphorus-doped TiO_2 (P-TiO_2) with mesoporous structure via hydrothermal treatment. Tetrabutyl titanate was used as the titanium source, and phosphoric acid was chosen as the precursor of the dopant. The hydrothermal treatment was carried out in a Teflon-lined stainless steel autoclave at 200 °C for 12 h. For comparison, P-doped samples were also prepared by high temperature calcination of sol-gel process in a similar way. Photocatalytic activity of the prepared samples was evaluated using as test reactions the degradation of methylene blue (MB) and the decomposition of formaldehyde (HCHO) under artificial light from a xenon lamp. Well-crystallized mesoporous P-doped titania nanoparticles were obtained by both hydrothermal and sol-gel methods; however, the samples synthesized by hydrothermal treatment have small average pore size, uniform crystallite size, large surface area, and better photocatalytic activity compared to those prepared by sol-gel process or Degussa-P25 catalyst (Guo et al. 2013). Additionally, the hydrothermal treatment creates more lattice defects in the P-TiO_2 semiconductor which are active to capture photogenerated electrons and improve catalytic activity.

In another work (Biswas and Baeg 2013), WO_3 was prepared by hydrothermal method for photoelectrochemical solar water oxidation and for decomposition of an aqueous solution of rhodamine B (RhB) under visible light irradiation. WO_3 was synthesized using ammonium metatungstate hydrate (AMT), as a tungsten precursor in acidic medium (HCl), carrying out the hydrothermal reaction in a Teflon-lined stainless steel autoclave at 180 °C for 4 h without stirring. The hierarchical hydrated WO_3 assemblies were also synthesized using HNO_3 or H_2SO_4 instead of HCl. The as-synthesized powders were annealed at 500 °C for 30 min in air. For comparative purposes, aqueous solution of AMT was hydrothermally treated at 180 °C for 24 h in the absence of acid. A well-defined three-dimensional (3D) hierarchical architecture of WO_3 hydrates resembling a flowerlike morphology was produced in the presence of the different acids, without using any template or any organic medium. No distinct influence of the anions of the acids was observed on directing the morphology of the final products. However, when the reaction was carried out in the absence of acid, the product was obtained in small amount and found to have an agglomerated, irregular sphere-like morphology. Regarding the photocatalytic behavior on the organic pollutant degradation, WO_3 hydrates with hierarchical structure showed better photocatalytic activity than commercial WO_3 (used as bulk counterpart). 84 % RhB decomposition was achieved compared with 40 % dye degradation obtained with commercial oxide. Moreover, the synthesized WO_3 microflowers showed higher activity for solar light-driven water splitting when WO_3 film-coated FTO electrode was illuminated under simulated solar light (Biswas and Baeg 2013).

Alternatively, ternary Bi_2WO_6 photocatalyst has been prepared as microspheres via a combination of sol-gel-hydrothermal technique. The precursors were $Bi(NO_3)_3{\cdot}5H_2O$ and $(NH_4)_6W_7O_{24}\cdot 6H_2O$ solutions in the presence of EDTA. The produced gel at pH 3.0 was aged for a whole night and was transferred into an autoclave maintained at 220 °C for 24 h. For comparison, Bi_2WO_6 was also synthesized directly by hydrothermal reaction without the sol-gel process. The sol-gel-hydrothermal (SH) Bi_2WO_6 consisted of monodispersed hierarchical microspheres (Fig. 3.1b) which were hollow in the middle, while hydrothermal (H)Bi_2WO_6 exhibits irregular platelike structure.

The results of methylene blue (MB) degradation in the presence of the different Bi_2WO_6 catalysts showed that SH-Bi_2WO_6 exhibits superior photocatalytic activity than H-Bi_2WO_6. The enhanced photocatalytic activity of SH-Bi_2WO_6 was attributed to its special hierarchical structure produced by the combination of sol-gel-hydrothermal processes (Liu et al. 2013c).

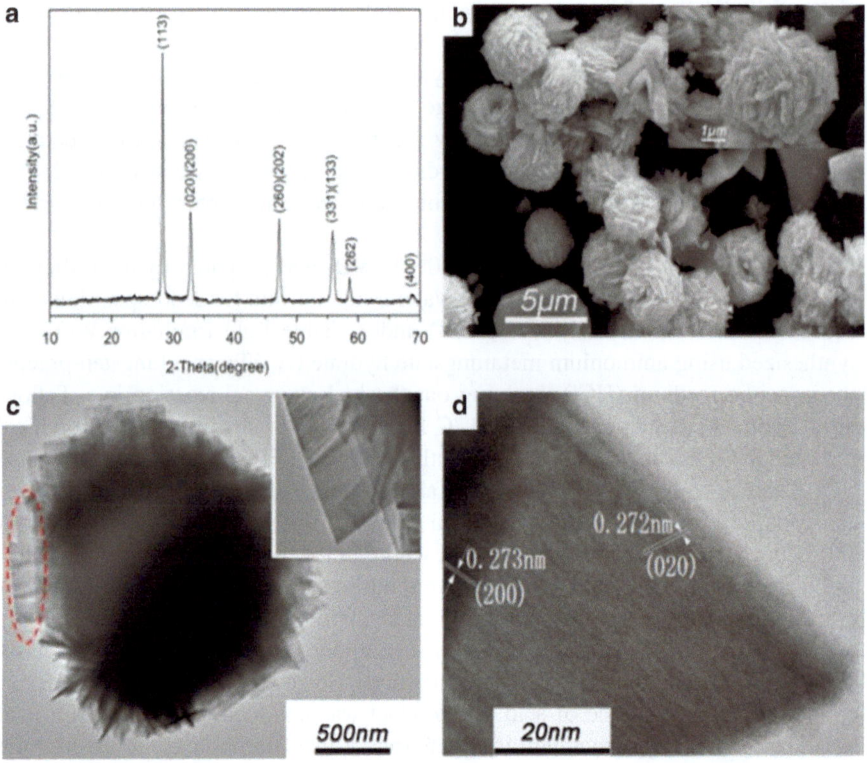

Fig. 3.1 (**a**) XRD pattern of Bi_2WO_6. (**b**) Low magnification SEM SH-Bi_2WO_6 (*inset* is the high magnification SEM). (**c**) TEM image of an individual hierarchical microsphere (*inset* is the detailed view of the outer edge of the 2D layer). (**d**) HRTEM image (From Liu et al. 2013c)

3.3 Solvothermal Technique

The solvothermal method is a derivation of the hydrothermal technique where the solvent is nonaqueous. The reaction temperature can be elevated to much higher levels than in the hydrothermal method, due to the variety of organic solvents with high boiling points that can be chosen. Generally, the semiconductor particle size, shape, distribution, and crystallinity can be better controlled using the solvothermal technique than with the hydrothermal method (Wu et al. 2013). The solvothermal method has been found to be a useful method for the synthesis of a variety of nanoparticles with narrow size distribution and dispersity and has been employed to synthesize TiO_2 nanoparticles and nanorods with or without the aid of surfactants. Otherwise, solvent plays an important role in determining the crystal morphology. Solvents with different physical and chemical properties can influence the solubility, reactivity, and diffusion behavior of the reactants; in particular, the polarity and coordinating ability of the solvent can influence the morphology and the crystallization behavior of the final products. The presence of ethanol at a high concentration not only can cause the polarity of the solvent to change but also strongly affects the ζ potential values of the reactant particles and increases solution viscosity. For example, in the absence of ethanol, short and wide flakelike structures of TiO_2 were obtained instead of nanowires. When chloroform is used, TiO_2 nanorods were obtained (Chen and Mao 2007).

Nonaqueous solvothermal routes usually provide better control over the size, crystallinity, and agglomeration behavior of the nanoparticles. For instance, mesoporous TiO_2 microspheres with rough surfaces generally have high specific surface area, which is an important feature for the semiconductor activity because more active sites are available for photocatalytic reactions. Several forms of mesoporous TiO_2 samples of different sizes have been prepared via solvothermal methods (He et al. 2014a, b; Tanasković et al. 2009; Zhu et al. 2013). In this context, the synthesis of hierarchical mesoporous TiO_2 microspheres with high crystallinity and high BET specific surface area was carried out using tetrabutyl titanate as a precursor into polyethylenimine solution mixed with absolute ethanol. The activity of obtained TiO_2 microspheres was evaluated in the degradation of methyl orange (MO) and phenol aqueous solutions under simulated solar light irradiation. The results showed that specific surface area was high (118.3 $m^2\ g^{-1}$) and pore-size distribution reveals a narrow distribution centered at 2.4 and 10.1 nm, respectively. The TiO_2 microspheres exhibit excellent photodegradation activities on both MO and phenol, comparing to P25 (Zhu et al. 2013).

In the solvothermal conditions, involving high pressure, the physicochemical behavior properties of the solvent are modified (i.e., the dielectric constant). In consequence, using this methodology, it is possible to obtain compounds which, under normal conditions of pressure and temperature, cannot be obtained. Such is the case of the synthesis of new photocatalytic semiconductors with activity in the visible region of the spectra. Particularly, bismuth oxyiodide (BiOI) has provoked great interest in heterogeneous photocatalysis because it shows high photocatalytic activity under visible radiation (Pingyu et al. 2009). BiOI microspheres have been synthesized by a solvothermal method as reported by Mera et al. (2014). In their experiment, ethylene glycol solution, containing 1 mmol of the ionic liquid 1-butyl-3-methylimidazolium iodide, was added slowly to 1 mmol of $Bi(NO_3)_3{\cdot}5H_2O$ dissolved in ethylene glycol. The solvothermal reaction was carried out in a Teflon-coated Parr® steel autoclave. The autoclave temperature was maintained constant at 120 °C and the reaction time was varied from 2 to 48 h. Structured microspheres were formed even at short periods of reaction (2 h), obtaining the most regular forms at 12 and 24 h. At higher reaction times (48 h), some disaggregation of the microspheres was observed. BiOI materials were constituted by spheres formed by irregular nanosheets connected between them and forming hierarchical microspheres. In Fig. 3.2 the composition of the material is confirmed by the EDS analysis and the nanosheets are clearly shown. The photocatalytic activity of the synthesized materials was evaluated using caffeic acid as model pollutant under UV-Vis radiation. The degradation profiles were compared to the catalyst Evonik P25.

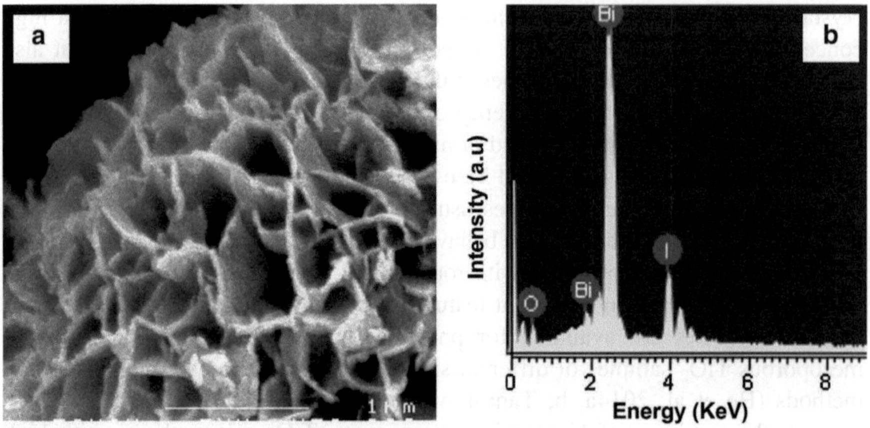

Fig. 3.2 (**a**) Expanded SEM image of the BiOI microspheres synthesized for 24 h. (**b**) EDS of BiOI microspheres synthesized for 24 h (From Mera et al. 2014)

All the materials prepared exhibit higher activity than Evonik P25. The reaction time in autoclave only affected the surface area value of the samples. The best results in catalytic activity correspond to the materials prepared at 12 and 24 h of reaction time which presented the highest BET surface area (54 and 58 m^2 g^{-1}, respectively). At times below 12 h, the microspheres are being formed, reaching the maximum growth between 12 and 24 h. At longer periods of reaction, the material can suffer sinterization, reducing the BET surface area (Mera et al. 2014).

3.4 Direct Oxidation Method

In the case of the synthesis of TiO_2 nanomaterials, it can be also obtained by direct oxidation of titanium metal using oxidants or under anodization. Crystalline TiO_2 nanorods have been obtained by oxidation of a titanium metal plate with hydrogen peroxide. TiO_2 nanorods on a Ti plate are obtained when a cleaned Ti plate is immersed in 50 mL of a 30 wt % H_2O_2 solution at 353 K for 72 h. The formation of crystalline TiO_2 occurs through a dissolution precipitation mechanism. By the addition of inorganic salts of NaX (X=F-, Cl-, and SO_4^{2-}), the crystalline phase of TiO_2 nanorods can be controlled. The addition of F^- and SO_4^{2-} helps the formation of pure anatase, while the addition of Cl^- favors the formation of rutile.

Anodic oxidation is a promising method for fabricating immobilized TiO_2-based photocatalysts, since it can lead to direct the growth of immobilized self-organized TiO_2 on differently shaped Ti substrates.

Anodic oxidation reaction is an electrochemical method for the production of an oxide film on a metallic substrate. It involves the application of an electrical bias at relatively low currents while the substrate is immersed in an acid bath. The films can be very dense and stable, with a variety of microstructural characteristics.

The phase formation, thickness, and pore size of the films strongly depend on the applied voltage and anodizing time (Abdullah and Sorrell 2007). This method is originally used to produce high-ordered nanoporous and nanotubular films (Wu et al. 2009). TiO_2 nanotubes with different morphologies can be obtained by anodic oxidation of titanium foil (Su et al. 2013; Wang et al. 2009b, 2011d). As an example, Ti plate was anodized in HF solution under 10–20 V for 10–30 min. Platinum was used as counter electrode. Crystallized TiO_2 nanotubes were obtained after the anodized Ti plate was annealed at 500 °C for 6 h in oxygen. The length and diameter of the TiO_2 nanotubes can be controlled over a wide range (diameter, 15–120 nm; length, 20 nm to 10 μm) with the applied potential between 1 and 25 V in optimized phosphate/HF electrolytes (Chen and Mao 2007).

In another work, three types of immobilized TiO_2 nanostructured films, nanowires (NW), irregular nanowires (IRNW), and nanotubes (NT), were prepared by potentiostatic anodization in a two-electrode electrochemical cell. For preparation, 0.1 mm thickness titanium foil (6 cm × 6 cm) was used as a working electrode connected to a DC power supply. NW and IRNW were prepared in ethylene glycol electrolyte solution that contained 0.2 wt % HF and 0.3 wt % water at 100 V for 3 h and 2 h, respectively. NT were prepared in ethylene glycol electrolyte solution that contained 0.08 wt % HF and 0.12 wt % water at100 V for 3 h.

The samples were tested in the photocatalytic degradation of toluene (200 mg m^{-3}) under UV irradiation. The best photocatalytic activity was shown by IRNW with mixed morphologies of irregular nanowires on top and nanotubes on bottom. The irregular nanowires were obtained with larger surface area, hollow interior wall, and hierarchical porous structures, which results in a faster diffusion of gas-phase species during the photocatalytic reaction (Wu et al. 2009).

As has been previously mentioned in Chap. 1, zinc oxide (ZnO) has significant importance in the field of the heterogeneous photocatalysis, and then, the physicochemical properties of ZnO and therefore its application also depend on its synthesis route. Although ZnO has been synthesized in the film form by a variety of methods, the electrochemical technique for the synthesis of nanosized metal oxide films has acquired a considerable interest because of its simplicity, low operating temperature, and viability for commercial production (Wang et al. 2011b; Chandrappa and Venkatesha 2012).

By direct oxidation, ZnO nanoparticles of varying sizes (20, 44, and 73 nm) have been successfully synthesized by electrochemical-thermal method using aqueous sodium bicarbonate electrolyte and sacrificial Zn anode and cathode in an undivided cell under galvanostatic mode. The nanoparticles morphology was similar for all the samples obtained at different current densities, and the size and shape did not depend on current density. The obtained samples were calcined at different temperatures (60, 300, and 600 °C) and tested in the photocatalytic decomposition of a methylene blue solution under UV light irradiation. The results showed that the as-prepared and calcined (600 °C) ZnO photocatalysts exhibit higher photocatalytic activity probably caused by the smaller particle size and higher crystallinity (Chandrappa and Venkatesha 2012).

3.5 Sonochemical Method

Over the last years, an increasing interest has been shown in power ultrasound potential for use over a much wider range of chemistry and processing which has been grouped together under the general title of sonochemistry. Most of these uses depend on the generation of acoustic cavitation in liquid media (Mason and Lorimer 2003). In this context, ultrasound has been very useful in the synthesis of a wide range of nanostructured materials, including high-surface-area transition metals, alloys, carbides, oxides, and colloids. The application of ultrasound under specific conditions allows the possibility of synthesizing nanocomposites in a short time, under mild conditions, in air, and without calcination.

The chemical effects of ultrasound do not come from a direct interaction with molecular species. Instead, sonochemistry arises from acoustic cavitation: the formation, growth, and implosive collapse of bubbles in a liquid. The collapse of cavitation bubbles generates localized hot spots with transient temperature of about 10,000 K, pressures of about 1,000 atm or more, and cooling rates in excess of 109 K s^{-1}. Under such extreme conditions, various chemical reactions and physical changes occur, and numerous nanostructured materials such as metals, alloys, oxides, and biomaterials can be effectively synthesized with required particle size distribution (Shirsath et al. 2013). The novelty of sonochemical method is the source that promotes the chemical reaction; it represents a sustainable alternative for saving energy during the synthesis of nanoparticles.

The synthesis of nanostructured TiO_2 can be carried out by template-free techniques among which hydrothermal and sonochemical methods are simple appropriated processes to obtain this kind of materials. Guo and coworkers (2009) prepared mesoporous TiO_2 nanorods using a novel template-free sonochemical approach from commercial Ti powder. The preparation was carried out at normal atmospheric pressure and at low temperature of 70 °C. Photocatalytic degradation of toluene was conducted under UV radiation to analyze the activity of the prepared samples. The mechanism for the formation of ultrasonic mesoporous TiO_2 nanorods was proposed and the photocatalytic activity was higher than P25 TiO_2. The superior activity of the sample can be ascribed to its higher surface BET of 91.4 m^2 g^{-1}, higher pore volume of 0.55 cm^3 g^{-1}, and mesoporous structure (Guo et al. 2009).

Various reports exist in the literature regarding the synthesis of metal- and nonmetal-doped TiO_2 by sonochemical methods (Del Ángel-Sanchez et al. 2013; Shirsath et al. 2013; Wang et al. 2011c; Yu et al. 2012). For example, Wang et al. (2011c) prepared N-doped TiO_2 nanocrystalline by sonication of the solution of tetraisopropyl titanium and urea in water and isopropyl alcohol at 80 °C for 150 min. The effect of sonication time on the formation of N-doped TiO_2 nanoparticles was investigated. The crystallinity of the sample increased at higher sonication time (60, 120, 150, and 180 min); however, 150 min was the optimal sonication time where the crystallization of TiO_2 was completed.

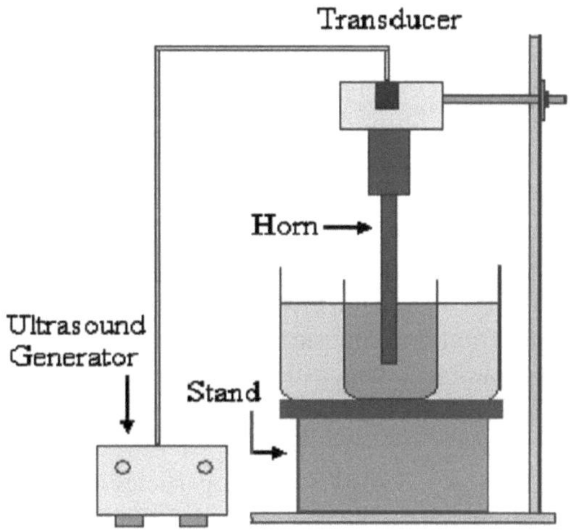

Fig. 3.3 Experimental setup for sonochemical synthesis (From Shirsath et al. 2013)

In another work (Shirsath et al. 2013), a detailed study about establishing the influence of ultrasound on the phase composition, structure, and performance of pure and doped TiO_2 nanocatalysts was investigated. Cerium and Fe-doped TiO_2 nanocatalysts with different amounts of doping elements were prepared by a single-step sonochemical method and conventional method at room temperature.

Cerium (III) nitrate, ferric nitrate, and titanium (IV) isopropoxide were used as precursors and 2-propanol as solvent in both conventional and sonochemical synthesis procedures. In the procedure involving ultrasound, the sonication was carried out by employing a direct immersion titanium horn in the sonication cell. Solutions of different concentrations of cerium/iron precursors were added into the ultrasound reactor to the titanium isopropoxide in 2-propanol solution, and the mixture was sonicated for further 30 min. The experimental setup for the sonochemical synthesis is shown in Fig. 3.3.

The reaction time required for the conventional synthesis was 4 h, while during sonochemical synthesis the precipitate was formed within 30 min. The reduction in the reaction time compared to conventional synthesis was due to the cavitational effect which results into accelerated hydrolysis reaction.

Also, the preparation method affects the optical properties of the doped samples. In this case, an increase in the absorption in the visible region which depends not only on the type of dopant but also on the method of preparation was observed. Hence, higher absorption was presented by the particles prepared by sonochemical method. In order to compare the photocatalytic activity of synthesized catalysts, crystal violet dye degradation studies were carried out under UV irradiation. The doped TiO_2 catalysts prepared by sonochemical method showed higher activity against catalyst prepared by the conventional method, attributed to the higher surface area for the reaction due to the lower particle size of the catalyst achieved with the sonochemical method. The increased high-velocity interparticle collisions among the particles can result in the fragmentation of the TiO_2 particles leading to lower size in the case of sonochemical method. Ultrasonic irradiation may accelerate the hydrolysis and formation of titania crystals. Furthermore, sonication method can evenly disperse the metal ions into the crystal lattice of TiO_2 (Huang et al. 2000, Shirsath et al. 2013).

Sonochemical method was also used for the synthesis of silica-coated ZnO nanoparticles. The ultrasound irradiation of a mixture of dispersion of ZnO, tetraethoxysilane (TEOS), and ammonia in an ethanol-water solution medium was carried out. The silica coating layer formed at the initial TEOS/ZnO loading of 0.8 for 60 min ultrasonic irradiation. The coating was uniform and extended up to 3 nm from the ZnO surface as revealed from HRTEM images. Silica-coated ZnO nanoparticles demonstrated a significant inhibition of photocatalytic activity against photodegradation of methylene blue dye in aqueous solution (Siddiquey et al. 2012).

Recently, TiO_2/WO_3 nanoparticles were synthesized at room temperature by means of the ultrasound method. Sodium tungstate dihydrate $[Na_2WO_4 \cdot 2H_2O]$ and titanium isopropoxide ($Ti(OC_3H_7)_4$) were used as a precursors (Anandan et al. 2014). The synthesis reaction was conducted at a high intensity in an ultrasonic horn in argon atmosphere for 2.5 h at room temperature. The TiO_2/WO_3 sample consisted of mixed square and hexagonal shape particles about 8–12 nm in diameter. The photocatalytic activity of TiO_2/WO_3 nanoparticles was tested for the degradation of methylene blue (MB) under visible light irradiation. The results indicated that TiO_2/WO_3 nanoparticles exhibit a higher degradation rate than bare TiO_2 nanoparticles under similar conditions.

3.6 Microwave Method

Microwave (MW) radiation has been widely explored as a means of activating energy to promote several organic reactions (Geyde 2002). In view of the advantages offered by this technique in the field of organic chemistry, such as reaction rate enhancement, increased yields, cleaner conditions (i.e., solvent-free reactions), and easier workup, the use of this approach has been extended to the production of several inorganic materials (Kitchen et al. 2014).

Despite the many advantages with the use of MW as a means of activating energy for chemical reactions, there still exists some controversy on the reasons that contribute to the enhancement of reactions by MW radiation. The debate centers on the impact of the thermal and nonthermal microwave effects (Kappe 2013). Microwave frequencies range from 0.3 to 300 GHz; the reactors that are designed to conduct chemical reactions operate at 2.45 GHz, a frequency region assigned for heating applications. It is known that MW irradiation is not sufficiently energetic to break hydrogen or covalent bonds; thus, it cannot induce chemical reactions since no bond breaking can occur by direct absorption of the MW energy.

In microwave chemistry, all components of the reaction (solvents, reagents, vessels, etc.) are capable of interacting with or perturbing the electromagnetic field. MW-assisted chemical reactions depend on the ability of the reaction mixture to efficiently absorb MW energy, which often depends on the choice of solvents for the reaction. The ability of a specific solvent or material to convert MW energy into heat is determined by the so-called loss tangent (δ). Compounds (solvents, molecules) that possess high tan δ values show good MW absorbing capacity and thus are expected that these chemicals will be favored by thermal effects of MW radiation.

Lately, Kappe (2013) conducted a systematic study aimed to elucidate the influence of the electromagnetic field on a chemical transformation, outside of the simple macroscopic change in bulk reaction temperature. In order to suppress the influence of the electromagnetic field, a silicon carbide (SiC) vessel (which presents a high MW absorptivity) was used to conduct chemical reactions. For comparison, a non-absorbing (of MW radiation) vessel was used to conduct the same reactions under identical reaction conditions. The obtained results allowed the research group to conclude that the electromagnetic field has no direct influence on the reaction pathway; it is the overheating phenomenon that drives the enhancements in MW chemistry (Kappe 2013).

Titanium dioxide materials with different morphologies (powders, thin films, nanotubes) have been synthesized using microwaves as heating source in order to achieve formation or crystallization of the material. Besides the energy and time savings, it has been encountered that the microwave hydrothermal method is superior to conventional hydrothermal process in at least three aspects: (a) the temperature of reaction is quickly achieved and keeps uniform within the time of reaction; (b) the kinetics of crystallization is increased by one-to-two orders of magnitude in comparison to conventional heat treatment; and (c) different crystalline phases can be produced (Komarneni et al. 1999, 2014; Baldassari et al. 2005; Li et al. 2009).

The synthesis of nanocrystalline TiO_2 (single-phase rutile) with diverse morphologies was reported using a microwave hydrothermal method (Chen et al. 2008). Mesoporous TiO_2 materials were easily prepared using a microwave-assisted esterification method (Li et al. 2009). It is important to remark that the use of MW led to the fast (~1 min) preparation of mesoporous TiO_2 materials with high specific surface area (210 $m^2\ g^{-1}$), without using any surfactants or templates. In another report, nanostructured crystalline TiO_2 powders with high specific surface areas (>200 $m^2\ g^{-1}$) were obtained using a two-step procedure (ball-milling followed by MW hydrothermal) and $K_2Ti_4O_9$ as Ti precursor (Zhang et al. 2009). Later, the use of an ionic liquid-like titanium precursor and MW radiation was explored for the synthesis of anatase (Estruga et al. 2010). The effect of MW radiation (compared to oil-bath heating conditions) on the physicochemical properties and photocatalytic activity of La^{+3} and Zr^{+4} co-doped nanocrystalline TiO_2 was reported by Shojaie and Loghmani (2010). The nano-TiO_2 samples prepared by MW method showed a smaller particle size and higher photoactivity than conventional heated produced samples. More recently, the photoreduction of CO_2 under simulated solar radiation and a MW-hydrothermally prepared zinc phtalocyanine/TiO_2 composite catalyst was reported (Wang et al. 2012).

Supported TiO_2 materials have also been prepared by means of MW irradiation (Liu et al. 2007; Mohamad-Alosfur et al. 2014). For instance, Horikoshi et al. (2013) prepared TiO_2/activated carbon (AC) particles using an aqueous titanium oxysulfate solution, AC, and MW irradiation. For comparison, the TiO_2/AC composites were also prepared by conventional heating (using an oil bath). The effect of reaction temperature and heating source is clearly illustrated in Fig. 3.4. SEM analysis suggests that the optimal reaction conditions were 90 °C and MW irradiation. This result was in accordance to the surface area and photocatalytic activity encountered for the materials. In the search for greener synthetic approaches to AC-supported materials, a combination of MW irradiation and ionic liquids was explored in order to obtain TiO_2/AC composites at low temperatures and ambient pressure. This approach rendered composites with good electrochemical properties as electrode material for capacitive deionization (Liu et al. 2013b).

Examples of MW-assisted synthesis of metal and metal-nonmetal-doped TiO_2 have recently appeared in literature: (S)Fe/TiO_2 (Esquivel et al. 2013), Ag-TiO_2 (Li et al. 2010), Zn_2TiO_4 (Arin et al. 2013), N-TiO_2/g-C_3N_4 (Wang et al. 2013b), and C-TiO_2 (Raji and Palanivelu 2011).

In the last decade, MW irradiation has been widely used for the production of TiO_2 nanotubes (Wu et al. 2005, 2007b; Li et al. 2011; Cui et al. 2012; Liu et al. 2013a). The nanotubes (NTs) are generally obtained by modifying TiO_2 nanoparticles using thermal MW radiation. TiO_2 powders suspended in a basic solution are introduced into the MW reactor and submitted to MW irradiation under different conditions (temperature, time, pressure). Comparative studies show that ordered arrays of TiO_2 NTs outperform TiO_2 nanoparticles for several applications. This enhanced activity of TiO_2-NTs can be justified due to the improved light harvesting and charge separation performed by these materials; thus, considerable effort is being devoted to the preparation of these complex surface structures (Grimes and Mor 2009). MW technique offers a simple approach to synthesize TiO_2-NTs.

Besides bare and doped TiO_2 materials, some other photocatalytic semiconductors have been prepared by using the MW-assisted solvothermal method. Bilecka and coworkers (2008) have published some papers regarding the synthesis of binary and ternary metal oxide (ZnO, Fe_3O_4, $BaTiO_3$) photocatalytic materials using non-hydrolytic colloidal method activated by MW radiation (Bilecka et al 2008; Bilecka et al 2009). It has been demonstrated that by the use of non-hydrolytic liquid-phase routes, size and shape control can be achieved easily. For instance, ZnO has been synthesized using this approach, rendering materials with diverse morphologies such as spheres, nanowires, flowerlike, nanorods, nano-needles, and hollow structures (Chen and Lo 2011; Zhu and Chen 2014). Some other examples of binary materials include $Bi_6O_6(OH)_3(NO_3)\cdot 1.5H_2O$ nanosheets (Xie et al. 2012) and ZnS that was prepared by a combination of MW and single-source precursor approach (Sun et al 2008). A recent review illustrates numerous examples on the MW synthesis of binary and ternary nanostructured photocatalytic materials. Some of these materials are ZrO_2, WO_3, CeO_2, $NiWO_4$, $CoWO_4$, $MnWO_4$, Bi_2WO_6,

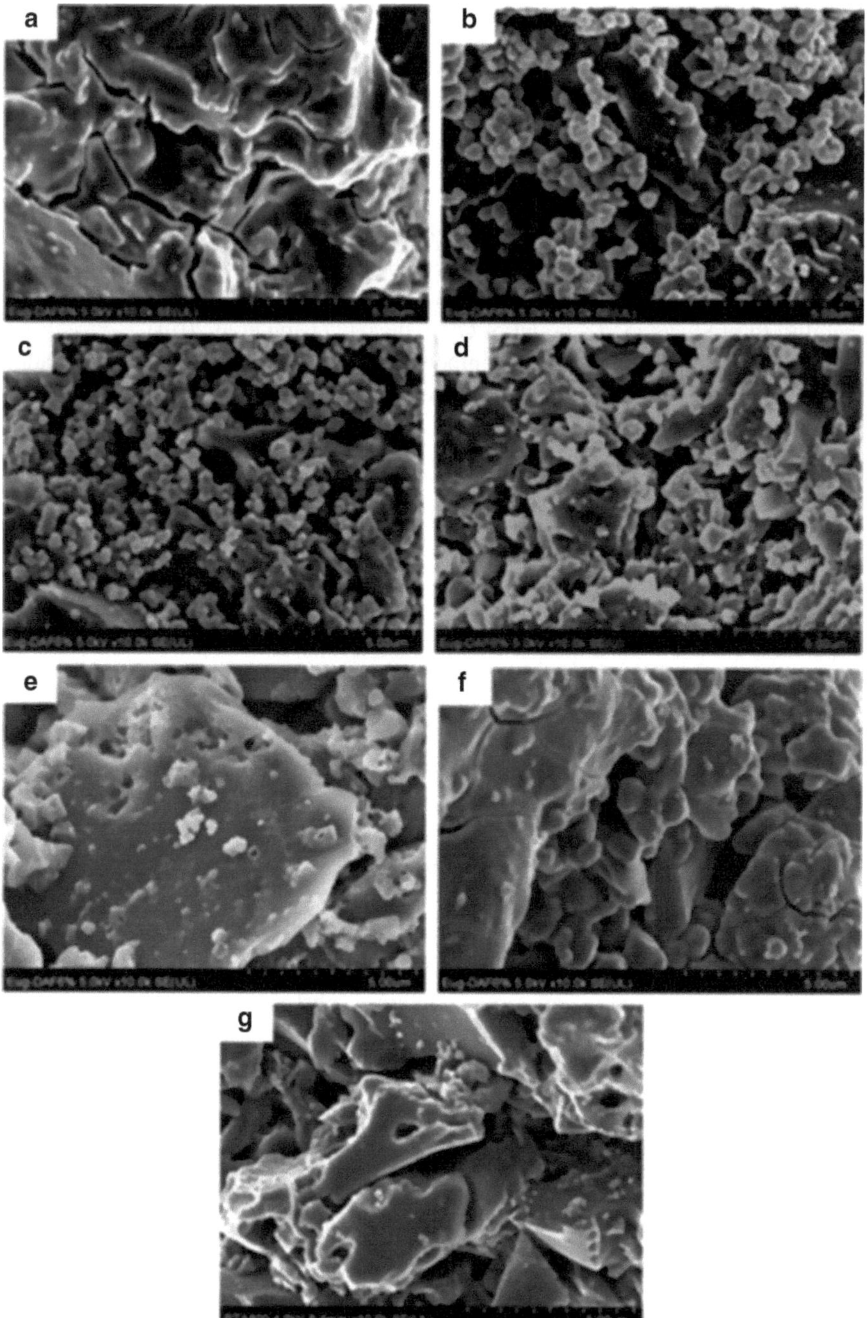

Fig. 3.4 Scanning electron micrographs of TiO_2/AC particles obtained under different synthetic conditions: reaction temperatures were (**a**) 70 °C, (**b**) 80 °C, (**c**) 90 °C, and (**d**) 120 °C under MW heating conditions; using the oil-bath heating method, the reaction temperatures were (**e**) 80 °C and (**f**) 90 °C; (**g**) naked AC as the control (with permission from Horikoshi et al. 2013)

$ZnFe_2O_4$, $NiFe_2O_4$, $MnFe_2O_4$, $CoFe_2O_4$, $BaTiO_3$, $SrTiO_3$, $NaTaO_3$, $BiVO_4$, PbS, CuS, CdS, ZnS, Ag_2S, Bi_2S_3, HgS, $AgInS_2$, etc. (Zhu and Chen 2014).

Cupric oxide (CuO), a p-type semiconductor with a narrow band gap (1.2 eV in bulk), is another example of a photocatalytic material that has been synthesized within minutes, using MW technique. Diverse CuO nanostructures such as nanorods and shuttle-like, flowerlike, platelike, leaflike, dandelion-like, and hollow structures have been prepared by means of this method. Organic dyes can be degraded using CuO as a catalyst and visible light; however, in order to achieve degradation of these compounds, hydrogen peroxide should be added since in the absence of this chemical, CuO is not an effective photocatalyst because of its inability to produce good amount of OH radicals (Zhang et al 2014).

Reports on the preparation of composite-ternary materials by MW irradiation have also appeared in the literature. For example, (RGO)-$BiVO_4$ (RGO: Reduced grapheme oxide) composite was prepared and its photocatalytic activity evaluated for the degradation of ciprofloxacin (Yan et al 2013). Another example relates to the synthesis, characterization, and photoactivity evaluation of Ag@AgBr-intercalated $K_4Nb_6O_{17}$ material (Cui et al 2013). Rare earth composites ($Ce_{1-x}Sm_xO_{2-\delta}$) have also been reported lately (Li et al 2013).

In summary, MW hydrothermal method is considered as a clean, faster, cost-effective, and convenient pathway for the production of photocatalytic materials. Also, the physicochemical properties of the materials can be easily improved with slight modifications of the reaction variables. Photocatalytic materials have been synthesized in various sizes, shapes, and morphologies. The applications of these materials are numerous.

3.7 Chemical Vapor Deposition

Chemical vapor deposition (CVD) is another technique that has been exploited for the preparation of supported titanium dioxide materials. Thin films (supported in glass, metallic substrates, or semiconducting substrates), composite materials (SiO_2/TiO_2, activated carbon/TiO_2, amine polymer/TiO_2,) mesoporous materials, and titanium dioxide nanorods, among others, have been prepared by CVD. Chemical vapor deposition is a technique that is at the boundary of chemical and physical techniques for material production. Chemical methods are the most popular techniques of manufacturing titanium dioxide materials due to relative simplicity of control over synthesis, narrow nanoparticle size distribution, low cost, and reliable stabilization of nanoparticles in the system. On the other hand, physical methods have been less explored, especially for large-scale production of materials since highly specialized equipment and operators are needed for material development. CVD technique allows the production of high purity materials; besides, it does not need the post-heat-treatment to improve the crystallinity of formed material. However, ultrahigh vacuum and precursors with high vapor pressure are required in order to attain material's deposition. Nowadays, a number of techniques have been

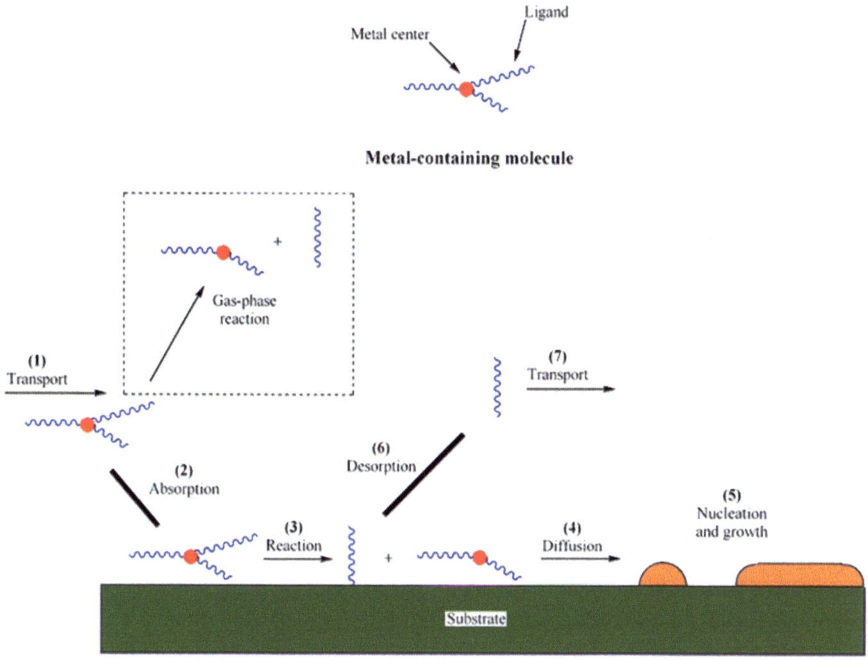

Fig. 3.5 Fabrication of thin film nanostructured materials by CVD technique (Medina-Ramírez 2005)

developed, based on gas deposition principles that present subtle differences, such as the type of precursor employed, the type of substrate, the desired degree of thin film uniformity, and the reaction conditions (Shan et al. 2010). Some examples of these techniques include atmospheric pressure chemical vapor deposition (APCVD), plasma-enhanced chemical vapor deposition (PECVD), metal-organic chemical vapor deposition (MOCVD) which is based on metal-organic precursors, and hybrid physical chemical vapor deposition (HPCVD).

CVD is considered as one of the most precise techniques when conformal deposition of the material is required. In this sense, CVD offers numerous advantages in the deposition of ordered nanostructured thin film of varied compositions (Medina-Ramírez 2005). CVD requires volatile molecular precursors as the source for film growth. The precursor molecules are transported in a stream of carrier gas which may be either chemically inert or reactive. They are converted by chemical reactions in the gas phase near the surface itself, into a thin solid film of the desired material. In CVD processes, heated carrier gases drive the deposition reaction. The main chemical events involved in thin film formation by means of CVD technique are illustrated Fig. 3.5 (Medina-Ramírez 2005).

Sarantopoulos et al. (2009a) fabricated TiO_2 thin films in an isothermal, isobaric, horizontal low-pressure-CVD reactor using borosilicate glass and Si wafers as substrates and titanium isopropoxide as molecular precursor. Besides the different substrates employed, deposition time, temperature, and amount of precursors were varied during deposition of the films. Different morphologies of the films were obtained due to differences in the experimental conditions. SEM micrographs of films grown at 400 °C are represented in Fig. 3.6. It was observed that for thin films (less than 400 nm) deposited on flat substrates, a compact structure is obtained whereas columnar morphology is observed if the thickness of the films is increased. On the other hand, non-flat substrates (glass microfiber) originate columnar growth independent of film thickness. The films were evaluated for the degradation of pollutants in aqueous solution (malic acid, imazapyr, and orange G) and gas phase (toluene). Higher photocatalytic efficiencies were achieved for films deposited on fibrous substrates since they exhibit a high porosity, a columnar morphology, and a high specific surface area.

The fabrication of nonmetal (B, C, N)-doped TiO_2 materials has been reported using CVD. Boron-doped TiO_2 nanotubes with photocatalytic activity under visible light were fabricated. TiO_2 nanotube synthesis was conducted using an electrochemical anodization method. After anodization, TiO_2 nanotube electrode was annealed in boron-containing vapor through CVD to carry out boron doping. Boric acid was selected as the source of boron (Su et al. 2008). Boron doping resulted in structural damage of the nanotubular layer (Fig. 3.7); however, B-doped samples displayed stronger absorption in both UV and visible ranges. The B-doped TiO_2 nanotubes electrode showed good photoelectrocatalytic efficiency and good stability (more than 10 times repeatability) in methyl orange degradation.

A high level of attention has been lately paid to the production of supported visible light active TiO_2 materials. Taking this into account, Sn^{+4}-TiO_2 (Chua et al. 2013), CNT/TiO_2 (Ma et al. 2012), TiO_2/Ag (Piszczek et al. 2013), carbon-doped TiO_2 (Wu et al. 2007a), and nitrogen-doped TiO_2 (Sarantopoulos et al. 2009b) were fabricated, and their environmental applications were explored. From the abovementioned results and some other examples encountered in the literature, it can be concluded that the higher catalytic activity of CVD-prepared TiO_2 materials is a result of the high-quality TiO_2 (bare or doped) film (uniform and regular surface and composition). In addition, the CVD methods can produce single-layer, multilayer, composite, nanostructured, and functionally graded coating materials with well-controlled dimension and unique structures at low deposition temperatures and in a short time (Jin et al. 2013).

Foster and coworkers (2010) reported on the growth of TiO_2/Ag and TiO_2/CuO thin films by atmospheric pressure thermal CVD. The films showed good microbicidal activity even against MRSA (methicillin-resistant *Staphylococcus aureus*) and some other pathogens (Foster et al. 2010). V_2O_5/TiO_2 nanocomposite particles and V_2O_5/TiO_2 nanocomposite films were synthesized using a modified CVD process. The materials were fully characterized and evaluated for the degradation of methylene blue (Song et al 2012a, b).

ZnO and some other chalcogenide films have been deposited by CVD. Afzaal and coworkers (2007) have reviewed the theme of deposition of zinc-containing

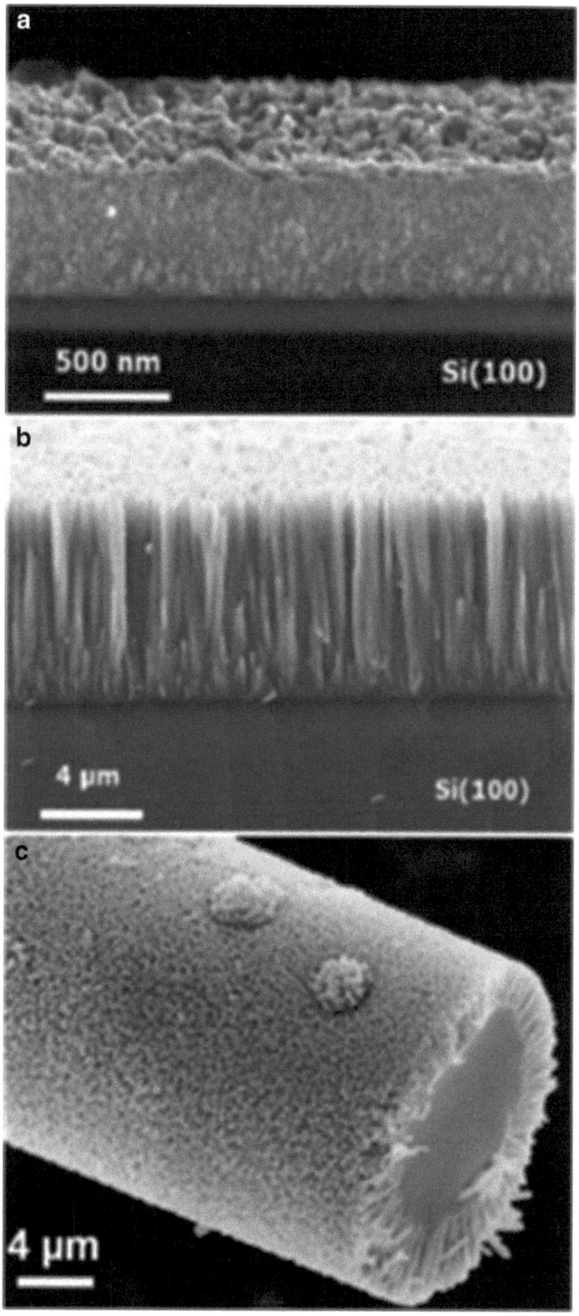

Fig. 3.6 SEM micrographs of TiO_2 films deposited at 400 °C: (**a**) thin film (500 nm) on Si (100) before columnar growth be developed; (**b**) thick film (~8 μm) on Si(100); (**c**) thick film on glass fibers (total pressure 20 Torr; TTIP mole fraction 7×10^{-6}) (with permission from Sarantopoulos et al. 2009a)

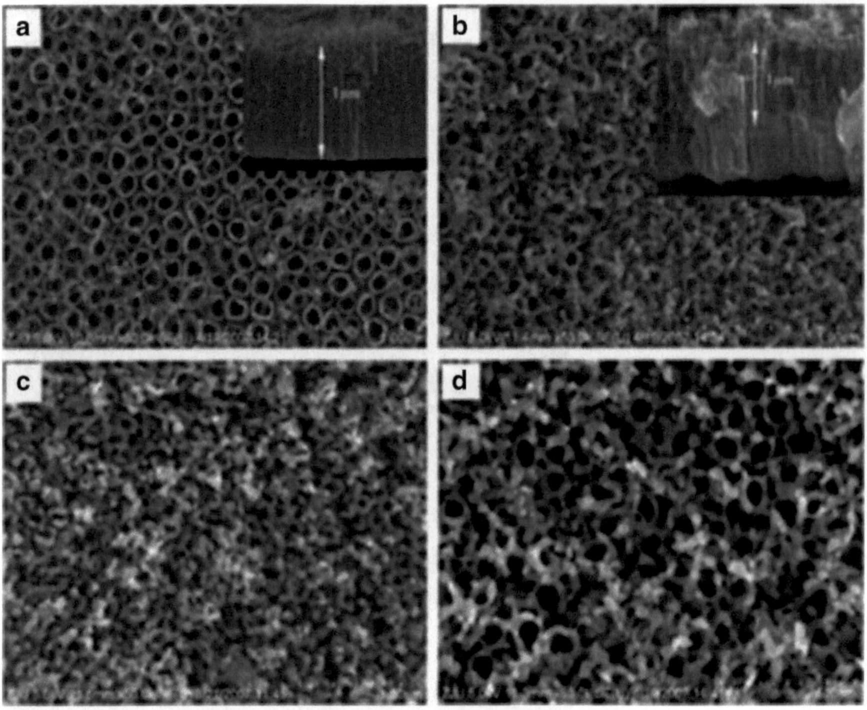

Fig. 3.7 Effect of B doping on the tubular structure of TiO_2. SEM top-view images of (**a**) undoped sample (700 °C), (**b**) B-doped sample (400 °C), (**c**) B-doped sample (550 °C), and (**d**) B-doped sample (700 °C). The cross-sectional images of the sample show the length is approximately 1 μm (with permission from Su et al 2008)

materials based on the use of single-source precursors. Depending upon the physicochemical properties of these compounds, MOCVD and AACVD are frequently used for the deposition of ZnO, ZnS, and ZnSe thin films (Afzaal et al. 2007). Among these materials, ZnO is the one that has encountered applications for environmental degradation; ZnS and ZnSe have been less explored due to low chemical stability.

In summary, CVD is an effective technique for preparing several photocatalytic materials in the form of thin films on a range of substrates. The use of organometallic precursors allows deposition of the materials at atmospheric pressure. Although the cost of large-scale production of photocatalytic materials by means of CVD might be higher in comparison to other fabrication techniques, the physicochemical properties of the deposited films by CVD might counterbalance production investment.

3.8 Physical Vapor Deposition (PVD)

PVD is another deposition technique that can be used for the fabrication of bare and doped titanium dioxide materials. In contrast to CVD process, in PVD, thin films are formed from the gas phase, but here without a chemical transition from precursor to product (Carp et al. 2004). The basic PVD processes fall into two general categories: sputtering and evaporation. Very high deposition rates have been achieved with the advent of electron-beam heated sources. Several PVD techniques are available for the deposition of thin films. Among them are cathodic arc vapor (plasma or arc ion plating) deposition, magnetron sputtering (or sputter ion plating), and combined magnetron and arc processes. These PVD techniques differ with respect to the type of evaporation of the precursors and the plasma conditions employed during the deposition process. In the case of arc evaporation, a small limited cathodic area is evaporated with a very high-energy arc that quickly moves over a spot on the metal surface to be evaporated. The plasma generated consists of highly ionized metal vapor. In the case of sputtering, atoms are ejected mechanically from a target by the impact of ions or energetic neutral atoms (Fig. 3.8). Advances in sputtering techniques have allowed the deposition of different thin films at low temperatures; with this, the use of polymeric substrates is now widely explored for industrial applications.

Yang (2004) and collaborators prepared N-doped TiO_2 films by ion-assisted electron-beam evaporation, using rutile and nitrogen ion bombardment at various nitrogen partial pressures. The photoactivity of the films was evaluated for the degradation of methylene blue. Despite the hydrophilicity, uniform composition, and nanometric nature of the films, the photocatalytic degradation process was lengthy (60 h). Iron-doped TiO_2 films were deposited on glass and polycarbonate substrates by dc reactive magnetron sputtering (Carneiro et al. 2005, 2007). The photodegradation of rhodamine B was evaluated, and it was encountered that degradation rates were higher for the films deposited on polymeric substrates. Later, the synthesis, characterization, and photoelectrochemical properties of

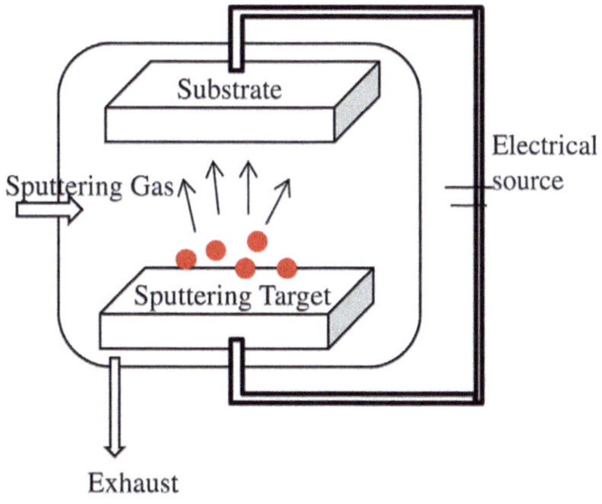

Fig. 3.8 Schematic representation of physical vapor deposition process (with permission from Ângelo et al. 2013)

W-doped TiO_2 thin films were reported. Tungsten-doped TiO_2 is of industrial interest because depending upon the doping content, the material can be useful as a photocatalyst or as a catalyst support in proton-exchange membrane fuel cells. The films were deposited by magnetron sputtering on unheated (001) Si substrates and fluorinated tin oxide-coated glass, using metallic Ti and W targets. The films were annealed in air at 550 °C for 2 h in order to obtain oxidized, crystalline films. The solubility limit of tungsten was found to be ~ 33 at. %; higher doping induced the formation of WO_3 segregated crystalline phase. The films presenting $Ti_{0.92}W_{0.08}O_2$ composition showed the best photoelectrochemical performance for H_2 production (Abadias et al. 2011). N-TiO_2 films were deposited on stainless steel (SS) substrates by oxidative annealing of sputtered TiN coatings (Wang et al. 2011a). Composite material N-TiO_2@SS shows improved wear resistance making it a promising candidate for environmental and biomedical applications. Titanium oxynitride films were prepared by reactive magnetron sputtering (Rawal et al. 2012). An optimization of the deposition variables (such as nitrogen flow rate, deposition time, and sputtering pressure) led to the fabrication of hydrophobic films that show promise for protective, wear, and corrosion resistance applications. Lately, the preparation of sputtered Ag:TiO_2 coatings was reported (Adochite et al. 2012). Interestingly, it was found that the presence of silver delayed the crystallization of TiO_2 matrix. Annealing of the films (~500 °C) led to the formation of anatase materials with improved tribological properties. The use of the PVD method and cryogenic cooling of substrates was explored for the formation of ZnO and TiO_2 films (Bruncko et al. 2013). Analysis of the ZnO and TiO_2 samples that were deposited on cryogenically cooled Si(100) and sapphire (001) substrates demonstrated their amorphous nature. For ZnO, deposition at room temperature resulted in the formation of crystalline films with wurtzite structure. In the case of TiO_2 films, the material was amorphous and with ordered surface morphology, independent of the deposition temperature (Fig. 3.9). Annealing of the films (600 °C) led to the formation of pure anatase materials (Fig. 3.9). Mc Donnell and coworkers implemented a simple, quick, and cost-effective powder coating technique, known as microblasting, for the deposition of TiO_2 material. It is important to remark that the use of this method allows the deposition of thin films at ambient pressure and temperature using compressed air as carrier gas and the deposition can be conducted on a range of substrates (such as polymers, metals, and fluorine tin oxide-coated glass).

Heterostructures (thin films, tilted nanorods, and vertical nanorods) of WO_3/TiO_2 were fabricated by electron-beam deposition. The materials were annealed at different temperatures (300 and 400 °C). The materials annealed at 400 °C exhibited orthorhombic (WO_3) and anatase (TiO_2) crystalline structures, whereas annealing at 300 °C renders WO_3 material with amorphous structure (TiO_2 in anatase structure). The material that showed better photocatalytic performance for the degradation of methylene blue was found to be the two-layer vertical TiO_2/WO_3 nanorods annealed at 300 °C, which had a photodegradation rate about 10 times higher than a single-layer TiO_2 vertical nanorod array. The explanation for this enhanced photoactivity is that the amorphous WO_3 has its conduction band level

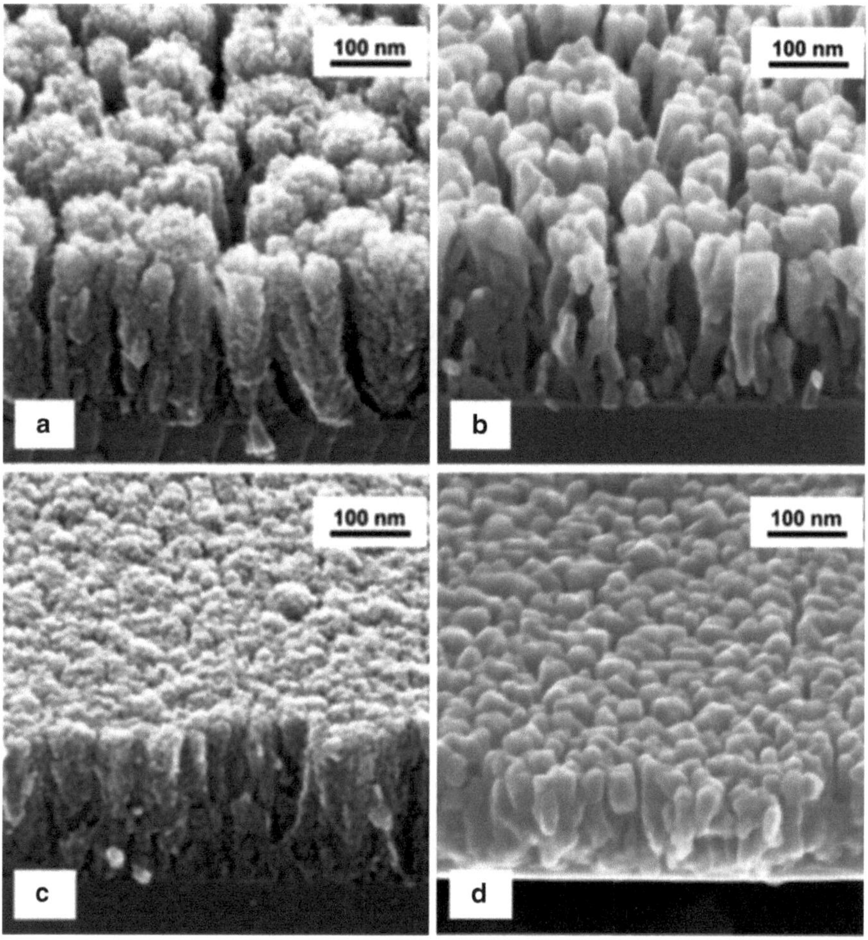

Fig. 3.9 SEM images that illustrate the thermal effect on the morphologies of TiO_2 films deposited at different temperatures on a Si(100) substrate. *Left*: (**a**) substrate near 77 K; (**b**) substrate at R.T. *Right*: after annealing at 600 °C (**a**) substrate near 77 K; (**b**) substrate at R.T. (with permission from Bruncko et al. 2013)

closer to that of TiO_2 and thus allow for easier charge-carrier transfer (Smith and Zhao 2008).

Lately, reactive ballistic deposition (RBD), a method derived from PVD, was reported. By using this technique, dramatic changes in film morphology and film composition can be achieved by modifying a single deposition parameter. During RBD, surface diffusion is limited either by cryogenically cooling the substrate or by the presence of strong interactions between substrate and deposited molecules (atoms) permitting the formation of kinetically trapped surfaces. The material must be vaporized by electron-beam bombardment or pulsed laser sources. High vacuum is necessary so that the evaporant mean free path is greater than the distance between the source and deposition substrate. Several photocatalytic materials have been prepared by this method: TiO_2, α-Fe_2O_3, Sn- and Ti-doped α-Fe_2O_3, $BiVO_4$, S- and C-doped TiO_2, and Mo- and W-doped $BiVO_4$ (Flaherty et al 2012).

Summarizing, PVD techniques have been widely used to deposit a variety of bare and doped TiO_2 thin films for various applications (such as environmental remediation, development of self-cleaning materials, hydrogen production, and wear resistance). Among the different PVD techniques, sputtering is broadly employed for large area deposition of the films. Also, several authors have reported that the use of sputtering method allows for precise control of the thickness, grain size, composition, and microstructure of the film (Ângelo et al. 2013). In the past, the use of sputtering techniques was limited due to its elevated cost and the detailed control of reaction conditions (working pressure, temperature, sputtering gas, reactive gas, time of reaction, among others) that must be kept in order to obtain films with the same composition and thickness; nevertheless, in the past five years, some research groups have been able to abolish the aforementioned problems associated to this technology being able to prepare films at ambient pressure and temperature. In contrast, the large area production of thin films by means of the pulsed laser deposition method is limited due to contamination of the films resulting from particulate formation within the course of film deposition.

3.9 Electrochemical Deposition

Electrochemical synthesis is used for the deposition of particles on metallic or conducting substrates; it may be employed to prepare advanced thin films such as epitaxial, superlattice, quantum dot, and nanoporous ones. In general, a powder material is dispersed in a solvent where two electrodes are immersed, the conducting substrate and the counter electrode, both arranged in parallel. Then, an electric potential is applied to the system and particles move toward the conducting substrate to make the film (Fig. 3.10). Also, varying electrolysis parameters like potential, current density, temperature, and pH can easily control the characteristic states of the films. Although electrodeposition of TiO_2 films by various Ti compounds such as $TiCl_3$ (An et al. 2005, Wang et al. 2013a), TiF_4 (Zhu et al. 2009), and $(NH_4)_2TiO(C_2O_4)_2$ (Ishikawa and Matsumoto 2001) is reported, the use of titanium inorganic salts in aqueous solutions is always accompanied by difficulties, due to the high tendency of the salts to hydrolyze. Therefore, electrolysis requires both an acidic medium and an oxygen-free environment (Chen et al. 2013). Nonaqueous solutions represent an option to overcome this problem (Hu et al. 2011). Numerous reports indicate the fabrication of bare, doped, and composite supported materials of TiO_2 with different morphologies (nanoparticles, nanotubes, nanowires), produced by means of electrochemical methods. For instance, Amadelli et al. (2009) reported on the fabrication of composite PbO_2-TiO_2 films. The materials showed good photoelectrocatalytic activity for the degradation of oxalic acid and benzyl alcohol.

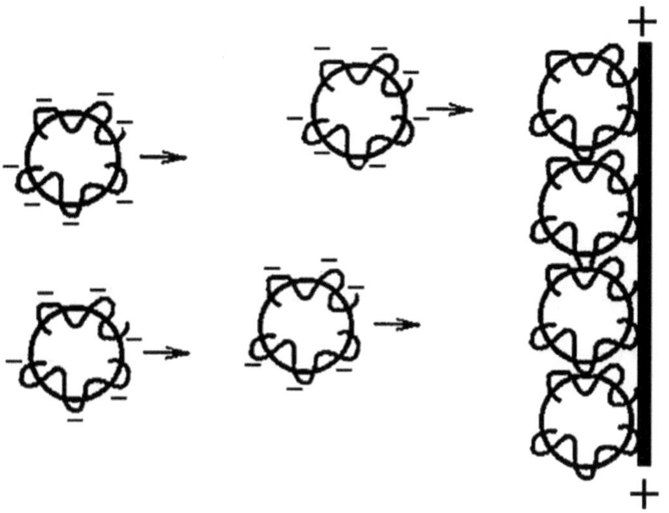

Fig. 3.10 Representation of thin film deposition by electrochemical methods (with permission from Amadelli et al. 2009)

Electrochemical deposition methods have been particularly useful for the fabrication of one-dimensional TiO_2 nanotube and nanowire arrays. In the last decade, these materials have attracted the attention of several research groups since comparative studies show that ordered TiO_2 nanotube arrays present improved characteristics than colloidal TiO_2 for numerous applications (such as sensing, electrochemical water splitting, environmental remediation, dye-sensitized solar cells, drug delivery, and other biomedical applications). Despite the fact that TiO_2 nanotubes have been fabricated by different synthetic approaches, anodization of titanium in a fluoride-based electrolyte renders precise control over the morphology and dimensions of the produced materials (Grimes and Mor 2009). By modification and optimization of electrochemical deposition variables, the production of TiO_2 nanotube arrays presenting different morphologies and properties has been accomplished. In Fig. 3.11, different morphologies of nanotube arrays fabricated by Ti anodization are illustrated. It can clearly be seen that variations of deposition conditions greatly affect the architecture of the material going from well-separated, stand-alone nanotubes to densely packed arrays. A typical nanotube's wall thickness ranges from 5 to 30 nm, the pore size from 20 to 350 nm, and the length from 0.2 to 1,000 mm (Grimes and Mor 2009).

Titanium dioxide-supported materials have been used for the fabrication of the working electrode of dye-sensitized solar cells; however, sintering of the material decreases its electrochemical properties. It has been demonstrated that electrodeposition provides a simple and inexpensive alternative for the fabrication of TiO_2 films at low temperature on different substrates (flat or non-flat materials). Nanocrystalline TiO_2 films were electrochemically deposited on SnO_2:F conductive glass substrates by cyclic voltammetry. The films were fabricated using $TiCl_4$ and

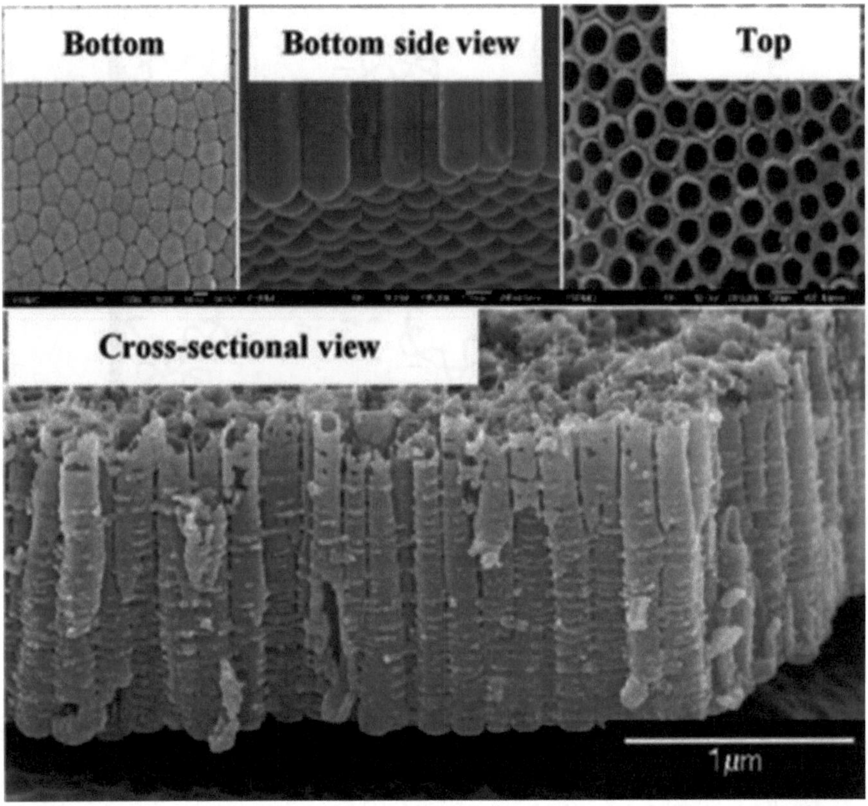

Fig. 3.11 Nanotube arrays deposited by Ti anodization. The effect of the solvent in the architecture of TiO_2 nanotube arrays is evaluated. The *upper* correspond to Ti anodization in an ethylene glycol bath; the *lower* image presents the material prepared in an aqueous KF bath (with permission from Grimes and Mor 2009)

$TiCl_3$ as the starting material and CTAB as the deposition-promoting agent (An et al. 2005). Also, Nguyen et al. (2007) reported on the preparation of TiO_2/SiO_2 nanocomposite electrode thin films. It is well known that the recombination of electrons and holes generated by band gap excitation diminishes the efficiency of TiO_2 materials. Incorporation of metal nanoparticles into the TiO_2 matrix has been used, since these particles can act as electron acceptors reducing the recombination reaction. Taking this into account, composites of nickel nanoparticle chains embedded in TiO_2 nanotube arrays (Fig. 3.12) were prepared using electrochemical method (Zhu et al. 2009). Another report indicates the synthesis of highly flexible composites of TiO_2 nanotubes with different conducting materials: polymers (polypyrrole, poly(3-hexylthiophene)), inorganic semiconducting materials (CdS), and metals (Ni and Au).

Different types of nanocomposites were fabricated by electrodeposition of a number of materials (conducting polymers, semiconducting materials, and metals)

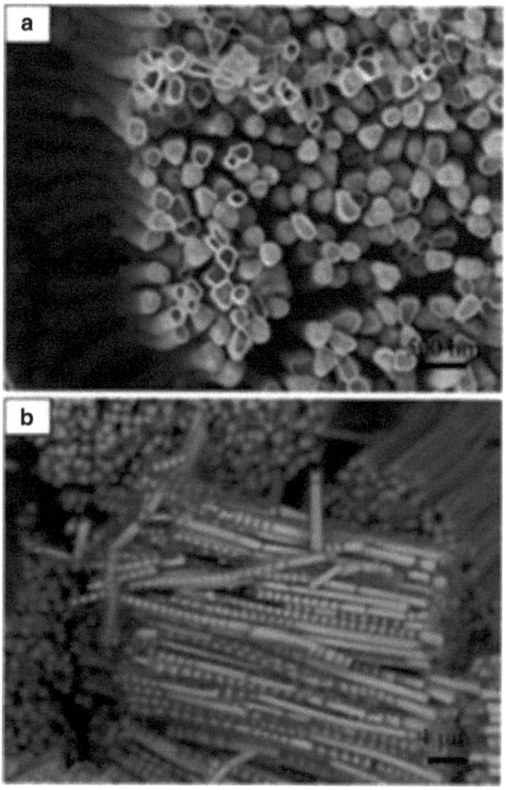

Fig. 3.12 Microstructures of Ni-TiO_2 composites fabricated by electrochemical method using TiF_4 and $NiCl_2{\cdot}6H_2O$ as precursors. (**a**) *Top view* secondary electron image of the nanotube arrays; (**b**) side view backscattered electron image of the nickel nanoparticle chains in nanotubes (with permission from Zhu et al. 2009)

into top-porous TiO_2 nanotubes. For instance, flexible TiO_2-NT-P3HT [poly (3-hexylthiophene)], TiO_2-NT-CdS, and TiO_2-NT-Ni nanohybrids can be prepared. These materials are of great importance for the development of solar cells (Wang et al 2009a).

Concluding Remarks

In this chapter, an extensive variety of synthetic methodologies for the preparation of photocatalytic materials has been presented. A brief description of each technique was provided, and several examples were enumerated in order to demonstrate the influence of the preparation method in the properties of the catalysts and, hence, in their photocatalytic activity. Materials can be prepared as powders or supported as films with different morphologies, surface textures, particles size, or crystalline phases, depending on the preparation method and the synthesis conditions. These features play an important role in the photocatalytic performance of the semiconducting

(continued)

materials; thus, election of the synthetic technique will depend on the infrastructure availability, the cost, and the properties of the catalyst that demand its application. We intended to show some of the several synthesis methods within this fascinating research field as well as their versatility to fabricate complex materials with tailor-made properties.

References

Abadias G, Gago AS, Alonso-Vante N (2011) Structural and photoelectrochemical properties of $Ti_{1-x}W_xO_2$ thin films deposited by magnetron sputtering. Surf Coat Technol 205:S265–S270

Abdullah HZ, Sorrell CC (2007) TiO_2 thick films by anodic oxidation. J Aust Ceram Soc 43:125–130

Adochite RC, Munteanu D, Torrell M, Cunha L, Alves E, Barradas NP, Cavaleiro A, Riviere P, Le Bourhis E, Eyidi D, Vaz F (2012) The influence of annealing treatments on the properties of Ag:TiO_2 nanocomposite films prepared by magnetron sputtering. Appl Surf Sci 258:4028–4034

Afzaal M, Malik M, O'Brien P (2007) Preparation of zinc containing materials. New J Chem 31:2029–2040

Amadelli R, Samiolo L, Velichenko AB, Knysh VA, Luk'yanenko TV, Danilov FI (2009) Composite PbO_2–TiO_2 materials deposited from colloidal electrolyte: Electrosynthesis, and physicochemical properties. Electrochim Acta 54:5239–5245

An HJ, Jang SR, Vittal R, Lee J, Kim KJ (2005) Cationic surfactant promoted reductive electrodeposition of nanocrystalline anatase TiO_2 for application to dye-sensitized solar cells. Electrochim Acta 50:2713–2718

Anandan S, Sivasankar T, Lana-Villarreal T (2014) Synthesis of TiO_2/WO_3 nanoparticles via sonochemical approach for the photocatalytic degradation of methylene blue under visible light illumination. Ultrason Sonochem 21:1964–1968

Ângelo J, Andrade L, Madeira LM, Mendes A (2013) An overview of photocatalysis phenomena applied to NOx abatement. J Environ Manage 129:522–539

Arin J, Thongtem S, Thongtem T (2013) Single-step synthesis of ZnO/TiO_2 nanocomposites by microwave radiation and their photocatalytic activities. Mater Lett 96:78–81

Baldassari S, Komarneni S, Mariani E, Villa C (2005) Microwave-hydrothermal process for the synthesis of rutile. Mater Res Bull 40:2014–2020

Bilecka I, Djerdj I, Niederberger M (2008) One-minute synthesis of crystalline binary and ternary metal oxide nanoparticles. Chem Commun 2008:886–888

Bilecka I, Elser P, Niederberger M (2009) Kinetic and thermodynamic aspects in the microwave-assisted synthesis of ZnO nanoparticles in benzyl alcohol. ACS Nano 3(2):467–477

Biswas SK, Baeg JO (2013) A facile one-step synthesis of single crystalline hierarchical WO_3 with enhanced activity for photoelectrochemical solar water oxidation. Int J Hydrogen Energy 38:3177–3188

Brinker CJ, Scherer GW (1990) Sol-gel science. Academic, San Diego

Bruncko J, Netrvalova M, Vincze A, Sutta P, Miroslav Michalka M, Uherek F (2013) Pulsed laser deposition of thin films on actively cooled substrates. Vacuum 98:56–62

Byrappa K, Yoshimura M (2001) Handbook of hydrothermal technology. A Technology for Crystal Growth and Materials Processing. Noyes, New York

Carneiro JO, Teixeira V, Portinha AL, Dupák L, Magalhaes A, Coutinho P (2005) Study of the deposition parameters and Fe-dopant effect in the photocatalytic activity of TiO_2 films prepared by dc reactive magnetron sputtering. Vacuum 78:37–46

Carneiro JO, Teixeira V, Portinha AL, Magalhaes A, Coutinho P, Tavares CJ, Newton R (2007) Iron-doped photocatalytic TiO_2 sputtered coatings on plastics for self-cleaning applications. Mater Sci Eng B 138:144–150

Carp O, Huisman CL, Reller A (2004) Photoinduced reactivity of titanium dioxide. Prog Solid State Chem 32:33–177

Chandrappa KG, Venkatesha TV (2012) Electrochemical synthesis and photocatalytic property of zinc oxide nanoparticles. Nano-Micro Lett 4:14–24

Chen YC, Lo SL (2011) Effects of operational conditions of microwave-assisted synthesis on morphology and photocatalytic capability of zinc oxide. Chem Eng J 170:411–418

Chen X, Mao SS (2007) Titanium dioxide nanomaterials: Synthesis, properties, modifications, and applications. Chem Rev 107:2891–2959

Chen Z, Li W, Zeng W, Li M, Xiang J, Zhou Z, Huang J (2008) Microwave hydrothermal synthesis of nanocrystalline rutile. Mater Lett 62:4343–4344

Chen B, Hou J, Lu K (2013) Formation mechanism of TiO_2 nanotubes and their applications in photoelectrochemical water splitting and super-capacitors. Langmuir 29:5911–5919

Choi J, Cho SH, Kim TH, Lee SW (2011) Comparison of sonochemistry method and sol-gel method for the fabrication of TiO_2 powder. Mater Sci Forum 695:109–112

Chua C, Tan OK, Tse MS, Ding X (2013) Photocatalytic activity of tin-doped TiO_2 film deposited via aerosol assisted chemical vapor deposition. Thin Solid Films 544:571–575

Cui L, Hui KN, Hui KS, Lee SK, Zhou W, Wan ZP, Thuc CNH (2012) Facile microwave-assisted hydrothermal synthesis of TiO_2 nanotubes. Mater Lett 75:175–178

Cui W, Wang H, Liang Y, Han B, Liu L, Hu J (2013) Microwave-assisted synthesis of Ag@AgBr-intercalated $K_4Nb_6O_{17}$ composite and enhanced photocatalytic degradation of Rhodamine B under visible light. Chem Eng J 230:10–18

Del Ángel-Sanchez K, Vázquez-Cuchillo O, Aguilar-Elguezabal A, Cruz-López A, Herrera-Gómez A (2013) Photocatalytic degradation of 2,4-dichlorophenoxyacetic acid under visible light: Effect of synthesis route. Mater Chem Phys 139:423–430

Esquivel K, Nava R, Zamudio-Méndez A, Vega González M, Jaime-Acuña OE, Escobar-Alarcón L, Peralta-Hernández JM, Pawelec B, Fierro JLG (2013) Microwave-assisted synthesis of (S)Fe/TiO_2 systems: Effects of synthesis conditions and dopant concentration on photoactivity. Appl Catal B Environ 140–141:213–224

Estruga M, Domingo C, Ayllón JA (2010) Microwave radiation as heating method in the synthesis of titanium dioxide nanoparticles from hexafluorotitanate-organic salts. Mater Res Bull 45:1224–1229

Flaherty DW, Hahn NT, May RA, Berglund SP, Lin YM, Stevenson KJ, Dohnalek Z, Kay BD, Mullins B (2012) Reactive ballistic deposition of nanostructured model materials for electrochemical energy conversion and storage. Acc Chem Res 45(3):434–443

Foster HA, Sheel DW, Sheel P, Evans P, Varghese S, Rutscke N, Yates HM (2010) Antimicrobial activity of titania/Silver and titania/Copper films prepared by CVD. J Photochem Photobiol A 216:283–289

Geyde RN (2002) Organic synthesis using microwaves in homogeneous media. In: Loupy A (ed) Microwaves in organic synthesis. Wiley-VCH, Weinheim, pp 115–146

Gionco C, Battiato A, Vittone E, Paganini MC, Giamello E (2013) Structural and spectroscopic properties of high temperature prepared ZrO_2-TiO_2 mixed oxides. J Solid State Chem 201:222–228

Grimes CA, Mor GK (2009) Fabrication of TiO_2 nanotube arrays by electrochemical anodization: four synthesis generations. In: Grimes CA, Mor GK (eds) TiO_2 nanotube arrays. Synthesis, properties, and applications. Springer, New York, pp 1–66

Guo S, Wu Z, Wang H, Dong F (2009) Synthesis of mesoporous TiO_2 nanorods via a mild template-free sonochemical route and their photocatalytic performances. Catal Commun 10:1766–1770

Guo S, Han S, Haifeng M, Zeng C, Sun Y, Chi B, Pu J, Li J (2013) Synthesis of phosphorus-doped titania with mesoporous structure and excellent photocatalytic activity. Mater Res Bull 48:3032–3036

Han C, Luque R, Dionysiou DD (2012) Facile preparation of controllable size monodisperse anatase titania nanoparticles. Chem Commun 48:1860–1862

He F, Li J, Li T, Li G (2014a) Solvothermal synthesis of mesoporous TiO_2: The effect of morphology, size and calcination progress on photocatalytic activity in the degradation of gaseous benzene. Chem Eng J 237:312–321

He F, Ma F, Li J, Li T, Li G (2014b) Effect of calcination temperature on the structural properties and photocatalytic activities of solvothermal synthesized TiO_2 hollow nanoparticles. Ceram Int 40:6441–6446

Hernández A, Maya L, Sánchez-Mora E, Sánchez EM (2007) Sol-gel synthesis, characterization and photocatalytic activity of mixed oxide ZnO-Fe_2O_3. J Sol-Gel Sci Technol 42:71–78

Horikoshi S, Sakamoto S, Serpone N (2013) Formation and efficacy of TiO_2/AC composites prepared under microwave irradiation in the photoinduced transformation of the 2-propanol VOC pollutant in air. Appl Catal B Environ 140–141:646–665

Hu L, Huo K, Chen R, Gao B, Fu J, Chu PK (2011) Recyclable and high-sensitivity electrochemical biosensing platform composed of carbon-doped TiO_2 nanotube arrays. Anal Chem 83:8138–8144

Huang W, Tang X, Wang Y, Koltypin Y, Gedanken A (2000) Selective synthesis of anatase and rutile via ultrasound irradiation. Chem Commun 15:1415–1416

Ishikawa Y, Matsumuto Y (2001) Electrodeposition of TiO_2 photo-catalyst into nano-pores of hard alumite. Electrochim Acta 46:2819–2824

Jin N, Yang Y, Luo X, Xia Z (2013) Development of CVD Ti-containing films. Prog Mater Sci 58:1490–1533

Jung YS, Kim KH, Jang TY, Tak Y, Baeck SH (2011) Enhancement of photocatalytic properties of Cr_2O_3-TiO_2 mixed oxides prepared by sol-gel method. Curr Appl Phys 11:358–361

Kappe CO (2013) Unraveling the mysteries of microwave chemistry using silicon carbide reactor technology. Acc Chem Res 46(7):1579–1587

Kitchen HJ, Vallance SR, Kennedy JL, Tapia-Ruiz N, Carassiti L, Harrison A, Whittaker AG, Drysdale TD, Kingman SW, Gregory DH (2014) Modern microwave methods in solid-state inorganic materials chemistry: from fundamentals to manufacturing. Chem Rev 114:1170–1206

Komarneni S, Rajha RK, Katsuki H (1999) Microwave-hydrothermal processing of titanium dioxide. Mater Chem Phys 61:50–54

Komarneni S, Esquivel S, Noh YD, Sitthisang S, Tantirungrotechai J, Li H, Yin S, Sato T, Katsuki H (2014) Novel synthesis of nanophase anatase under conventional-and microwave-hydrothermal conditions : DeNOx properties. Ceram Int 40:2097–2102

Li Y, Li H, Li T, Li G, Cao R (2009) Facile synthesis of mesoporous titanium dioxide nanocomposites with controllable phase compositions by micro-wave-assisted esterification. Microporous Mesoporous Mater 117:444–449

Li X, Wang L, Lu X (2010) Preparation of silver-modified TiO_2 via microwave-assisted method and its photocatalytic activity for toluene degradation. J Hazard Mater 177:639–647

Li L, Qin X, Wang G, Qi L, Du G, Hu Z (2011) Synthesis of anatase TiO_2 nanowires by modifying TiO_2 nanoparticles using the microwave heating method. Appl Surf Sci 257:8006–8012

Li X, Hu Z, Zhao X, Lu X (2013) $Ce_{1-x}Sm_xO_{2-\delta}$-attapulgite nanocomposites: synthesis via simple microwave approach and investigation of its catalytic activity. J Rare Earths 31(12):1157–1162

Linnik O, Smirnova N, Korduban O, Eremenko A (2013) Gold nanoparticles into Ti-xZnxO2 films: Synthesis, structure and application. Mater Chem Phys 142:318–324

Liu Y, Yang S, Hong J, Sun C (2007) Low-temperature preparation and microwave photocatalytic activity study of TiO_2-mounted activated carbon. J Hazard Mater 142:208–215

Liu M, Piao LY, Wang WJ (2011) Hierarchical TiO_2 spheres: Facile fabrication and enhanced photocatalysis. Rare Metals 30:153–156

Liu N, Chen X, Zhang J, Schwank JW (2013a) A review on TiO_2-based nanotubes synthesized via hydrothermal method: Formation mechanism, structure modification, and photocatalytic applications. Catal Today 225:34–51

Liu PI, Chung LC, Shao H, Liang TM, Horng RY, Ma CCM, Chang MC (2013b) Microwave-assisted ionothermal synthesis of nanostructured anatase titanium dioxide/activated carbon composite as electrode material for capacitive deionization. Electrochim Acta 96:173–179

Liu Y, Li Z, Lv H, Tang H, Xing X (2013c) Synthesis of hierarchical Bi_2WO_6 microspheres with high visible-light-driven photocatalytic activities by sol-gel-hydrothermal route. Mater Lett 108:84–87

Liu N, Chen X, Zhang J, Schwank JW (2014) A review on TiO_2-based nanotubes synthesized via hydrothermal method: Formation mechanism, structure modification, and photocatalytic applications. Catal Today 225:34–51

Livraghi S, Pelaez M, Biedrzycki J, Corazzari I, Giamello E, Dionysiou DD (2013) Influence of the chemical synthesis on the physicochemical properties of N-TiO_2 nanoparticles. Catal Today 209:54–59

Ma L, Chen A, Zhang Z, Lu J, He H, Li C (2012) In-situ fabrication of CNT/TiO_2 interpenetrating network film on nickel substrate by chemical vapour deposition and application in photoassisted water electrolysis. Catal Commun 21:27–31

Mason TJ, Lorimer JP (2003) Front matter and subject index. In applied sonochemistry: uses of power ultrasound in chemistry and processing. Wiley-VCH Verlag GmbH & Co. KGaA, Weinheim, FRG. doi:10.1002/352760054X.fmatter_indsub

Maya-Treviño ML, Guzmán-Mar JL, Hinojosa-Reyes L, Ramos-Delgado NA, Maldonado MI, Hernández-Ramírez A (2014) Activity of the ZnO–Fe_2O_3 catalyst on the degradation of Dicamba and 2,4-D herbicides using simulated solar light. Ceram Int 40:8701–8708

Medina-Ramírez I (2005) Chalcopyrite CVD Precursors containing Silylated Ligands. Ph.D. dissertation. Tulane University

Mera AC, Moreno Y, Pivan JY, Peña O, Mansilla HD (2014) Solvothermal synthesis of BiOI microspheres: Effect of the reaction time on the morphology and photocatalytic activity. J Photochem Photobiol A 289:7–13

Mohamad-Alosfur FK, Jumali MHH, Radiman S, Ridha NJ, Umar AA (2014) In-situ dressing of MWCNTs with TiO_2 nanoparticles: Microwave assisted synthesis towards water treatment. Procedia Technol 12:264–270

Nguyen TV, Lee HC, Khan MA, Yang OB (2007) Electrodeposition of TiO_2/SiO_2 nanocomposite for dye-sensitized solar cell. Sol Energy 81:529–534

Niederberger M, Pinna N (2009) Metal oxide nanoparticles in organic solvents. Synthesis, formation, assembly and application. Springer, London

Palanisamy B, Babu CM, Sundaravel B, Anandan S, Murugesan V (2013) Sol-gel synthesis of mesoporous mixed Fe_2O_3/TiO_2 photocatalyst: Application for degradation of 4-chlorophenol. J Hazard Mater 252–253:233–242

Pingyu W, Qinglin Y, Lin G (2009) Bismuth oxyhalide compounds as photocatalysts. Prog Chem 21:1734–1741

Piszczek P, Muchewicz Z, Radtke A, Gryglas M, Dahm H, Różycki H (2013) CVD of TiO_2 and TiO_2/Ag antimicrobial layers: deposition from the hexanuclear μ-oxo Ti(IV) complex as a precursor, and the characterization. Surf Coat Technol 222:38–43

Raji JR, Palanivelu K (2011) Sunlight-induced photocatalytic degradation of organic pollutants by carbon-modified nanotitania with vegetable oil as precursor. Ind Eng Chem Res 50:3130–3138

Ramos-Delgado NA, Hinojosa-Reyes L, Guzman-Mar JL, Gracia-Pinilla MA, Hernández-Ramírez A (2013) Synthesis by sol-gel of WO_3/TiO_2 for solar photocatalytic degradation of malathion pesticide. Catal Today 209:35–40

Rawal SK, Chawla AK, Jayaganthan R, Chandra R (2012) Structural, wettability and optical investigation of titanium oxynitride coatings: effect of various sputtering parameters. J Mater Sci Technol 28(6):512–523

Saragiotto Colpini LMS, Alves HJ, Andreo dos Santos OA, Macedo Costa CM (2008) Discoloration and degradation of textile dye aqueous solutions with titanium oxide catalysts obtained by the sol-gel method. Dyes Pigm 76:525–529

Sarantopoulos C, Puzenat E, Guillard C, Herrmann JM, Gleizes AN, Maury F (2009a) Microfibrous TiO_2 supported photocatalysts prepared by metal-organic chemical vapor infiltration for indoor air and waste water purification. Appl Catal B Environ 91:225–233

Sarantopoulos C, Gleizes A, Maury F (2009b) Chemical vapor deposition and characterization of nitrogen doped TiO_2 thin films on glass substrates. Thin Solid Films 518(4):1299–1303

Shan AY, Ghazi IM, Rashid SA (2010) Immobilisation of titanium dioxide onto supporting materials in heterogeneous photocatalysis: A review. Appl Catal A 389:1–8

Shirsath SR, Pinjari DV, Gogate PR, Sonawane SH, Pandit AB (2013) Ultrasound assisted synthesis of doped TiO_2 nano-particles: characterization and comparison of effectiveness for photocatalytic oxidation of dyestuff effluent. Ultrason Sonochem 20:277–286

Shojaie AF, Loghmani MH (2010) La^{3+} and Zr^{4+} co-doped anatase nano TiO_2 by sol-microwave method. Chem Eng J 157:263–269

Siddiquey IA, Furusawa T, Sato M, Bahadur NM, Alam MM, Suzuki N (2012) Sonochemical synthesis, photocatalytic activity and optical properties of silica coated ZnO nanoparticles. Ultrason Sonochem 19:750–755

Smith W, Zhao Y (2008) Enhanced photocatalytic activity by aligned WO_3/TiO_2 two-layer nanorod arrays. J Phys Chem C 112:19635–19641

Song MY, Chin S, Jurng J, Park Y (2012a) One step simultaneous synthesis of modified-CVD-made V_2O_5/TiO_2 nanocomposite particle. Ceram Int 38:2613–2618

Song MY, Park Y, Jurng J (2012b) Direct coating of V_2O_5/TiO_2 nano-particles onto glass beads by chemical vapor deposition. Powder Technol 231:135–140

Sreethawong T, Ngamsinlapasathian S, Yoshikawa S (2012) Surfactant-aided sol-gel synthesis of mesoporous-assembled TiO_2-NiO mixed oxide nanocrystals and their photocatalytic azo dye degradation activity. Chem Eng J 192:292–300

Su C, Hong BY, Tseng CM (2004) Sol-gel preparation and photocatalysis of titanium dioxide. Catal Today 96:119–126

Su Y, Han S, Zhang X, Chen X, Lei L (2008) Preparation and visible-light-driven photoelectrocatalytic properties of boron-doped TiO_2 nanotubes. Mater Chem Phys 110:239–246

Su Z, Zhang L, Jiang F, Hong M (2013) Formation of crystalline TiO_2 by anodic oxidation of titanium. Prog Nat Sci 23:294–301

Sun JQ, Shen XP, Chen KM, Liu Q, Liu W (2008) Low-temperature synthesis of hexagonal ZnS nanoparticles by a facile microwave-assisted single-source method. Solid State Commun 147:501–504

Tanasković N, Radovanović Ž, Dokić V, Krstić J, Drmanić S, Janaćković D, Petrović R (2009) Synthesis of mesoporous nanocrystalline titania powders by nonhydrolitic sol-gel method. Superlattices Microstruct 46:217–222

Tang H, Zhang D, Tang G, Ji X, Li W, Li C, Yang X (2013) Hydrothermal synthesis and visible-light photocatalytic activity of α-Fe_2O_3/TiO_2 composite hollow microspheres. Ceram Int 39:8633–8640

Tobaldi DM, Pullar RC, Gualtieri AF, Seabra MP, Labrincha JA (2013) Sol–gel synthesis, characterisation and photocatalytic activity of pure, W-, Ag- and W/Ag co-doped TiO_2 nanopowders. Chem Eng J 214:364–375

Wang D, Liu Y, Wang C, Zhou F, Weimin Liu W (2009a) Highly flexible coaxial nanohybrids made from porous TiO_2 nanotubes. ACS Nano 3(5):1249–1257

Wang N, Li X, Wang Y, Quan X, Chen G (2009b) Evaluation of bias potential enhanced photocatalytic degradation of 4-chlorophenol with TiO_2 nanotube fabricated by anodic oxidation method. Chem Eng J 146:30–35

Wang H-F, Tang B, Li X-Y (2011a) Microstructure and wear resistance of N-doped TiO_2, coatings grown on stainless steel by plasma surface alloying technology. J Iron Steel Res Int 18(7):73–78

Wang HJ, Sun YY, Cao Y, Yu XH, Ji XM, Yang L (2011b) Porous zinc oxide films: controlled synthesis, cytotoxicity and photocatalytic activity. Chem Eng J 178:8–14

Wang XK, Wang C, Guo WL, Wang JG (2011c) A novel single-step synthesis of N-doped TiO_2 via a sonochemical method. Mater Res Bull 46:2041–2044

Wang Y, Wu Y, Qin Y, Xu G, Hu X, Cui J, Zheng H, Hong Y, Zhang X (2011d) Rapid anodic oxidation of highly ordered TiO_2 nanotube arrays. J Alloys Compd 509:L157–L160

Wang Q, Wu W, Chen J, Chu G, Ma K, Zou H (2012) Novel synthesis of ZnPc/TiO_2 composite particles and carbon dioxide photo-catalytic reduction efficiency study under simulated solar radiation conditions. Colloids Surf A 409:118–125

Wang H, Song Y, Liu W, Yao S, Zhang W (2013a) Template synthesis and characterization of TiO_2 nanotube arrays by the electrodeposition method. Mater Lett 93:319–321

Wang XJ, Yang WY, Li FT, Xue YB, Liu RH, Hao YJ (2013b) In situ microwave-assisted synthesis of porous N-TiO_2/g-C_3N_4 heterojunctions with enhanced visible-light photocatalytic properties. Ind Eng Chem Res 52:17140–17150

Wu X, Jiang QZ, Ma ZF, Fu M, Shangguan WF (2005) Synthesis of titania nanotubes by microwave irradiation. Solid State Commun 136:513–517

Wu G, Nishikawa T, Ohtani B, Chen A (2007a) Synthesis and characterization of carbon-doped TiO_2 nanostructures with enhanced visible light response. Chem Mater 19:4530–4537

Wu X, Jiang QZ, Ma ZF, Shangguan WF (2007b) Tile overlapping model for synthesizing TiO_2 nanotubes by microwave irradiation. Solid State Commun 143:343–347

Wu Z, Guo S, Wang H, Liu Y (2009) Synthesis of immobilized TiO_2 nanowires by anodic oxidation and their gas phase photocatalytic properties. Electrochem Commun 11:1692–1695

Wu Z, Wu Q, Du L, Jiang C, Piao L (2013) Progress in the synthesis and applications of hierarchical flower-like TiO_2 nanostructures. Particuology. doi:10.1016/j.partic.2013.04.003

Xie L, Wang J, Hu Y, Zheng Z, Weng S, Liu P, Shi X, Wang D (2012) Template-free microwave-assisted hydrothermal synthesis and photocatalytic performance of $Bi_6O_6(OH)_3(NO_3)_3{\cdot}1.5H_2O$ nanosheets. Mater Chem Phys 136:309–312

Xu J, Ao Y, Chen M, Fu D (2009) Low-temperature preparation of Boron-doped titania by hydrothermal method and its photocatalytic activity. J Alloys Compd 484:73–79

Yan Y, Sun S, Song Y, Yan X, Guan W, Liu X, Shi W (2013) Microwave-assisted in situ synthesis of reduced graphene oxide-$BiVO_4$ composite photocatalysts and their enhanced photocatalytic performance for the degradation of ciprofloxacin. J Hazard Mater 250–251:106–114

Yang MC, Yang TS, Wong MS (2004) Nitrogen-doped titanium oxide films as visible light photocatalyst by vapor deposition. Thin Solid Films 469–470:1–5

Yang D, Liu H, Zheng Z, Yuan Y, Zhao JC, Waclawik ER, Ke X, Zhu H (2009) An efficient photocatalyst structure: TiO_2(B) nanofibers with a shell of anatase nanocrystals. J Am Chem Soc 131:17885–17893

Yu C, Fan Q, Xie Y, Chen J, Shu Q, Yu JC (2012) Sonochemical fabrication of novel square-shaped F doped TiO_2 nanocrystals with enhanced performance in photocatalytic degradation of phenol. J Hazard Mater 237–238:38–45

Zaleska A (2008) Doped TiO_2: a review. Recent Patents Eng 2:157–164

Zhang P, Yin S, Sato T (2009) Synthesis of high-activity TiO_2 photocatalyst via environmentally friendly and novel microwave assisted hydrothermal process. Appl Catal B Environ 89:118–122

Zhang Q, Zhang K, Xu D, Yang G, Huang H, Nie F, Liu C, Yang S (2014) CuO nanostructures: synthesis, characterization, growth mechanisms, fundamental properties, and applications. Prog Mater Sci 60:208–337
Zhu Y, Chen F (2014) Microwave-assisted preparation of inorganic nanostructures in liquid phase. Chem Rev 114:6462–6555
Zhu W, Wang G, Hong X, Shen X, Li D, Xie X (2009) Metal nanoparticle chains embedded in TiO_2 nanotubes prepared by one-step electrodeposition. Electrochim Acta 55:480–484
Zhu L, Liu K, Li H, Sun Y, Qiu M (2013) Solvothermal synthesis of mesoporous TiO_2 microspheres and their excellent photocatalytic performance under simulated sunlight irradiation. Solid State Sci 20:8–14

Chapter 4
Physicochemical Characterization of Photocatalytic Materials

Monserrat Bizarro and Sandra E. Rodil

Abstract A wide range of analytical techniques has been employed to obtain the physical–chemical properties of photocatalytic semiconductors prepared as powders or thin films. The photocatalytic activity of a semiconductor material depends on both surface and structural properties. For bulk materials, the composition, crystalline structure, and electronic properties of the materials are intrinsically correlated; however, surface characteristic such as surface area, particle size distribution, and porosity can be independently modified. On the other hand, for nanomaterials, the size becomes an important feature affecting the physical properties, such as the optical band gap. Moreover, the particle size is of primary importance in heterogeneous photocatalysis, because it is directly related to the efficiency of a catalyst through the enhancement of its specific surface area.

The techniques described in this chapter were divided into five major topics: elemental composition, structure and topography, surface area and porosity, and vibrational and optical properties. The most common techniques are briefly described giving some examples about their use for the analysis of a photocatalytic material.

4.1 Elemental Composition

The analytical methods used to measure the composition of solids can be divided in terms of the probe used to analyze the material: ions, electrons, or photons. The spatial resolution for each technique is different and dependent on the instrumentation but also on the type of the detected beam. Some techniques are destructive, while others are nondestructive, as well as surface sensitivity or bulky. This is shown in Table 4.1, where the incident and detected beams are described, as well as the standard spatial resolutions (Egerton 2011).

M. Bizarro (✉) • S.E. Rodil
Instituto de Investigaciones en Materiales, Universidad Nacional Autónoma de México, Circuito Exterior sn Ciudad Universitaria 04510 México D.F., México
e-mail: monserrat@iim.unam.mx

A. Hernández-Ramírez, I. Medina-Ramírez (eds.), *Photocatalytic Semiconductors*,
DOI 10.1007/978-3-319-10999-2_4

Table 4.1 Classification of the analytical characterization techniques in terms of the probe and detected signal

Incident beam	Detected beam	Name	Spatial resolution (nm)	Destructive	Surface sensitive
Ions	Ions	Rutherford backscattering (RBS)	1,000	No	No
		Secondary ion mass spectroscopy (SIMS)	40	Yes	Yes
	Photons	Proton-induced X-ray emission (PIXE)	500	No	Yes
Electrons	Electrons	Electron energy loss spectroscopy (EELS)	<1	No	No
		Auger electron spectroscopy (AES)	2	No	Yes
	Photons	Energy dispersive spectroscopy (EDS)	2–10	No	No
Photons	Photons	X-ray fluorescence spectroscopy (XRF)		No	No
		X-ray absorption spectroscopy (XAS)	20	No	No
	Electrons	Ultraviolet photoelectron spectroscopy (UPS)	1,000	No	Yes
		X-ray photoelectron spectroscopy (XPS)	5–10	No	Yes

4.1.1 Ion Beam Methods

4.1.1.1 Rutherford Backscattering Spectroscopy, RBS

Rutherford backscattering spectroscopy is an ion scattering technique used for the analysis of the surface of solids, and it has become one of the most frequently used techniques for quantitative analysis of composition, thickness, and depth profiles of thin solid films. In RBS a beam of accelerated ions in the MeV range (0.5–4 MeV) impinges on the sample to be analyzed. During the bombardment, a small fraction of the incident particles undergo a direct collision with the nucleus of one of the atoms in the upper few micrometers of the sample. The combination and interplay of three basic concepts allow RBS analysis to detect masses and composition as a function of depth. The first concept is based upon elastic collision of ions with the target nuclei, causing the ions to backscatter and carry information about the mass of the atom they backscatter from. The second concept is the probability of backscattering from different elements, which is highly predictable and proportional to the square of the atomic number, allowing the determination of the sample composition. The third concept is depth analysis, which is possible because backscattered ions are slowed down due to the inelastic energy losses before reaching the surface. The energy measured for a particle backscattering at a given angle depends upon the material's stopping power and the masses of the projectile to target atoms. The number of backscattering events that occurs for a given element in a sample depends upon two factors: the concentration of the element and the effective size of its nucleus. The probability that a material will cause a collision is called its scattering cross section.

By measuring the heights of the peaks and normalizing them by the scattering cross section for the respective element, the elemental ratio can be obtained at any given depth in the film. Therefore, RBS allows the quantitative determination of the composition of a material without the need of a standard, as well as depth profiling of individual elements with a depth resolution in the nanometer range (Fig. 4.1). Another useful feature of RBS is that it provides an accurate concentration of the total atoms/cm^2, which is present in a film. If the thickness of the film can be measured by another method, then the density of the film can be easily calculated (Tesmer and Nastasi 1995).

4.1.1.2 Elastic Recoil Detection Analysis, ERDA

RBS has a very good sensitivity for heavy elements in the ppm range but is not sensitive to light masses. In order to determine the heavy and the light components of a matrix, RBS and elastic recoil detection analysis (ERDA) are used simultaneously. ERDA uses the same incident beam as RBS but detects the energy of the recoiled atoms. An absorber foil prevents scattered particles from the sample entering the detector. As a result, only the lighter recoiled particles (hydrogen or deuterium) will be measured (Tesmer and Nastasi 1995). Nowadays, a new configuration has been designed, heavy incident ion-ERDA (HI-ERDA), where heavier ions are used (^{35}Cl, ^{63}Cu, ^{127}I, and ^{197}Au). This configuration coupled with elemental mass-sensitive detectors allows the identification of all atoms from the target sample with masses lighter than the projectile ion. No absorber is used in this case, but the recoiled projectile can be also identified. The most common detectors are magnetic spectrographs, gas ionization detectors, or time-of-flight-ERDA (TOF-ERDA); in the latter the elements are identified by the simultaneous detection of velocity and energy (Siketic et al. 2010). A great improvement has been obtained in terms of depth resolution, which is in the 1–2 nm range, allowing depth profile analysis with high precision and atomic resolution better than 0.1 at.% (Schiettekatte and Chicoine 2004).

4.1.1.3 Particle-Induced X-Ray Emission or Proton-Induced X-Ray Emission PIXE

In PIXE, the elemental composition is obtained by measuring the energy of the X-rays emitted by the sample after bombardment with energetic ions. Since the incident protons (typically of energy 1–5 MeV) are only slightly deflected by the nuclear field, the bremsstrahlung background radiation characteristic of the

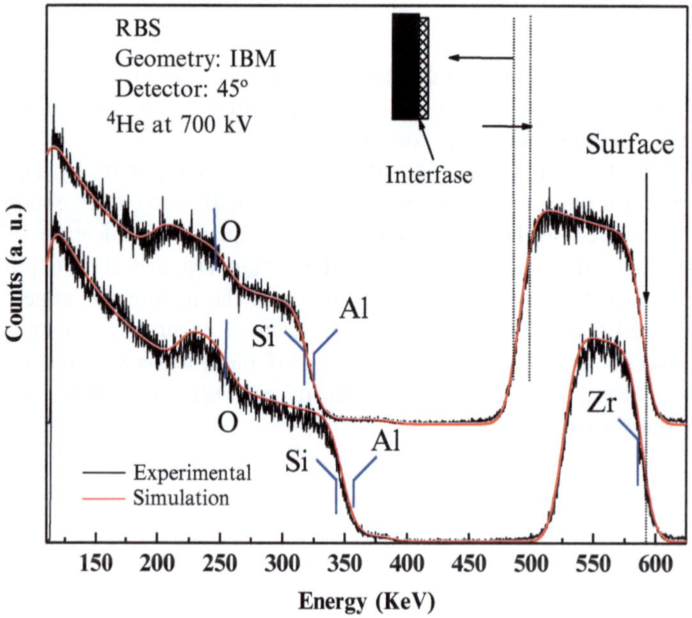

Fig. 4.1 Example of an RBS spectra for Al-doped ZrO_2 thin-film samples having different thicknesses. The ions used are indicated, as well as the specific sample–detector geometry (IBM)

X-ray peaks is lower than when using incident electrons, and then sensitivities as low as 0.1–10 ppm can be attained. Besides, the spatial resolution can be improved up to 1 μm using tightly focused ion beams (Johansson et al. 1995).

4.1.1.4 Secondary Ion Mass Spectrometry, SIMS

In SIMS, the sample is bombarded with a focused ion beam 1–20 keV ions (O^{2+}, O^-, Cs^+, Ar^+, Ga^+, or neutrals); this causes the sputtering of surface atoms (secondary ions), whose mass/energy ratio are analyzed by passing them through a mass spectrometer. The mass analyzer may be quadrupole mass analyzers or magnetic sector mass analyzers, and more recently, time-of-flight (TOF) analyzers have been used providing substantially higher sensitivity and mass resolution, as well as a much greater mass range. The secondary ions come from a surface layer of 1–2 nm thick; typically one monolayer is analyzed in static mode SIMS, while in dynamic mode, bulk composition and depth distribution of trace elements are investigated with a depth resolution ranging from sub-nanometer to tens of nanometer. One example of the profile composition for light elements from a thin film (amorphous hydrogenated carbon nitride) deposited on silicon is shown in Fig. 4.2. All elements from hydrogen to above are detectable, and the technique is capable of detecting impurity elements present in a surface layer at <1 ppm concentration and bulk concentrations of impurities of around 1 ppb. Moreover, chemical images are

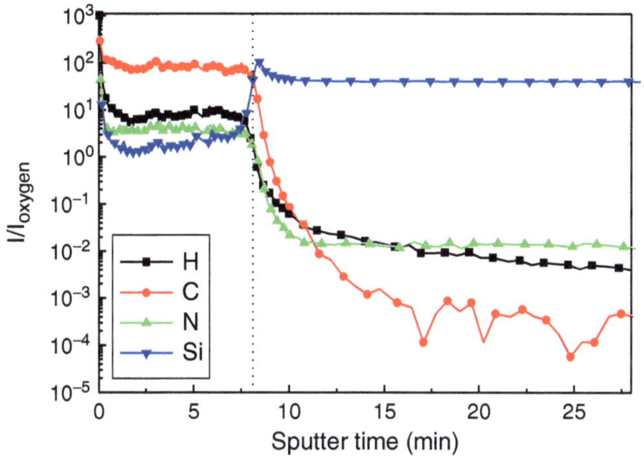

Fig. 4.2 Example of a depth profile for a hydrogenated carbon nitride thin film deposited on a silicon substrate. Note the clear change in the composition at the film–substrate interface

possible by collecting a mass spectrum at every pixel as the primary ion beam is scanned across the sample surface. However, quantification in SIMS is complicated by the fact that sputtering gives rise to both neutral atoms and ions, but most of the spectrometers detect only the ions. Therefore, for quantitative analysis it is necessary to use standard samples with similar matrix composition. The flux of sputtered ions is very low and strongly depends on the surface composition (matrix effects). While the probability of sputtering neutral atoms is much higher (90–99.9999 %) and less sensitive to the surface composition, its detection allows a much better quantitative estimation of the sample stoichiometry. This has led to the design of a different method of detection for a better quantification, namely, secondary neutral mass spectroscopy SNMS, that ionizes the neutral atoms (by an electron gun, laser, or radio-frequency cavity) before detection (Briggs and Sheah 1996).

4.1.2 Electron Beam Methods

In this group, we consider all those techniques where the elemental composition can be obtained by analyzing the interaction of an electron beam with core electrons; such interactions bring information about the binding energy of the core electrons, and then the elemental composition as well as the chemical bonding and electronic structure of the material can be obtained. The most common techniques are Auger electron spectroscopy (AES), electron energy loss spectroscopy (EELS) and X-ray emission spectroscopy, most commonly known as energy dispersive X-ray spectroscopy (EDX) or electron probe microscopy analysis (EPMA). The fundamental physical phenomena used to estimate the chemical composition for the different techniques are schematically shown in Fig. 4.3.

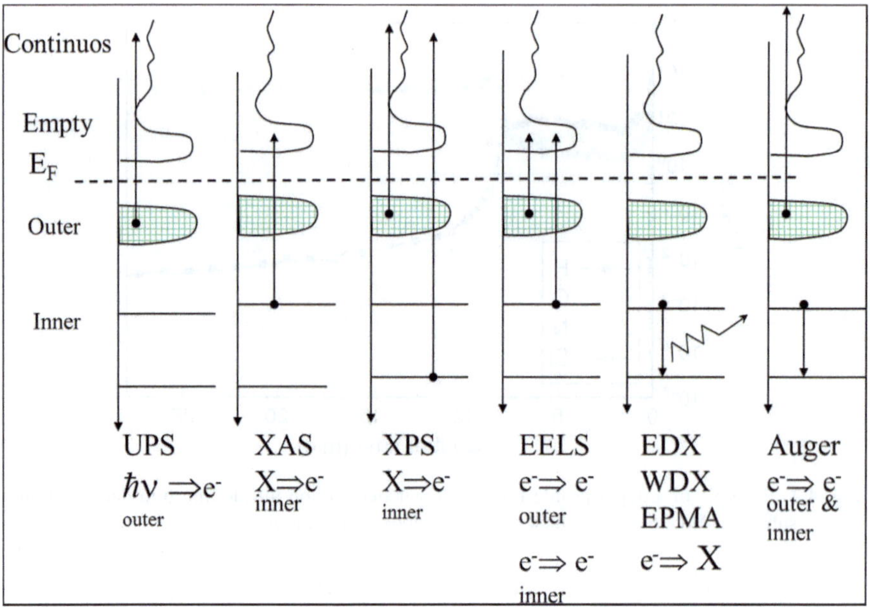

Fig. 4.3 Schematic representation of the electronic transitions and the de-excitation mechanisms used for the chemical characterization of solid

The energy of the incoming electrons have to be large enough to remove one electron from a core level, leaving a hole in the energy level, which is later filled through a decay mechanism, such as the emission of an X-ray (EDX) or the emission of a secondary electron (AES), where the energy of the emitted beam is directly measured and correlated to the corresponding energy level. Meanwhile in EELS, the energy loss suffered by the incident electron due to the interaction with the core electrons is measured.

4.1.2.1 X-Ray Emission Spectroscopy, XES

As its name indicates, in this technique the energy of X-rays emitted by the sample after interaction with a beam of energetic electrons is measured (Goldstein et al. 2003). The energy or wavelength of the emitted X-rays is analyzed using either a wavelength spectrometer (WDX) or an energy spectrometer (EDX). The resolution of the WDX is much larger (10 eV), allowing a better identification of the elements; however, the acquisition time is much larger than using energy spectrometers (130 eV energy resolution). As a method of elemental bulk analysis, the technique is known as EPMA, and it has been improved to give elemental accuracy of about 1–2 % using appropriate standards and corrections for atomic number, absorption, and fluorescence effects. On the other hand, EDX has been attached to either transmission or scanning electron microscopes, leading to a high spatial resolution. In any case, no H or He can be detected, and detection of light elements (B, C, N, O) within heavy element matrices is always controversial, since the probability of excitation decay by X-ray emission is much lower for the light elements. Moreover, in order to keep the required vacuum in the microscope column, a window between the chamber and the detector has to be used, which filters the low energy photons. Windowless equipment allows the detection and quantification of elements above boron at high sensitivity with detection limits between 0.01 and 0.1 %. One interesting feature that has been extensively improved in the last years is the X-ray mapping, where nanometric segregations of certain compounds or elements within the matrix can be easily detected. EDX is nowadays the most common technique used to analyze the composition of bulk samples. It is important to consider that it is not a surface science technique, since the X-rays are generated in a region about 2 μm in depth and quantitative analysis is mainly done by software, instead of using appropriate standard samples.

4.1.2.2 Electron Energy Loss Spectroscopy, EELS

Electron energy loss spectroscopy is another microanalytical technique used to obtain information about the elemental and chemical composition, as well as the electronic structure of materials. However, since EELS measures the energy loss of transmitted electrons, it has to be coupled to a transmission or scanning transmission electron microscope. It has very good spatial resolution (0.1–1 nm) but requires very thin samples, either nanoparticles or thin films (<50 nm). Imaging is also possible, providing the highest resolution compositional maps.

The loss spectroscopy consists of the analysis of the energy distribution of initially monoenergetic electrons that have interacted inelastically with the specimen. Provided the incident energy is high enough and the film sufficiently thin, practically all of the incident electrons are transmitted without reflection or absorption. The elastic and inelastic processes occurring when electrons bombard solid samples are shown in Fig. 4.4. Inelastic interaction with core electrons takes place inside the specimen, and information about internal structure can be obtained by passing the transmitted beam into a spectrometer (Egerton 2011).

Energy loss implies inelastic collision but the spectrum will also contain electrons that have interacted elastically with the specimen (Froitzheim 1977). Thus, the EELS spectra consist of three principal regions, as shown in Fig. 4.5:

- The zero-loss region consists primarily of elastic forward scattered electrons but also contains electrons that have suffered minor (unresolvable) energy losses.
- The low-loss region up to an energy loss of ~50 eV contains electrons which have interacted with the weakly bound outer-shell electrons of the atoms in the specimen or by the excitation of vibrational transitions. Such low energy losses can be detected in high-resolution EELS, which is used to study the adsorption of molecules on surfaces.

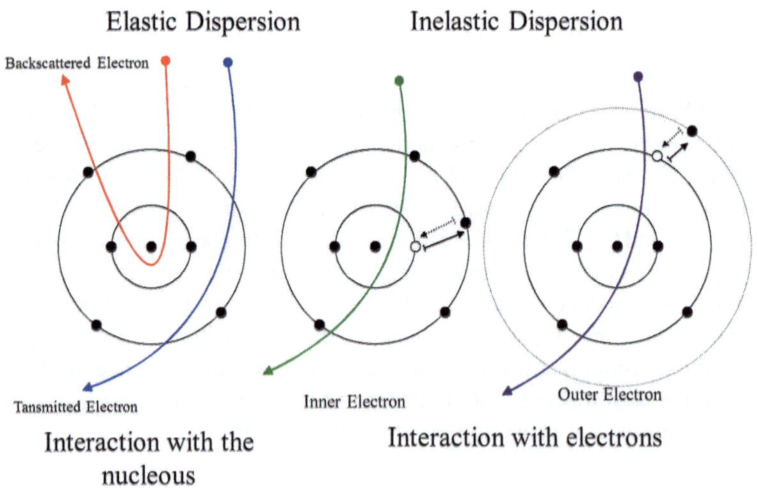

Fig. 4.4 Representation of the inelastic and elastic interaction of the incidence electrons with the atoms of a solid sample

Fig. 4.5 Typical energy loss spectra, after background subtraction and plural scattering removal, where C and N edges are observed in the high-loss region

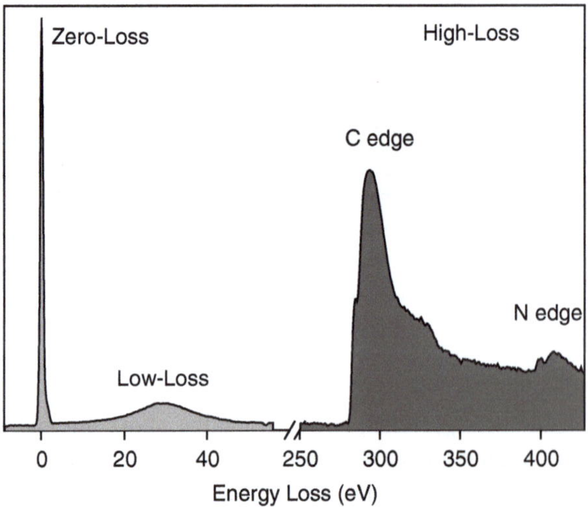

- The high-loss region contains electrons that have interacted with the more tightly bound inner-shell or "core" electrons.

Each region gives different information; the zero-loss peak defines the energy resolution and is used to calibrate the energy scale of the spectrum.

The low-loss region is due to inelastic interaction between the incident electrons and the outer-shell electrons of the specimen. There are two types of interactions, single electron excitation in which a valence electron makes an interband transition across the energy gap for a semiconductor or in the case of a metal a conduction electron makes a transition to a higher state within the same energy band. The analysis of this part of the spectra allows the determination of the optical band gap and its character, as well as inter-intraband transitions. The other interaction is a collective electron excitation in which the scattering involves many atoms of a solid. This collective effect is known as a plasma resonance and takes the form of a longitudinal traveling wave. The excitation can also be described in terms of the creation of a pseudoparticle, the plasmon, whose energy is given by $E_p = \hbar w_p$, where w_p is the frequency (rad/s) that is proportional to the square root of the valence electron density, as shown in Eq. (4.1):

$$E_p = \hbar \left(\frac{n_e e^2}{\varepsilon_o m^*} \right)^{1/2} \tag{4.1}$$

where e and m^* are the electron charge and effective mass, ε_o is the permittivity of free space, and n_e is the valence electron density. Even though E_p is proportional to the density of free electrons, there are also plasmon-like spectra from insulator materials. A simple explanation for the occurrence of plasmon effects in insulators is that electrons that receive energy in excess of their binding energy are released to take part in collective oscillations. Thus, analysis of the low-loss region can provide information about the inter-intraband transitions, optical gap, and surface and bulk plasmons of the sample.

The high-loss regions comprised the chemical information, composition and chemical bonding. When the beam of electrons transfer more energy than the attractive fields of the nucleus to K, L, M, N, or O shell electrons, the electrons are transferred to an unoccupied electron state. Because the total energy is conserved at each collision, the fast incident electron loses an amount of energy comparable to or greater than the core electron binding energy. A specific minimum energy transfer is required to overcome the binding energy of the electron to the nucleus and ionize the atom. This minimum energy constitutes the ionization threshold E_c characteristic of the given inner shell (K, L, or M). The ionization loss electrons have an energy distribution that ideally shows a sharp rise to a maximum E_c, followed by a slowly decreasing intensity above E_c back toward the background due to electrons that have undergone random plural inelastic scattering events (steplike peaks). The edge may also contain fine structure around E_c that is due to the bonding effects, and its analysis is termed electron loss near edge structure (ELNES). More than ~50 eV after the edge, small intensity oscillations may be detectable due to diffraction effects from the atoms surrounding the ionized atoms (extended energy loss fine structure, EXELFS).

Therefore, elemental analysis by EELS is enabled by the inner-shell edges. The position and intensity of which can be used to identify and quantify atomic species within the sample, while the analysis of the fine structure gives information about the chemical bonding (π, σ bonding character), most suitable for atoms with $Z < 33$. The quantification of the EEL spectra requires the knowledge of the partial scattering cross section, i.e., differential cross section integrated over angle and energy.

This cross section can be determined either theoretically or experimentally by using standards. Such calculation and the lack of extended databases have limited the use of the technique as a widespread characterization tool (Keast et al. 2001).

4.1.2.3 Auger Electron Spectroscopy, AES

Auger spectroscopy owns its name to Pierre Auger, who discovered that the emission of a lower binding energy electron was also a relaxation process after the ionization of a core electron by the bombardment with electrons or photons with enough energy. For AES, electrons of energy 3–20 keV are incident upon a conducting sample. These electrons cause core electrons from atoms contained in the sample to be ejected resulting in a photoelectron and an atom with a core hole. The atom then relaxes by the transition of an electron with a lower binding energy dropping into the core hole. The energy thus released can be converted into an X-ray (as in EDX) or emit an Auger electron (Fig. 4.6). After the emission of the Auger electron, the atom is left in a doubly ionized state, and the energy of the Auger electron is characteristic of the element that emitted it and can thus be used to identify the element (Davis 1996; Briggs and Sheah 1990).

The short inelastic mean free path (IMFP) of Auger electrons in solids ensures the surface sensitivity of AES. For the Auger process, there are always three energy levels involved with well-defined selection rules for the transition, allowing elemental identification. The energy level of the core hole will often determine which transition types will be favored. For single energy levels, i.e., K, transitions can occur from the L levels, giving rise to strong KLL type peaks in an Auger spectrum. Higher level transitions can also occur but are less probable. For other shells, transitions are available from higher energy orbital, such as LMM and LLM. As the atomic number Z increases, so does the number of potential Auger transitions. Fortunately, the strongest electron–electron interactions are between levels that are close together, giving rise to characteristic peaks in an Auger spectrum of the KLL and LMM type. Quantitative analysis is possible by the use of standards or relying on Auger databases (semiquantitative) where the yield of Auger electrons has been reported for different elements. The depth analysis of AES is of a few nanometers, 2–5 nm, and can detect elements from Li and onward in the periodic table. Due to the possibility of focusing the incoming electrons, it has a better spatial resolution (10 nm) than XPS, so the analyzed area becomes more specific. The acquisition of Auger spectra has changed in the last years due to the improved sensitivity of the electron analyzers. During the initial years, derivative spectra were obtained to enhance the small signals, but nowadays direct spectra can be acquired and this has changed the method to analyze AES data. Chemical imaging is also possible taking advantage of the focused electron beam, which can be scanned on the surface; this has been named as scanning Auger microscopy (SAM). The major limitation of AES is that the surface of the sample is charged, which leads to peak shifting and loss of resolution, and so conductive samples have to be used or a secondary process has to be implemented to avoid charging effects. Note that the Auger signals can

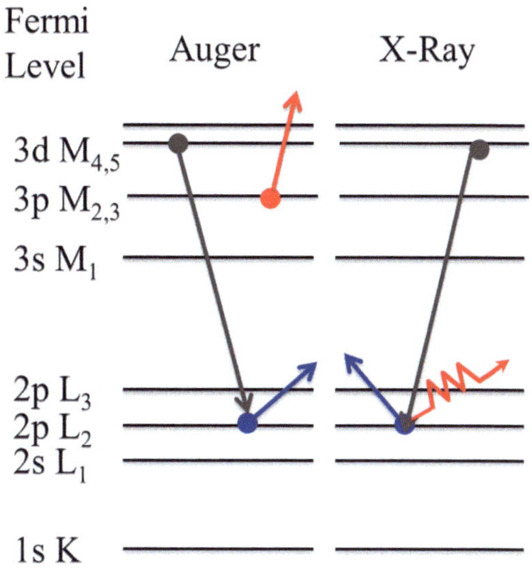

Fig. 4.6 Schematic representation of the energy levels involved in an Auger event in comparison to X-ray emission

also be detected in any of the other analytical techniques, when the electrons constitute the detected beam, as in the XPS or NEXAFS.

4.1.3 X-Ray Beam Methods

Another group of elemental composition techniques use photons as the incident beam. The energy of the photons (X-ray range) has to be enough to cause ionization of the inner or core-shell electrons, and the emitted electrons (XPS) or photons (XAS) are detected.

4.1.3.1 X-Ray Photoelectron Spectroscopy, XPS

The phenomenon used for photoelectron spectroscopy is based on the photoelectric effect outlined by Einstein where the concept of the photon was used to describe the ejection of electrons from a surface when photons impinge upon it.

For XPS, also known as electron spectroscopy for chemical analysis (ESCA), the photon energies of choice are the fundamental emissions from Al-κα (1,486.6 eV) or Mg-κα (1,253.6 eV) cathodes. The XPS technique is highly surface specific due to the short IMFP of the photoelectrons that are excited from the solid. The energy of the photoelectrons leaving the sample is characteristic of each element.

Each peak area is proportional to the number of atoms being present in the material; the shape of the peak and the binding energy (BE) can be slightly altered by the chemical site of the emitting atom; hence XPS can provide chemical bonding information as well. The fundamental equation that relates the BE of the emitted photoelectron with the measured electron kinetic energy (KE) is given by

$$BE = h\nu - KE - \phi \tag{4.2}$$

where ($h\nu$) is the incident photon energy and (ϕ) is the work function, which has to be well defined for each equipment in order to determine precisely the BE that is usually measured in reference to the material's Fermi level.

The analysis of the XPS spectrum (counts per binding energy) allows us to obtain the surface elemental composition by comparing to standard materials (quantitative) or using XPS databases (semiquantitative). The depth of analysis in XPS is about 1–10 nm depending on the incidence energy of the photons, the kinetic energy of the outgoing electrons, and the angle of collection (angle resolved XPS, ARPES). The XPS technique requires ultra-high vacuum conditions (UHV, pressure in a range from 10^{-8} to 10^{-10} mbar), and so the samples have to be evaluated always in dry conditions. The XPS is essentially a nondestructive technique but is very sensitive to surface contamination; so in some cases, an in situ cleaning process is applied using an ion gun (Watts and Wolstenholme 2003; Briggs and Sheah 1990).

4.1.3.2 X-Ray Absorption Spectroscopy, XAS

Absorption of X-rays by a solid sample can also give information about the chemical composition and structure of the sample (Bunker 2010). X-ray absorption includes different techniques: near edge X-ray fine structure (NEXAFS), X-ray absorption near edge structure (XANES), and extended X-ray absorption fine structure (EXAFS), schematically shown in Fig. 4.7. The term NEXAFS is typically used for soft X-ray absorption spectra (Stohr 2003), where the edge (similar to EELS) corresponding to elemental absorption is observed, whereas XANES is used for hard X-ray spectra and EXAFS refers to the analysis of the oscillations due to the interaction of the electron wave with the rest of the atoms in the sample. In general, a monochromatic X-ray beam is directed at the sample, and the photon energy of the X-rays is gradually increased. Below the absorption edge, the photons cannot excite the electrons of the relevant atomic level, and thus absorption is low. However, when the photon energy is sufficient to excite the core electrons into empty states, then a large increase in absorption occurs, known as the absorption edge. A few electron volts above the absorption edge, the spectra shows a series of oscillations due to interaction of the outcoming photoelectrons with the atoms surrounding the emitting atom, but also because the probability of backscattering is dependent on both the energy of the photoelectrons and the photon energy. These oscillations can be used to determine the atomic number, distance, and coordination number of the atoms surrounding the element whose absorption edge is being examined. The necessity to sweep the photon energy (200–35,000 eV) requires the use of synchrotron radiation.

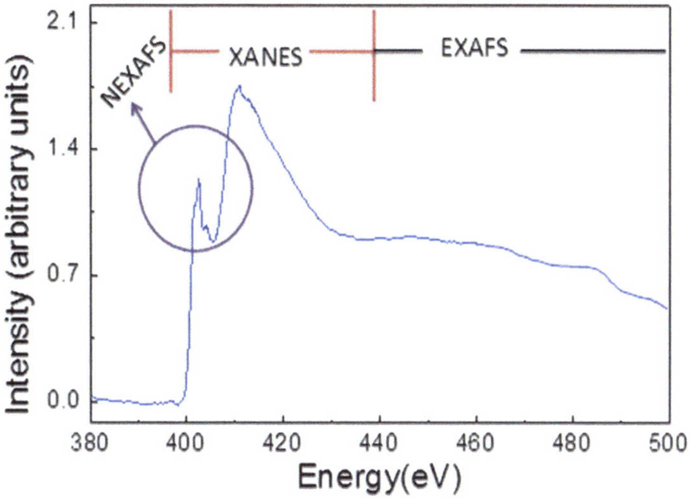

Fig. 4.7 Schematic representation of X-ray absorption spectra showing the different region of analysis

In NEXAFS, the X-ray energy is scanned and the absorbed X-ray intensity is measured, by detecting the emitted photoelectrons, the Auger electrons, or the fluorescent photons. NEXAFS is very sensitive to the surface, and it is particularly important to study chemisorbed molecules on surfaces. For XANES experiments, analysis of the near edge region (up to 40 eV above the edge) can provide information about vacant orbitals, electronic configuration, and site symmetry of the absorbing atom. The absolute position of the edge contains information about the oxidation state of the absorbing atom. These techniques are very sensitive to directional bonds by the use of linearly polarized X-rays and therefore are well suited for covalent systems like low-Z molecules, macromolecules, and polymers (Haw 2002).

4.1.4 Examples

One interesting example where different analytical techniques were used to understand the electronic–catalytic response of ceria-doped titania samples is given by Kundu et al. (2012). The interest about Ce–TiO_2 comes from theoretical calculations done by Catlow et al. (2010) that suggest that the presence of reduced cerium cations will enhance the photocatalytic activity of TiO_2. However, none of the reported works (Li et al. 2009; Xie et al. 2010) have investigated the oxidation states of the cerium. For such purpose, the authors (Kundu et al. 2012) produced Ce-modified TiO_2 samples by wet impregnation of TiO_2 using two different Ce concentrations: 6 and 15 wt %. Then, HRTEM, EELS, NEXAFS, and XANES were used to determine the oxidation state of the cerium and their structural distribution. The HRTEM images showed that CeO_2 particles were generally located at a TiO_2 corner site, and they have a truncated pyramidal shape. Electron energy loss spectroscopy was employed to measure the Ce^{3+}/Ce^{4+} ratio with high spatial resolution within and around the ceria nanoparticles. The presence of Ce^{3+} was observed mainly at the CeO_2–TiO_2 interface, as indicated by the NEXAFS forming a mixed oxide at the interface. The XANES spectra showed that the incorporation of the CeO_2 onto the TiO_2 surface changed the charge transfer between oxygen and Ce, probably due to a Ce–O–Ti interaction. The authors also analyze the Pt/CeO_2/TiO_2 systems showing that without the presence of Pt, the pure CeO_2/TiO_2 mixed metal oxide showed a very small activity toward water splitting in the visible light. This result indicates that the existence of Ce^{3+} drastically enhances the formation of electron–hole pairs, but it does not help to suppress their recombination. The presence of a small amount of Pt is needed in order to act as an electron trap and prolong the lifetime of electron–hole pairs. However in this particular system, Pt not only acts as an electron-trapping center, but also increases the amount of Ce^{3+} in the support itself, thus enhancing the photon absorption efficiency of the catalyst in the visible region.

McDermott et al. (2013) have used a combination of X-ray absorption and electron energy loss spectroscopies along with theoretical calculations to investigate the electronic structure of poly(triazine imine) (PTI), a graphitic carbon nitride-type material with LiCl intercalation. The PTI/LiCl is aimed to improve the water-splitting photocatalytic reaction, since Li and Cl ions can donate or withdraw charge from the polymeric matrix. It was observed that the addition of LiCl reduced the band gap (~2.2 eV) with respect to polymeric carbon nitride (2.7 eV).

Qi et al. (2012) have produced transition metal-doped TiO_2 transparent sols via a simple chemical route. The sol samples were analyzed using HRTEM, XRD, EELS, and XPS, verifying that most metal ions have been doped with TiO_2 forming Ti–O–metal bonds. Analyses of XRD, EELS, and XPS experiments allow them to verify that most of the dopant metal ions substitute the Ti^{4+} sites, forming metal-doped TiO_2 sols instead of just mechanically mixed phases. These doped TiO_2 sols exhibited good photocatalytic activities under visible light irradiation and were stable in sealed conditions for several months without formation of precipitates.

Oxide supported noble metal systems have been studied intensively due to their extreme importance in catalysis; therefore, gathering information about the interface of the noble metal with the support is always of great interest, but not easily done. Wang et al. (2013) prepared thin iron oxide films of FeO(111), Fe_3O_4(111), and Fe_2O_3(0001) on Mo(110) substrate. The growth and electronic structures of silver on these films with different morphologies were investigated by using various surface analytical techniques including low energy electron diffraction (LEED), AES, UPS, and EELS. The analysis of the valence band structure and surface plasmons showed that silver grows on FeO(111) and Fe_2O_3(0001) as three-dimensional (3D) clusters but forms a wetting layer on Fe_3O_4(111). This difference on growth will have an impact of the photocatalytic response of the noble metal–iron oxide system, although it was not evaluated in the paper.

Nitrogen doping of TiO_2 samples has been extensively investigated as a method to enhance the photocatalytic effect of TiO_2 under visible or white light irradiation. However, it is not trivial to detect nitrogen at very low levels in order to determine the doping characteristics. Doping concentrations as low as 1.19 at.% were shown by Tavares et al. (2009) to improve the photocatalytic degradation of an organic dye by TiO_2-anatase films deposited by RF magnetron sputtering. Such low concentration levels were accurately determined using HI-ERDA. The effect of nitrogen and sulfur doping in commercial TiO_2 powders was also investigated by Rengifo-Herrera et al. (2008), where a careful analysis of the XPS peak assignments for both N and S was presented. The authors were able to identify the type of doping, interstitial or substitutional, through XPS analysis, a subject that is important to understand the energy position of the gap level states introduced in the structure by the dopants. Sério et al. (2012) used RBS and ERDA to study the effect of the nitrogen doping on the growth, structure, crystallinity, and optical properties of TiO_2 thin films deposited by DC-reactive magnetron sputtering. Similarly, Moser et al. (2013) studied the effect of N and C doping on TiO_2 films and demonstrated that the high activity of the films was associated to the presence of N and C acting as dopants, which increases the absorption in the near-UV region. The dopant level was demonstrated by RBS, ERDA, and XPS.

4.2 Structure and Topography

4.2.1 Structure

The photocatalytic activity of semiconductor materials mainly depends on its intrinsic properties, such as the composition and crystalline structure. In this regard, it is important to know the active crystalline phase or mixture of phases in the photocatalyst. The most important techniques used to identify the crystalline phases in materials are X-ray diffraction (XRD) and transmission electron microscopy (TEM).

4.2.1.1 X-Ray Diffraction, XRD

X-ray diffraction involves the measurement of the intensity of X-rays scattered from crystals. Waves scattered at atoms at different positions arrive at the detector with a relative phase shift, producing constructive and destructive interference patterns depending on the angle and the relative atomic positions (Suryanarayana and Norton 1998; Waseda et al. 2011). The easiest access to the structural information in crystals is using the Bragg equation, which describes the principle of X-ray diffraction in terms of a reflection of X-rays by sets of lattice planes:

$$n\lambda = 2d \sin \theta \tag{4.3}$$

where n is called the order of diffraction and is equal to the number of wavelengths in the path difference between rays scattered by adjacent planes, λ is the wavelength of the incident photon, d is the interplanar distance, and θ is the diffraction angle.

Lattice planes are crystallographic planes, characterized by the index triplet (*hkl*), the so-called Miller indices. Parallel planes have the same indices and are equally spaced, separated by the distance d_{hkl}. Bragg analysis treats the X-rays as a mirror-like reflection at the lattice planes. As the X-rays penetrate deep inside the material, multiple reflections occur at thousands of consecutive parallel planes.

For an infinite stack of lattice planes, Bragg's equation gives the position of delta-function Bragg peaks. In practice, finite size crystallites give rise to Bragg peaks of finite width. This size broadening is described by the Scherrer equation given by

$$L_{hkl} = \frac{K\lambda}{\beta_{hkl} \cos \theta} \tag{4.4}$$

where K depends on the shape of the grains; for perfect spheres $K = 0.89$, and $K = 0.94$ for cubic shaped grains. Besides the size broadening of the diffraction peaks described by the Scherrer equation, the width is also affected by other microstructural features like the crystallite shape, the shape distribution, and any distortion of the crystal structure like microstrain, dislocations, twin planes, and stacking faults.

The X-ray diffraction technique is a powerful nondestructive tool for phase identification of crystalline samples. Unlike other techniques to obtain the chemical composition, XRD does not give elemental composition but can distinguish different packing ways of the same set of elements. In the case of TiO_2, it is known that it is composed of titanium and oxygen atoms, but XRD allows determining the crystalline arrangement in which they are disposed. In the case of TiO_2 we can find a tetragonal structure (anatase or rutile) or even an orthorhombic structure (brookite), as shown in Fig. 4.8 for P25 Degussa TiO_2 particles.

Interpretation of XRD patterns have become easy due to the large amount of crystalline structures that have been already identified and classified into proper databases (more than 50,000 validated patterns for inorganic materials). Nevertheless for unknown materials, the interpretation is rather more complicated. Fortunately, about 45 years ago, Rietveld (1969) provides us with a quantitative tool to analyze the X-ray powder diffraction patterns, known as the Rietveld method (RM). The Rietveld method is a standard-less method that allows the accurate determination of a wide range of structural parameters. It is a computing-demanding method, and different open-source or commercial software have been developed. The most common information obtained is:

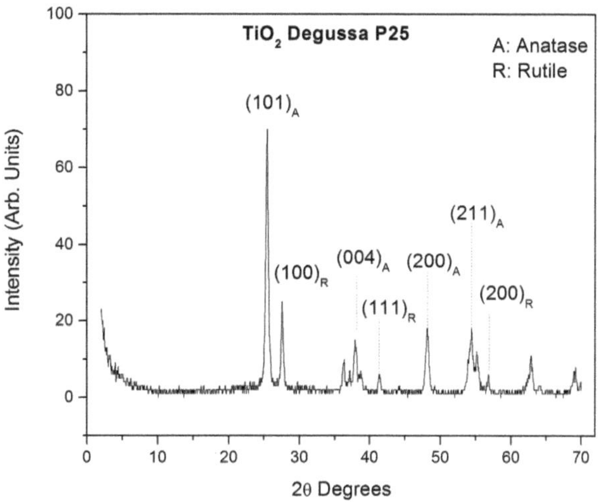

Fig. 4.8 XRD pattern for standard P25 TiO_2 particles (Degussa) showing the two crystalline phases: anatase and rutile

- Quantitative determination of the percentage of different phases present in one sample (quantitative phase analysis) (Gualtieri 2000)
- Percentage of crystallinity or amount of disorder (Chipera and Bish 2013)
- Atomic resolution for solid solution materials
- Unit cell determination
- Crystalline size
- Global residual stress.

Therefore, the percentage of anatase and rutile in the XRD pattern shown in Fig. 4.8 could be accurately determined by a Rietveld refinement of the structure. On the other hand, for the calculation of this fraction where only two polymorphs are present, semiquantitative methods can also be applied. In such case, the integrated intensity of the strongest peak for each phase is compared to the integrated intensity of the pure phase (Guillén et al. 2014). However, such calculations are prompt to errors due to the presence of not identified phases, overlapping peaks, or preferred orientation of individual phases. An alternative quantitative method that is more precise is the matrix flushing method or normalized RIR (Reference Intensity Ratio) method (Chung 1974). In this case, a least square fitting of the full experimental pattern is performed to identify the phases present in the mixture. Then, the weight fraction of each phase can be determined using the RIR values commonly listed in the databases such as the ICDD. The RIR values contain the intensity ratio between the target material/phase and a known standard, usually corundum. However, there is some criticism about the use of the RIR method, and other methods combining both RIR and Rietveld methods have also been proposed (Chipera and Bish 2002, 2013).

Different modes have been attached to the XRD systems; the most commonly used is the Bragg–Brentano (BB) configuration suitable for powders and polycrystalline bulk samples. In the BB mode, the penetration depth of the X-rays is often found in the 10–100 μm range, causing than when thin films (less than 1 μm thick) are studied, the substrate peaks dominate the pattern. Thus, for the analysis of thin films, X-ray diffraction techniques have been developed for which the primary beam enters the sample under very small angles of incidence. In its simplest variant this configuration is denoted by GIXRD that stands for grazing incidence X-ray diffraction (Ulyanenkov 1998). More recently, the In-Plane mode has been introduced to study the crystal structure of very thin films in the direction perpendicular to the substrate. Information about the texture, i.e., preferred crystallographic orientations in thin films and polycrystalline samples, can also be obtained by XRD using a mode called XRD Pole figures. In thin films, it is common that a certain crystallographic direction (*hkl*) is preferentially oriented with respect to the sample reference frame (powders) due to the directional growth. This anisotropy of crystallite orientation has important effects on different physical properties of both the thin films and polycrystals, such as oxidation, diffusion of atoms, phase transformations, as well as mechanical, electrical, and magnetic properties.

Other interesting variation of the XRD technique is the small angle X-ray scattering (SAXS) configuration (Glatter and Kratky 1982), which is typically done using synchrotron radiation or dedicated SAXS equipment, but recently has been also incorporated into standard XRD systems (GISAXS). This is a very powerful technique for the structural characterization of nanomaterials and surfaces. In this configuration the sample is irradiated with a monochromatic X-ray beam at very low angles, typically 0.1–10°. This angular range contains information about the shape and size distribution of nanoparticles (1–100 nm), characteristic distances of partially ordered materials, pore sizes, and other data. SAXS is capable of delivering structural information of macromolecules between 5 and 25 nm, of repeat distances in partially ordered systems of up to 150 nm.

4.2.1.2 Transmission Electron Microscopy, TEM

The transmission electron microscope uses a high energy electron beam that is transmitted through a very thin sample to image and analyze the microstructure of materials with atomic scale resolution (Williams and Carter 1996; Reimer and Kohl 2008). The TEM is an electron–electron technique; the incident electrons are accelerated at high voltages (100–1,000 kV) to a velocity approaching the speed of light (0.6–0.9*c*). The electrons travel through the sample suffering both elastic and inelastic dispersion (as shown in Fig. 4.3). The elastically dispersed electrons are used to form a diffraction pattern which follows the Bragg law and therefore contains information about the crystalline structure. Meanwhile, the inelastically dispersed electrons are those typically detected in the EELS experiments described above. High-resolution TEM images are formed focusing the transmitted electrons to a plane where both the CCD camera and/or the fluorescence screen are placed; there the electron density is converted into light-optical images. The great advantage in imaging microscopy is that the associated wavelengths are five orders of magnitude smaller than the visible light wavelength, about 0.04–0.008 Å, depending on the voltage of acceleration. This would mean a nearly infinite resolution. However, the resolution is limited by aberrations inherent to the electromagnetic lenses, to about 1–2 Å. Nevertheless this is good enough to obtain nanometric imaging and structure determination at the atomic level.

Besides imaging mode, the next most useful mode is diffraction mode, also called selected area diffraction, SAD, where the crystal lattice of a material is used as an interference pattern between the transmitted and diffracted beams, allowing the direct observation of the crystalline structure, as well as defects such as planar and line defects, grain boundaries, interfaces, etc., with atomic scale resolution. The brightfield/darkfield imaging modes of the microscope, which operate at intermediate magnification, combined with electron diffraction, are also invaluable for giving information about the morphology, crystal phases, and defects in a material. Another mode of the TEM is called microdiffraction, where the incident electron beam is focused (2 nm) to obtain the diffraction patterns from very small particles, allowing the identification of crystalline nanoparticles.

Figure 4.9 shows an example of the information that can be obtained from a HRTEM analysis. In that figure, a zinc sulfide (ZnS) thin film (<100 nm) deposited by spray pyrolysis was scratched from the substrate, and the particles were analyzed using a JEOL JEM2010 TEM working at 200 kV. The TEM image shows the morphology of the particles that resemble a rodlike structure. Moreover, the HRTEM images can be used to determine the interplanar distance of the crystalline structure (right corner), and, finally, the crystalline planes are further identified in the SAED pattern allowing the identification of the hexagonal ZnS structure using the ICDD diffraction database.

The addition of analytical methods, such as EDX or EELS, to obtain the composition of the samples at the microscale is also a great advantage. The main disadvantage of the TEM is the sample preparation process, since, as mentioned above, the sample must be thin enough to allow the transmission of the electrons with a minimum of inelastic scattering; ion beam erosion methods are usually required previous to the TEM analysis. For nanoparticles (NPs), the process is easier, and just a good distribution of dried NPs needs to be dispersed on appropriate microscopy grids.

4.2.2 *Surface Topography*

Surface topography or morphology can be observed by contacting or non-contacting techniques. The most commonly used contacting techniques are atomic force microscopy (AFM) and profilometry, while the most common non-contacting technique is optical or scanning electron microscopy.

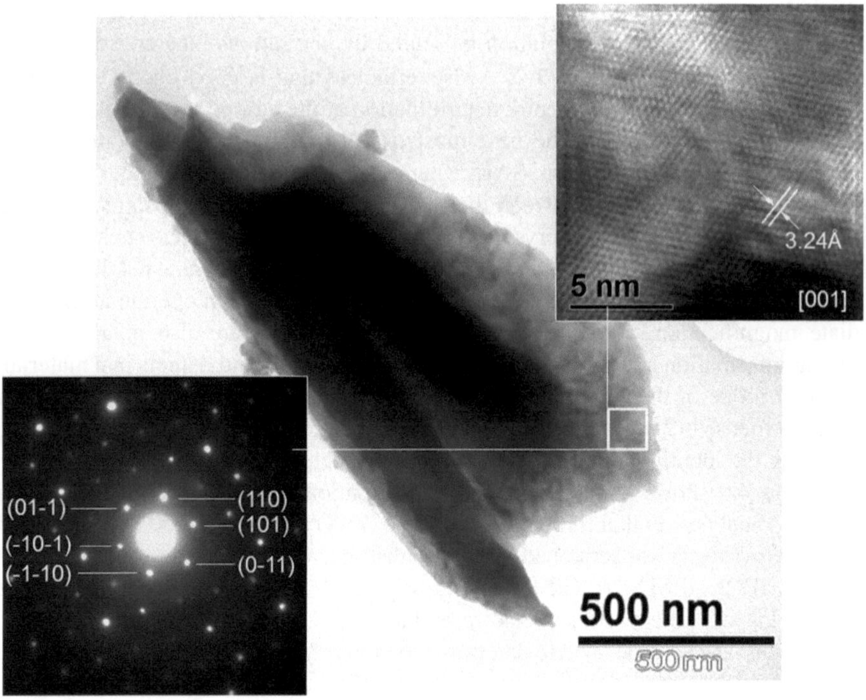

Fig. 4.9 TEM image of a particle obtained from a ZnS thin film. The HRTEM from the selected area and the microdiffraction pattern are also shown. Image provided by J. Santoyo, CINVESTAV, Mexico

4.2.2.1 Contact Techniques

Atomic force microscopy, AFM, is classified as a contact technique, although in some operation modes, no real contact occurs (Haugstad 2012; Eaton and West 2010; Moria et al. 2002). In the AFM, the probe consists of a sharp tip with radius of curvature of a few nanometers, normally made of Si or SiN, which is attached to a very sensitive cantilever. The tip is brought into close proximity to the surface of the sample, where the force between the tip and the sample leads to a deflection of the cantilever. The cantilever deflections are then measured with the aid of a laser and transformed into an analogous signal. The tip height (s) is regulated using a feedback mechanism and a piezoelectric piece above the tip, which allows the scanning of the sample without collisions. The tip height is plotted as a function of the sample dimensions (x, y) forming a topographic 3D image. The vertical resolution of the AFM is at the atomic level, about 0.1 nm, while the lateral resolution is about 1 nm. The area of the scans can be controlled from a molecular or atomic scale to about 100 μm^2, depending on the equipment. Other AFM modes measure different sample–tip interactions such as electrical (STM and Kelvin Probe), magnetic (MFM), piezoelectric (PFM), etc.

Profilometry is a real contacting technique where a stylus, to which a slight pressure is applied, is scanned across the surface. Profiles are formed by scanning the stylus along a direction keeping the stylus in contact with the surface at all moments; the displacements of the stylus in the vertical direction are measured and recorded. The stylus motion in the vertical direction is detected by piezoelectric or capacitors and is a direct measurement of the surface topography. The vertical resolution of the profilometry is lower (2 or 3 nm) than that of the AFM, but it can be used to obtain the surface roughness above the microscale, where AFM does not work.

Since both AFM and profilometry are sensitive to the vertical scale, surface topography can be quantified in terms of the surface roughness. Different definitions have been used to characterize the surface roughness; some of the most common are (1) the average roughness, R_a or S_a which are the arithmetic mean deviation of a lineal or a profile scan, respectively; (2) the RMS roughness, R_q or S_q which are the root mean square deviation of a lineal or a profile scan, respectively; and (3) the R_z or S_z and the R_t or S_t, which are a measure of maximum peak to valley distances in the lineal or profile scans.

4.2.2.2 Non-contact Techniques

Scanning electron microscopy (SEM) is a non-contacting technique where a beam of electrons, typically in the energy region of 0.2–40 keV, is focused on the sample's surface to form a spot of about 0.4–5 nm, which is scanned forming a 2D image of the sample (Goldstein et al. 2003; Goodhew et al. 2000; Amelinckx et al. 2008). The image can be formed by the secondary electrons (SEI), emitted from the valence and conduction bands of the material and whose intensity is proportional to the surface topography. The contrast difference produced from the emitted electrons at different heights (penetration field) produces a three-dimensional appearance in the image. However, images can also be formed by elastically scattered electrons (backscattered electrons, BSEI), and in such case the image contains information about the composition of the sample, since the contrast is due to electron–atom interaction, and so it is sensitive to the atomic mass. In modern microscopes, a composed SEI–BSEI image can be produced which contains both topography and composition information. The lateral resolution of the SEM can be as good as 1 nm when working at the highest incident electron energies; however, care has to be taken since the energy of the incident electron beam may damage the surface. The images obtained in the SEM are similar to the standard optical microscopes with a larger depth focus, which gives the impression of a 3D image. However, under normal operation condition, no quantitative information is obtained about the vertical scale.

For nanomaterials, high-resolution TEM (HRTEM) has also shown to be essential for studying the topography of the samples since high magnifications are required to really observe the detailed features, which might have a strong impact on the photocatalytic properties.

Finally, other methods to obtain information about the surface topography are the confocal optical microscopes and optical profilometers. However, they are not described here since they have not been extensively used for the photocatalytic samples.

4.2.3 Examples

Sridharan et al. (2013) reported the synthesis and visible light photocatalytic activity of graphitic C_3N_4–TiO_2 composites. In their synthesis, they obtained, in a first stage, microspheres of C- and N-codoped TiO_2 with stacked nanorods observed by SEM; however, the XRD pattern showed only the presence of pure tetragonal rutile phase of TiO_2. After pyrolysis in air at 300 and 500 °C, the XRD pattern corresponded again to pure rutile phase; however, XRD results alone were not conclusive since the diffraction peak formed at $2\theta = 27.4°$ matches both the (110) plane of rutile and the (002) plane of g-C_3N_4. Then, they used TEM to reveal the presence of two dimensional carbon nitride sheets with TiO_2 nanoparticles intercalated in between them, forming a TiCN composite, in the sample annealed at 300 °C. The spot diffraction pattern of the nanoparticle and its corresponding lattice resolved HRTEM image confirmed the single crystalline nature of the particles, while EDX spectra showed the presence of C, N, Ti, and O. The thermal treatment at 500 °C suppressed the formation of g-C_3N_4. The g-C_3N_4–TiO_2 composite showed an enhanced visible photocatalytic activity, attributed to the synergistic heterojunction formed between g-C_3N_4 nanosheets and doped TiO_2, that enhanced the electron–hole separation and the ability to absorb visible light.

Another interesting case of analysis is the one reported by Deng et al. (2012), who produced nanoporous ZnO microrods decorated with Ag nanoparticles. The XRD patterns showed the typical hexagonal wurtzite structure of ZnO. Additional peaks corresponding to the FCC Ag were identified, which increase their intensity as the Ag content increases. In addition, SEM images showed that the bare ZnO sample had rod shapes composed of nanoparticles. The microrods were about 90–150 nm in diameter and had a length of 0.5–3 μm. After the decoration with the Ag-NPs, the morphology looked the same, but further structural characterization with TEM showed that the sample had a typical nanoporous rodlike structure, and the 20–50 nm Ag-NPs were found attached onto the surfaces of the ZnO microrods (Fig. 4.10). HRTEM images revealed the interplanar spacing of about 0.28 nm, corresponding to the (100) plane of ZnO, while the interplanar spacing of 0.235 nm could be assigned to the (101) plane of Ag. Further composition analysis using XPS confirmed the presence of only Zn, O, Ag, and C, where the splitting of the 3d Ag peak indicated the metallic nature of silver, in agreement to other reported works (Lai et al. 2010; Lin et al. 2009). Additionally, they used Raman spectroscopy (RS) to evaluate the crystalline quality and the amount of defects as RS is sensitive to crystallization, structural disorder, and defects in micro- and nanostructures. The Raman spectra showed the Raman active modes for the hexagonal ZnO (A_1, E_1 and

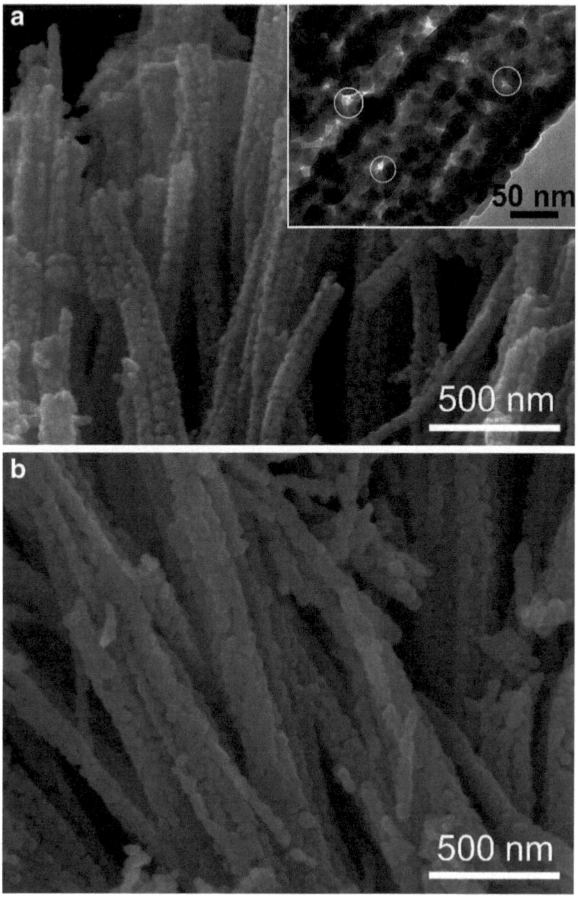

Fig. 4.10 (**a**) SEM image of the as-prepared n-ZnO microrods. The *inset* shows the high magnification view of TEM and (**b**) ZnO–Ag sample. From Deng et al. 2012. Copyright 2012 American Chemical Society

$2E_2$); the E_1 (longitudinal optical) mode, associated with the formation of oxygen deficiency, interstitial Zn, or the free charge carrier, was barely observed. Therefore, the appearance of a high intensity and dominated E_2 mode and almost no E_1 mode in the Raman spectrum indicated high quality crystalline specimens with little structural defects. The Ag-decorated ZnO microrods had a significant increase in the photocatalytic activity under solar light irradiation compared to the bare ZnO microrods, and the photostability was highly improved. This was attributed to the unique nano-/micro-configured structure achieved and to the Ag-NPs that acted as electron wells promoting the charge separation.

The effect of the topography on the wettability, the antifogging properties, and the photocatalytic activity of SiO_2–TiO_2 nanoparticle coatings was studied by Li and He (2013). The SEM and AFM images showed that hierarchically raspberry-like structures were produced showing superhydrophilic, antifogging, and photocatalytic activity toward organic pollutants, being the surface nanotopography the main responsible for the properties.

Graphene and graphene oxide have demonstrated to have extraordinary properties with potential applications in many fields, including photocatalysis. It is particularly attractive for the photo-induced water reduction to produce H_2, the fuel of the future. During the synthesis and functionalization of graphene sheets, AFM and optical microscopy are crucial characterization techniques as shown by Zhu et al. (2013) and Liu et al. (2013b), which used AFM to measure the number of graphene layers or the functionalization of the graphene.

As a final example of the effect of structure and topography in the photocatalytic properties, the ZnO–carbon nanotube (CNT) nanocomposites produced via a simple sol method are presented (Ahmad et al. 2013). The composites were produced at different concentrations of ZnO (0–20 wt %), and the photo-, sono-, and sonophotocatalytic degradation of the rhodamine blue was investigated, finding that the response was always higher for the combined process but was dependent on the ZnO concentration. In order to understand the role played by the CNTs, a wide range of characterization techniques were used. The SEM, TEM, and HRTEM showed the morphology and distribution of the ZnO nanoparticles attached to the multi-walled carbon nanotubes. Meanwhile, other techniques, such as BET, XPS, photoluminescence (PL), and DRS measurements, demonstrated that all the ZnO/CNT nanocomposites have stronger light absorption in the visible light region compared to pure ZnO, effect that acted synergistically with a larger specific surface area of the composites and the fact that the CNTs act as electron acceptors for ZnO, hindering the recombination of charge carriers.

4.3 Surface Area and Porosity

Surface area and porosity are important parameters for the photocatylic applications. The standard methods to determine these properties include direct observation by optical or electronic microscopy, gas adsorption, mercury porosimetry, as well as small angle scattering of X-rays (SAXS) or neutrons (SANS). Since the particle size for powders or nanopowder samples is directly related to the surface area, light scattering methods can also be used to determine indirectly the surface area.

4.3.1 Gas Adsorption

Gas adsorption methods are the most commonly used based on the adsorption of an inert gas, mainly nitrogen, on the surface of a solid material (Condon 2006; Lowell et al. 2004).

Gas adsorption occurs on the outer surface and, in case of porous materials, also on the surface of pores and allows the determination of the specific surface area (SSA). Several methods are used for SSA determination (i.e., Brunauer, Emmett, and Teller (BET) and de Boer *t*-Plot methods); nevertheless the BET method is the most commonly used, and it provides precise specific surface area evaluation by nitrogen multilayer adsorption, measured as a function of relative pressure using a fully automated analyzer (Brunauer et al. 1938).

Monolayer formation of gas molecules on the surface is used to determine the specific surface area, while the principle of capillary condensation can be applied to assess the presence of pores, pore volume, and pore size distribution. The technique encompasses external area and pore area evaluations to determine the total specific surface area in $m^2\ g^{-1}$.

According to IUPAC (Sing 1985), an adsorption isotherm is the function which relates the amount of adsorbed adsorbate to the pressure (or concentration) of the adsorbate in the fluid phase under equilibrium condition. IUPAC also classified adsorption isotherms into six types as shown in Fig. 4.11. The detailed discussion of the shape of adsorption and desorption isotherms can be found in Sing (1985) and Lowell and Shields 1984.

As can be seen in Fig. 4.10, type IV and V exhibit a hysteresis loop, i.e., the adsorption and desorption isotherms do not coincide over a certain region of external pressures. The type IV isotherm is typical for mesoporous adsorbents. Type V hysteresis loop is a typical sign of a weak fluid–wall interaction. It is less common but observed with certain porous adsorbents. Measuring sorption scanning curves helps to identify the underlying mechanism of hysteresis, which is crucial for obtaining an accurate and comprehensive pore size analysis of mesoporous and micro-mesoporous materials.

Pore width, pore shape, and the effective adsorption potential are the factors that determine pore filling. In the case of so-called micropores (pore width <2 nm, according to IUPAC classification), pore filling occurs in a continuous way, whereas in case of mesopores (pore widths in the range from 2 to 50 nm) pore filling occurs by pore condensation, which reflects a first-order gas–liquid phase transition. So-called classical macroscopic, thermodynamic concepts are based on the assumption of a certain pore-filling mechanism. Methods based on the Kelvin equation (e.g., Barrett–Joyner–Halenda, the BJH method) are linked to the pore condensation phenomena, i.e., they are applicable for mesopore size analysis, but they fail to describe the pore filling of micropores and even narrow mesopores in a correct way. In contrast to these macroscopic approaches, methods like the density functional theory (DFT) or methods of molecular simulation (Monte Carlo simulation methods (MC), molecular dynamics methods (MD)) provide not only a microscopic model of adsorption but also a more realistic description of the thermodynamic properties of the pore fluid (Thommes Quantachrome). The pore shape can further be obtained from the hysteresis loop from corresponding isotherm.

A detailed description of isotherm types, hysteresis loop, and methods related with gas adsorption in porous materials mentioned above can be found in several articles and books already published (Sing 1985; Lowell and Shields 1984; Brunauer et al. 1938; Barrett et al. 1951; Fagerlund 1973; Gauden et al. 2004; Kowalczyk et al. 2003; Lastoskie and Gubbins 2000; Dombrowski et al. 2001; Rouquerol et al. 2014).

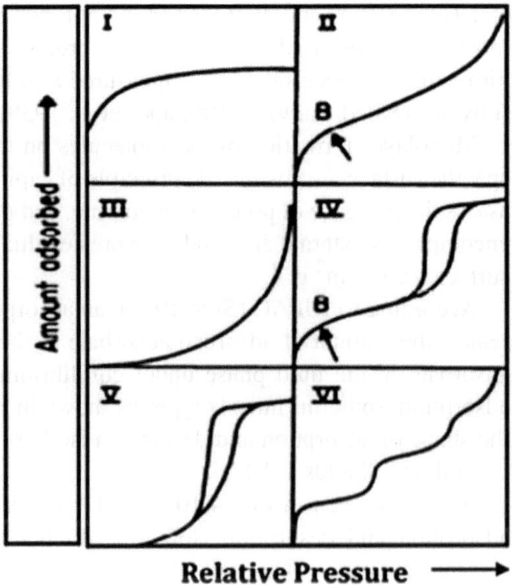

Fig. 4.11 Classification of adsorption isotherms (IUPAC, Sing 1985)

4.3.2 *Mercury Porosimetry*

Mercury porosimetry uses the non-wetting properties of mercury to gain information on the porosity, pore volume, pore size distribution, and density (Lowell et al. 2004). The technique consist of applying a high pressure (about 400 MPa for the smallest pores) to force the intrusion of mercury in the sample's small pores, whereas mercury intrusion in larger pores already occurs at low pressure. In this way a wide dynamic range of pore sizes can be measured, and the pore size distribution can be obtained starting from 4 nm up to approx. 800 μm. Thus, mercury porosimetry is extremely suitable for materials showing broad distributions of pore sizes or mainly macropores.

4.3.3 *Dynamic Light Scattering*

Light scattering techniques have become very popular these days to measure the size and size distribution of nanoparticles in a solution. Basically, the techniques consist of shining a monochromatic light beam, such as a laser, onto the solution that contains the spherical particles (Schartl 2007). The Brownian motion of the particles (movement of particles due to the random collision with the molecules of the liquid that surrounds the particle) causes a scattering of the light intensity, and such scattering depends on the particle velocity. For Brownian motion, small particles move quickly and large particles move more slowly. Therefore, the measurements of the intensity of the light by a photomultiplier placed at 90° as a function of the frequency leads to a frequency spectrum of the intensity of the scattered light. The spectra will have the form of a Lorentzian shaped line whose width depends on the diffusion coefficient, *D*, and the scattering angle. The assumption of pure Brownian movement and spherical particles allows the determination of size and size distribution of particles and molecules. This is possible, since the probability density function for Brownian movement is well defined and depends mainly in the diffusion coefficient, which according to the Stokes–Einstein equation is a function of the temperature, the viscosity of the solvent, and the size of the particle.

4.3.4 Examples

Porous TiO_2-coated magnetic core-shell nanocomposites were prepared by Liu et al. (2013a) in order to improve the separation and recovery of nanosized TiO_2 while maintaining good catalytic activity. The nanocomposite was formed by magnetic Fe_3O_4 particles covered with an interlayer of SiO_2 and an outer layer of porous anatase TiO_2 (TiO_2/SiO_2@Fe_3O_4 microspheres). The XRD pattern presented the characteristic peaks of anatase TiO_2 and the peaks of cubic Fe_3O_4. The N_2 adsorption–desorption isotherms were acquired. The samples exhibited a combination of type I and IV patterns with distinct H2 and H4 hysteretic loops, which confirmed the existence of micro- and mesoporous in the outer layer of anatase. Such multiple-pore structure can promote rapid diffusion of various reactants and products during the photocatalytic reaction and enhance the photocatalytic activity. The BET surface area of the 9 %TiO_2/6 %SiO_2@Fe_3O_4 microspheres was 373.5 m^2g^{-1}, and the BJH pore volume was 0.28 cm^3g^{-1}. This material was 20 % more efficient than P25 in the degradation of methylene blue and methyl orange. The high photocatalytic performance was maintained after six cycles. The samples presented a high degree of superparamagnetism; although the magnetization values were weak compared to pure Fe_3O_4, they are strong enough to guarantee a good recovery by magnetic separation.

Recently, ZnO nanostructures have gained a lot of attention; particularly, nanorods are potentially useful in photocatalysis due to a high surface area. Hafez (2012) used a simple hydrothermal method to obtain ZnO rodlike nanostructures with pointed-shape ends. TEM images showed straight and uniform rodlike structures with diameters between 30 and 50 nm and length between 400 and 650 nm. XRD pattern indicated the hexagonal wurtzite structure of ZnO, while EDS analysis confirmed the presence of only Zn and O without any other elemental contamination (Fig. 4.12). The N_2 adsorption/desorption measurements showed a type II isotherm with small type H3 adsorption hysteresis loop. The BET surface area of the nanorods was about 270 m^2g^{-1}. The pore size distribution presents two types of pores (small mesopores at 1–5 nm and large meso- and macropores centered at 10–100 nm) (Fig. 4.13). Additionally the hysteresis loop approached $P/P_o = 1$,

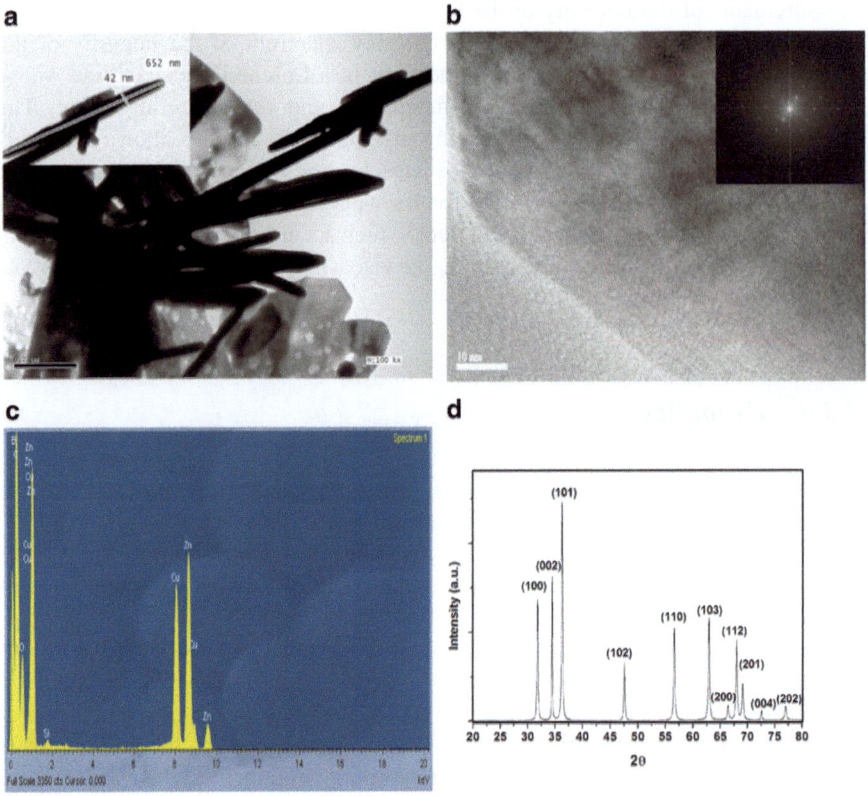

Fig. 4.12 (**a**) TEM image; (**b**) SAED; (**c**) EDX; (**d**) XRD patterns of the obtained ZnO rodlike nanostructures. From Hafez 2012

confirming the presence of macropores (>50 nm). In fact, the nanorod structures did not contain mesopores and macropores; the observed micro- and mesopores came from the aggregation of the nanorods which leads to formation of such porous structure and this high surface area. The sample exhibited good photocatalytic activity in the degradation of reactive yellow 15 dye that can be attributed to the high adsorption percentage (34.7 %) of the dye on the surface of the catalyst as a consequence of the high surface area.

4.4 Vibrational Spectroscopies

Vibrational spectroscopies are widely used techniques to characterize materials because they are nondestructive, noninvasive tools that provide information about the molecular composition and structure. These two analytical techniques, namely,

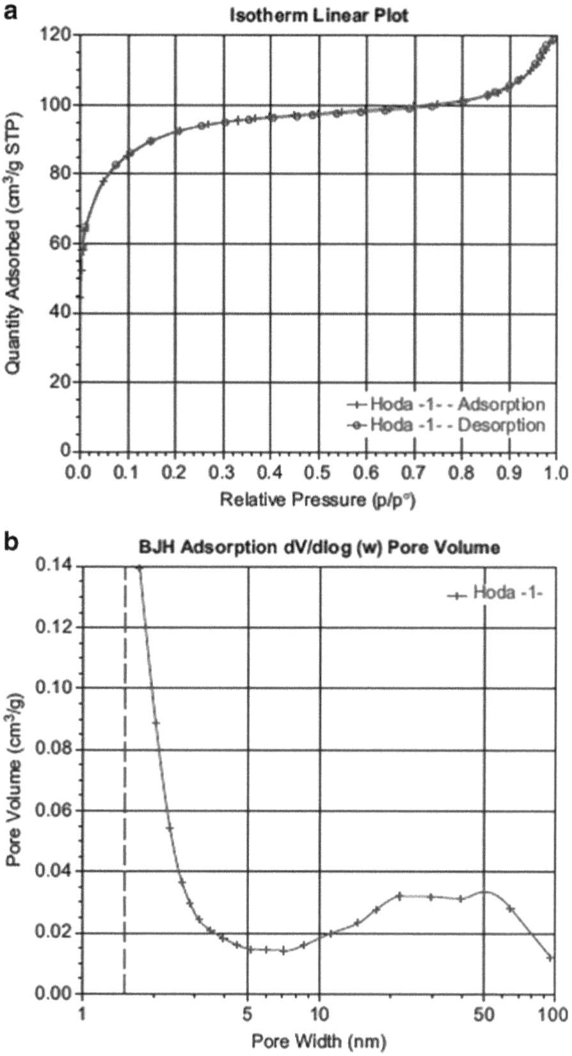

Fig. 4.13 (**a**) Nitrogen adsorption–desorption isotherm and (**b**) BJH pore size distribution of the ZnO rodlike nanomaterials. From Hafez 2012

infrared and Raman spectroscopy, measure vibrational energy levels which are associated with the chemical bonds in the sample. The spectrum is unique, like a fingerprint, and vibrational spectroscopy is used for identification, characterization, structure elucidation, reaction monitoring, and quality control. FTIR and Raman have been implemented for in situ characterization of catalytic reactions (Haw 2002).

4.4.1 Fourier Transform Infrared Spectroscopy, FTIR

Infrared spectroscopy is a technique based on the detection of the vibrations of the atoms of a molecule by measuring the absorption of electromagnetic light by the sample (Kuzmany 2010). An infrared spectrum is commonly obtained by passing infrared radiation through a sample (transmission) and determining the particular energy at which light is absorbed. The IR energy range (700 nm to 1 mm or 1.24 meV to 1.7 eV) coincides with the energies needed to excite fundamental modes of rotation and vibration in materials; thus absorption of the incident IR signal is due to the excitation between ground and excited vibrational states. The transmission or reflection (attenuated total reflectance (ATR) mode or diffuse reflectance) spectra show the specific vibrational modes, characteristic of each molecular structure that depends on the bond strength, structure, masses of the atoms, etc. IR spectroscopy is a nondestructive technique which can be used in any type of samples: gases, liquids, solids, powders, fibers, etc. It is very surface sensitive and can detect very small quantities of molecules such as 10^{-5} monolayers. Infrared light interacts only with those vibrations whose dipole moment (μ) periodically changes due to the oscillation of the atoms. If the oscillating electromagnetic field of the incident photon couples with the dipole oscillating at the same frequency, it is absorbed. Those vibrations that are not accompanied by a change in the dipole moment are IR-inactive, but may be Raman active if the polarizability (α) of the electron configuration changes in the course of an oscillation period.

4.4.2 Raman Spectroscopy

Raman spectroscopy provides information about molecular vibrations that can be used for sample chemical identification and quantitation. In Raman spectroscopy, the sample is excited by a monochromatic light source (i.e., laser in the visible, near-IR, or near-UV range, 1–12.4 eV), and the scattered light is detected (Ferraro et al. 2003). The majority of the scattered light is of the same frequency as the excitation source (Rayleigh or elastic scattering). A small fraction of the scattered light (ca. 5–10 % of the incident light intensity) is shifted in energy from the laser frequency due to interactions between the incident electromagnetic waves and the vibrational energy levels of the molecules in the sample. Plotting the intensity of the Raman shift (scattered minus Rayleigh energy) versus frequency results in a Raman spectrum of the sample, where the peak positions will lie at frequencies that correspond to the energy levels of different functional group vibrations, and the spectra can be analyzed similar to IR. The energy shift of the scattered photons is independent of the energy of the incident photons, except for resonance phenomena, and is characteristic of the chemical bonds, electronic configuration, and vibrational modes that were excited. In the same sense as in IR spectroscopy, it is possible to characterize the surface chemistry from the generated spectrum which is a fingerprint of the molecular structure. Although the physical process and instrumentation of Raman spectroscopy is more complex, it has the advantage of being more versatile due to the different incident energies; it is nondestructive and less affected by the environment. Moreover, the new Raman systems are highly sensitive, and chemical imaging has also been developed, as well as in situ analysis of surface processes (Lamberti et al. 2010).

4.4.3 Examples

FTIR technique can be very useful for the identification of chemical bonds present in the sample and thus confirm a specific structure, especially for complex compounds. Such is the case of photocatalytic graphitic carbon nitride compounds (Yan et al. 2009; Huang et al. 2013; Dante et al. 2013), which presents strong bands in the 1,200–1,650 cm^{-1} region that correspond to the typical stretching modes of CN heterocycles. Additionally, the characteristic breathing mode of the triazine units at 801 cm^{-1} is observed. The presence of a broad band at around 3,000–3,500 cm^{-1} is indicative of NH stretching vibration modes. In polymeric carbon nitride obtained from melamine cyanurate, this band can evolve toward a definition of four bands when the pyrolysis temperature is increased from 450 to 700 °C (Dante et al. 2013); the bands appearing at 3,248, 3,150, and 3,078 cm^{-1} are due to NH interacting via hydrogen bond, as observed in Fig. 4.14. FTIR spectra at 650 and 700 °C

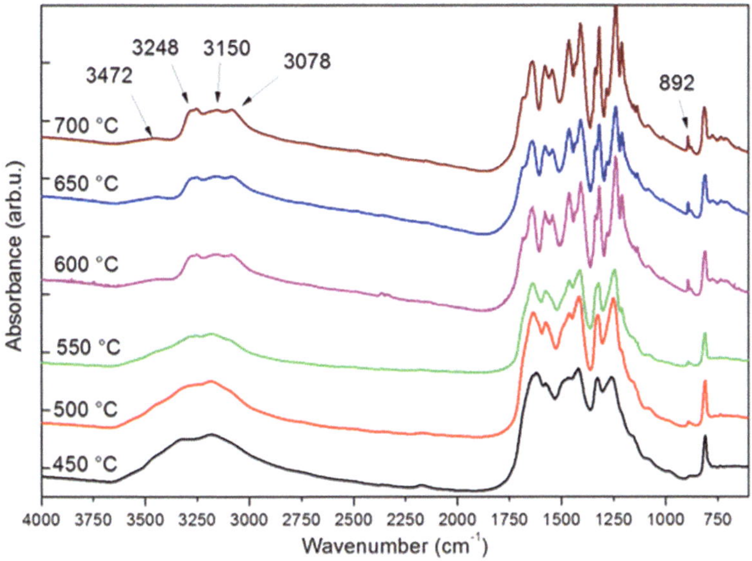

Fig. 4.14 FTIR spectra of polymeric carbon nitride obtained from melamine cyanurate at 650 and 700°

correspond substantially to a melon (polymeric carbon nitride) which is known to consist of heptazine units, observed at 892 cm^{-1} that increase with the temperature (Horvath-Bordon et al. 2004).

Liang et al. (2012) hybridized Ag_3PO_4 nanospheres with graphene oxide (GO) sheets and used FTIR and Raman spectroscopy to demonstrate the existence of strong charge interactions. They obtained the FTIR spectra of pure GO and pure Ag_3PO_4 and compared them with the spectrum of the GO–Ag_3PO_4 nanocomposite. The latter showed a shift in the peak associated with the carbonyl group due to charge interactions with the Ag_3PO_4 nanospheres. The characteristic peak assigned to the stretching vibration of the P–O–P group also shifted to a higher wavenumber, suggesting that Ag_3PO_4 successfully attached with GO sheets. Moreover, Raman spectroscopy, an important tool for investigating the detailed structure of graphitic materials, was applied to roughly evaluate the disorder degree in the hexagonal graphitic layers and the average size of the sp^2 domains of the graphite materials. They obtained the intensity ratio of the D and G bands for the GO sheets and the GO–Ag_3PO_4 nanocomposite, finding an increase in this ratio from 1.13 of GO to 1.21 of the composite, indicating a larger amount of defects formed in the graphitic layers after the incorporation of the Ag_3PO_4 nanospheres. The interactions are not only beneficial to improve the stability but also favorable to enhance the photocatalytic activity of Ag_3PO_4. In an interesting experiment, reflection absorption infrared spectroscopy (RAIRS) was used to compare the photocatalytic activity of TiO_2 polymorphs (Mingchun Xu et al. 2011); the anatase (101) surface was more active for the photo-oxidation of CO into CO_2 than the rutile (110). The authors explained the difference as a consequence of a one order of magnitude larger electron–hole lifetime for the anatase surface than rutile, information obtained from transient photoconductance experiments. This is a key experiment providing evidence to support the commonly observed different photocatalytic activities for the two polymorphs.

Raman spectroscopy, on the other hand, has been more used for either phase identification, temperature-induced phase transformation, determination of quality of the synthesized material, or detection of carbonaceous residues left during the synthesis (Hardcastle 2011; Beuvier et al. 2009). However, care has to be taken when analyzing nanocrystalline materials since it has been shown that Raman shift position and line width change in comparison to bulk materials. The changes in the Raman spectra of nanocrystalline TiO_2 have been interpreted as due to phonon confinement, non-stoichiometry, or internal stress (Balaji et al. 2006; Swamy et al. 2005 and references therein).

4.5 Optical Properties

Semiconductors and insulators have the fundamental band gap in the near infrared, visible, or ultraviolet spectral region (Fox 2010). Therefore, light absorption is zero below the absorption edge, and it increases rapidly once the energy of the light is enough to excite electrons across the optical gap, which is called interband transition and is related to the band structure of the material. The physical model for such process is based on the application of the quantum mechanical description of the light–matter interaction to the band states of solids. The optical interband transitions occur when a photon excites an electron from the filled valence band to the empty conduction band. The threshold for absorption occurs when the photon energy equals the band gap energy (E_g), and after there is a continuous absorption spectrum up to an energy which depends on the specific structure of the bands involved. The band structure also determines the interband absorption rate, since it depends on the population of the bands (density of states), as well as in other quantum-mechanically (QM) defined selection rules. The interband absorption creates a hole in the valence band and puts an electron in the conduction band, which is typically referred as the creation of an electron–hole pair. The frequency or energy dependence of the absorption coefficient, $\alpha(\omega)$, contains all the information concerning the light–matter interaction and therefore is QM determined by the probability of exciting an electron from an initial to a final quantum state by the absorption of a photon. This calculation leads to the following simplified equations: for any semiconductor,

$$\text{For } \hbar\omega < E_g, \quad \alpha = 0 \tag{4.5}$$

However, as the photon energy increases, for a direct gap semiconductor, as in Figure 4.15(a)

$$\text{For } \hbar\omega \geq E_g, \quad \alpha \propto \left(\hbar\omega - E_g\right)^{1/2} \tag{4.6}$$

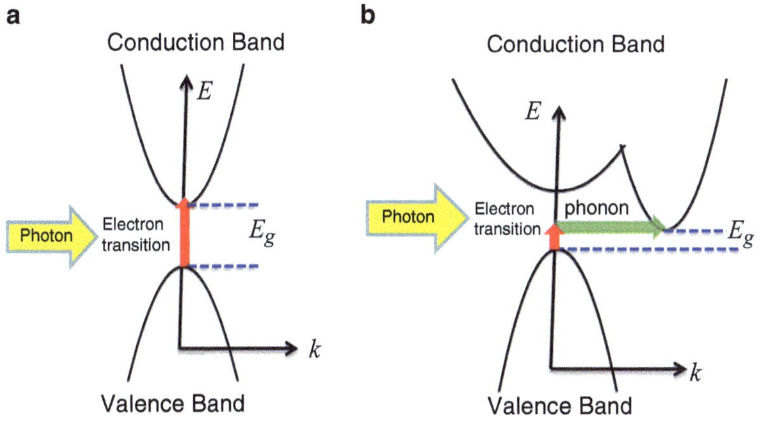

Fig. 4.15 Interband transition in solids. (**a**) Direct gap and (**b**) indirect gap. Since the phonon energies are much lower than the optical gap, it is very common to neglect it for an estimation of the gap

For an indirect gap semiconductor, as in Figure 4.15(b)

$$\text{For } \hbar\omega \geq E_g, \quad \alpha \propto \left(\hbar\omega - E_g \mp \hbar\Omega\right)^2 \tag{4.7}$$

where $\hbar\Omega$ is the phonon energy that assists the transition process, as shown in Fig. 4.15. Some authors used $\alpha\hbar\omega \sim (\hbar\omega - E_g)^p$ where the exponent is the same as before and the energy factor introduces only a small correction to the gap values.

Optical gap calculations for solids are usually done using one of the equations in Eq. (4.6) or (4.7). For this, the absorption coefficient has to be obtained experimentally by some of the methods described later, and the optical gap can be calculated by plotting

$$\alpha^2 \text{ vs. } \hbar\omega \text{ for a direct gap} \tag{4.8}$$

$$\alpha^{1/2} \text{ vs. } \hbar\omega \text{ for indirect gap} \tag{4.9}$$

From such plots, the E_g is calculated as the point at which a linear extrapolation of the relations (Eqs. 4.8 and 4.9) crosses the energy axis at $\alpha = 0$, as illustrated in Fig. 4.16.

The study of amorphous semiconductors brought a lot of attention during the 1960s, for the development of amorphous silicon and germanium and their possible applications in the newly born microelectronic industry (Tauc et al. 1966; Cody et al. 1982). Such studies proposed certain models to estimate the optical band gap for the amorphous semiconductors considering that the optical transitions conserve energy but not the momentum vector and that the densities of states near the band extreme have the same energy dependence as in the crystalline counterparts. The most known of such models is the Tauc-gap model (Tauc et al. 1966) which states that the absorption edge follows Eq. (4.10):

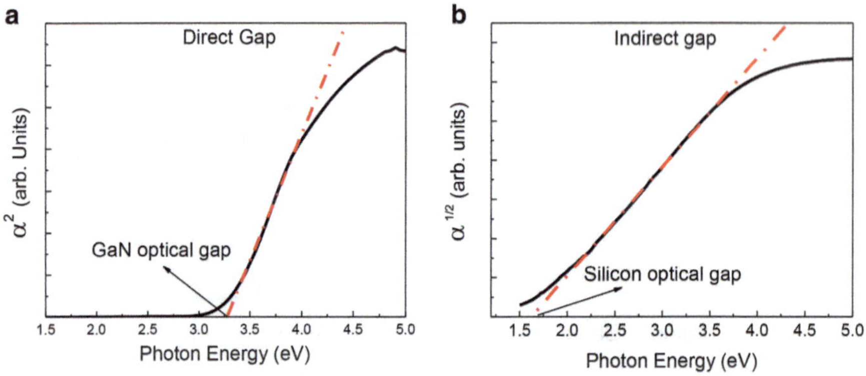

Fig. 4.16 Example of the calculation of the optical gap for standard semiconductors. (**a**) GaN direct gap and (**b**) Si indirect gap

$$\omega^2 \epsilon_2 \sim (\hbar\omega - E_g)^2 \tag{4.10}$$

where ϵ_2 is the complex dielectric function, related to the refractive index and the extinction coefficient as $\epsilon_2 = 2nk$. Using Eq. (4.11) that relates the absorption coefficient with the extinction coefficient,

$$\alpha(\lambda) = \frac{4\pi k(\lambda)}{\lambda} \tag{4.11}$$

where λ is the light wavelength, the Tauc-plot equation is given by

$$\sqrt{\alpha(\hbar\omega)} \sim (\hbar\omega - E_g) \tag{4.12}$$

And so the optical gap for amorphous semiconductors is estimated plotting $\sqrt{\alpha(\hbar\omega)}$ versus $\hbar\omega$, which is similar to Eqs. (4.8) and (4.9), but care has to be taken since for amorphous semiconductors the direct–indirect concept is no longer valid.

4.5.1 *Transmission and Reflection*

For low absorbing materials, the absorption coefficient can be calculated from the equation

$$T = (1 - R)^2 \exp(-\alpha d) \tag{4.13}$$

where T is the transmission, R is the reflectance, and d is the sample thickness (Fox 2010).

T and R are measured as the fraction of light intensity transmitted or reflected from an optical medium in comparison to the intensity of an incident light. The spectral dependence of T and R are measured with the aid of a monochromator.

As Eq. (4.13) shows, it is important to measure both T and R, although in many cases R is neglected and so the absorption coefficient is estimated only from the T spectra. However, it is important to understand that such model gives only approximate absorption coefficients, since reflection from semiconductor samples in the UV–Vis region can be as high as 30 %, as shown in Fig. 4.17 for different materials.

Another method to analyze the T and R spectra (particularly useful when the absorption is larger) which is valid for homogeneous materials and smooth surfaces, in which scattering losses are negligible, is to use the total transmission and reflection coefficients for normal incidence (Fresnel coefficients) and calculate the dispersion of both the refractive index (n) and the extinction coefficient (k). In this case, the absorption is calculated from Eq. (4.11).

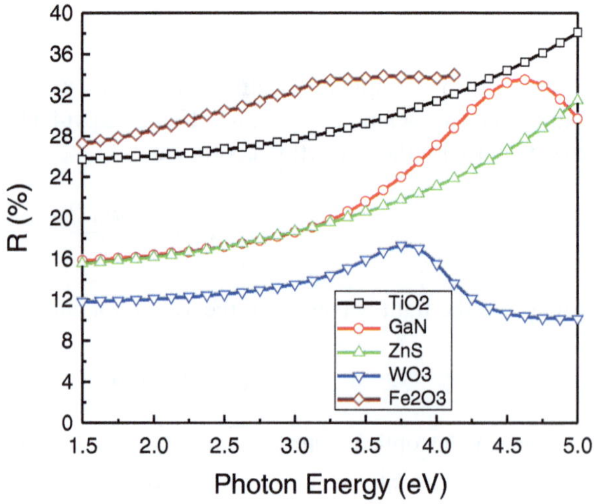

Fig. 4.17 Reflectance spectra of different semiconductor slabs estimated using their known optical properties (n, k)

Some specific software uses minimization algorithms to solve iteratively the system of equations (Eqs. 4.14 and 4.15), where the film thickness could be also a fitting parameter:

$$T_{\text{cal}}(n, k, \lambda) - T_{\text{exp}}(\lambda) = 0 \tag{4.14}$$

$$R_{\text{cal}}(n, k, \lambda) - R_{\text{exp}}(\lambda) = 0 \tag{4.15}$$

Multiple solutions are a usual problem in this type of calculation; however when R and T are measured under the same conditions, the multiple solutions for k do not deviate considerably from the true values, so that the problem is mainly restricted to the refractive index, a problem that can be reduced having previous knowledge of the film thickness.

4.5.2 Diffuse Reflectance

For the case of nonhomogeneous, opaque, or powdered materials, the determination of the band gap is not possible using either the optical transmission–reflection or the spectroscopic ellipsometric methods. Measurement of the diffuse reflectance instead is a more adequate technique (Murphy 2007a). Diffuse reflectance spectroscopy (DRS) is a suitable technique to measure optical properties of rough surfaces or powders. This technique also provides information about the oxidation state and coordination environment of transition metal ions in catalytic solids. If the surface of the material or sample is microscopically rough, the light rays will reflect and diffuse in many different directions. In this case, each individual ray follows the law of reflection. However, the roughness of the material makes that each

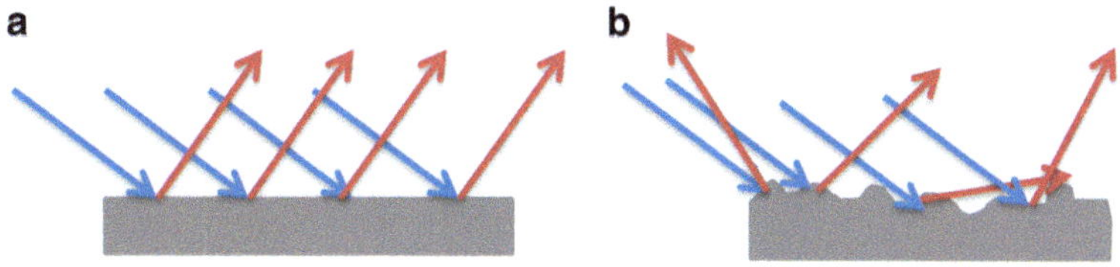

Fig. 4.18 Comparison between (**a**) specular reflectance (smooth surfaces) and (**b**) diffuse reflectance (rough surfaces)

individual ray meets a surface with different orientation. Subsequently, when the individual rays reflect off the rough surface, they scatter in different directions (see Fig. 4.18).

When electromagnetic radiation is incident upon a system which is able to scatter as well as absorb photons, Lambert–Beer's law is no longer useful due to two main reasons; first, the path length l is no longer well defined, and, second, the optical response is no longer determined by k but by $\hat{n} = n - ik$ (Delgass et al. 1979).

The diffuse reflectance phenomenon is complex, but in the simplest case of a nonabsorbing material ($k = 0$), it involves only scattering of photons. The scattering of light can be divided into three different regimes: (1) single scattering, (2) multiple scattering, and (3) dependent scattering (Weckhuysen et al. 2000). The first type arises when scattering centers are sufficiently apart that each is illuminated only by photons not previously scattered, and on average, there is no phase relationship between the photons scattered form the neighboring particles. In the case of multiple scattering, the scattering centers are still far apart that they may be treated as independent, but each center is now illuminated by photons scattered from adjacent particles. Dependent scattering arises when, in addition to multiple scattering, phase coherence exists between scattered photons from adjacent centers. Dependent scattering is present if the average distance (d) between scattering centers is less than three times the particle diameter.

The division of light scattering systems into the above three categories determines the theoretical approach necessary to describe the scattered light intensity. For single and multiple scattering, the absence of phase coherence ensures that the total scattered light intensity is merely the sum of the intensities of the individual scatters. Both types of scattering can be mathematically treated in a rather easy way. In contrast, when the particles become so tightly packed that phase coherence becomes important, amplitudes rather than intensities must be summed. This is the case for heterogeneous catalysts because they are always investigated in the form of densely packed powders. Then the radiation transfer theory has to be considered, leading to the well-known Kubelka–Munk theory. The Kubelka–Munk model uses an effective scattering coefficient S and an effective absorption coefficient K to describe the optical properties of a compact powder sample; a sufficiently thick sample so that all the incident light is absorbed or scattered, no light is considered.

Then, the Kubelka–Munk (K-M) function, $F(R_\infty)$, is related to the apparent absorption (K) and the apparent scattering coefficient (S) by Eq. (4.16):

$$F(R_\infty) = \frac{(1 - R_\infty)^2}{2R_\infty} = \frac{K}{S} \tag{4.16}$$

Where $\left(R_\infty = \frac{R_{\text{sample}}}{R_{\text{standard}}}\right)$, and it is the equivalent of the Lambert–Beer law for dilute species measured by diffuse reflectance experiments:

$$F(R_\infty) = \frac{\varepsilon C}{s} \tag{4.17}$$

Where ε is the molar absorptivity and C the concentration.

For such powder infinitely thick samples, the optical band gap can be easily calculated using Eqs. (4.8) and (4.9) but substituting α by $F(R_\infty)$

However, as discussed by Murphy (2007a), when a thin powder sample (<1 mm) or a rough thin coating on a substrate is measured using DRS, the reflection and transmission of the light through the different interfaces need to be taken into account to avoid the thickness dependence of the optical gap. For this, Murphy (2006, 2007b) proposed a modified K-M two-flux model to calculate the reflectance of the coating as a function of the refractive index, absorption coefficient, scattering coefficient, and thickness. Then a fitting procedure is used to correlate the calculated reflectance spectra to the measurements using a spectral projected gradient algorithm, and in this way, appropriate optical properties can be obtained.

DRS measurements are achieved using a UV–Vis (can also be used for infrared analysis) spectrophotometer, but with a special accessory called integrating sphere. This accessory provides the ability to collect a quantitative reflectance spectrum from highly scattering or irregularly shaped samples, but acquiring only diffuse light, not the specular signal and so the K-M theory can be applied.

4.5.3 Spectroscopic Ellipsometry

Optical ellipsometry is a common method used to determine dielectric functions or complex refractive index, thickness, and morphology of thin films of various natures, but requires flat surfaces. Ellipsometry measures the change of the light polarization upon reflection at an interface (Tompkins and Irene 2005; Fujiwara 2007). In ellipsometry, the amplitude ratio and the phase difference between the parallel (p) and the perpendicular (s) components of the reflected light polarized with respect to the plane of incidence are measured and expressed by two quantities, namely, the ellipsometric angles, psi Ψ (which measures the amplitude ratio) and delta Δ (which measures the relative phase change). These are given by

$$\overline{\rho} = \frac{\overline{r}_p}{\overline{r}_s} = \frac{|r_p|\exp(\delta_p)}{|r_s|\exp(\delta_s)} = \frac{|r_p|}{|r_s|}\exp(\delta_p - \delta_s) = \tan(\Psi)\exp(i\Delta) \tag{4.18}$$

where the tilde in r_p and r_s represents that they are the complex reflections coefficients for the p and s components of the electric field, respectively. For bulk samples, r_p and r_s are the Fresnel coefficients, and for thin films on a substrate, r_p and r_s are substituted by the total reflection coefficients R_p and R_s given by

$$\overline{R}_{p,s} = \frac{E^R_{p,s}}{E^0_{p,s}} = \frac{\overline{r}_{0,1p,s} + \overline{r}_{1,2p,s}e^{i2\beta}}{1 + \overline{r}_{0,1p,s}\overline{r}_{1,2p,s}e^{i2\beta}} \text{ and} \tag{4.19}$$

$$\beta = 2\pi\left(\frac{d}{\lambda}\right)\sqrt{\overline{n}_1^2 - n_o^2 \sin^2\theta} \tag{4.20}$$

where $\overline{r}_{0,1p,s}$ and $\overline{r}_{1,2p,s}$ are the Fresnel coefficients for the interfaces between media 0 and 1 and between media 1 and 2, respectively, d is the film thickness, λ is the wavelength, and θ the incidence angle, as shown in Fig. 4.19.

The advantage of ellipsometry over conventionally transmission–reflection measurements is that ellipsometry measures the ratio between two quantities, so it is potentially more precise and accurate. However, the disadvantage is the complexity of the optical system.

In the most general case the reflecting surface is made up of a stack of planar multilayer materials, and then polarization change, ρ, depends on the incidence angle, the radiation wavelength, and the thickness and complex refraction index of each layer. The thickness and complex refractive index of the layers are obtained from measurements of $\Delta(\lambda)$ and $\Psi(\lambda)$ by fitting the experimental data with a suitable optical model of the system under consideration.

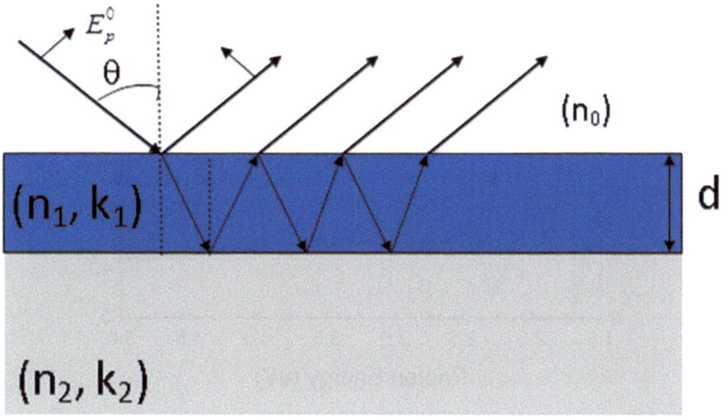

Fig. 4.19 The light pathway for a transparent film on an absorbing thick substrate

The development of digital light detectors together with the need for faster acquisition, having the capability of measuring psi and delta even if the sample was continuing changing (to study film growth), led to the development of various approaches for the determination of the polarization state of transverse electromagnetic plane waves, such as the rotating analyzer ellipsometer, the rotating polarizer ellipsometer, and the phase modulation ellipsometer. These dynamic measurements are based on the idea that one of the parameters that describes the optical component of the system is periodically varied with time and the detected signal is Fourier-analyzed. The different hardware configurations together with their advantages and disadvantages have been extensively discussed in many papers.

Ellipsometry does not directly measure the film thickness or optical constants; it measures psi and delta. To extract useful information about the sample, it is necessary to perform a model-dependent analysis of the ellipsometric data. Therefore, after the measurements are made, a model for the optical structure of the sample is constructed. This may include a substrate and a single film on top, or substrate plus film with roughness on top, or more complex multilayered structures. Assuming a sample structure and dispersion formulae for the optical constants of the different layers, the measured ellipsometric spectra are then fitted using a standard minimization algorithm; an example of the fitting for a Bi_2O_3 thin film deposited on a silicon substrate by magnetron sputtering is shown in Fig. 4.20.

The optical constants of the layers are simulated using dispersion formulae (Fujiwara 2007; Tompkins and Irene 2005) such as the following:

- The Classical Lorentz oscillator model, which allows a fairly good visualization of the atom–field interactions, considers the material medium as an assemblage

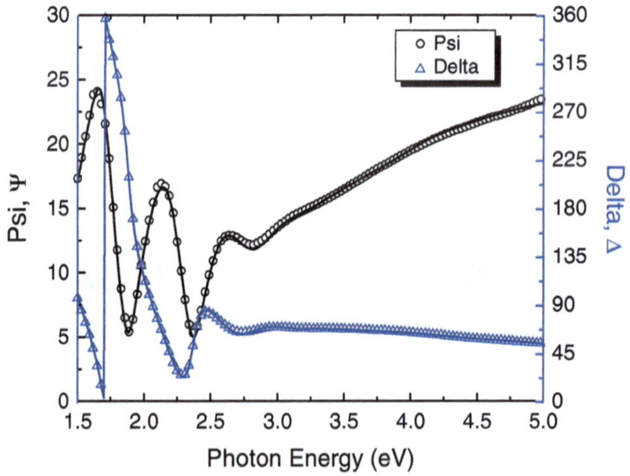

Fig. 4.20 Ellipsometric measurements are represented as *symbols* (some of them have been omitted for clarity), and the *lines* correspond to the fittings using a Tauc–Lorentz dispersion model for a nanometric Bi_2O_3 thin film

of many polarizable atoms with bound electrons that respond to the excitation with light as oscillators.

- The Drude model that takes into account the absorption from free electrons.
- The Lorentz–Drude model where both bound and free electrons are considered.
- The Sellmeier model which corresponds to the Lorentz model when the imaginary dielectric function (or the extinction coefficient) approaches to zero.
- The Cauchy model, which is an approximate function of the Sellmeier model obtained by doing a series expansion of the Sellmeier function.
- The Tauc–Lorentz model which was proposed for amorphous semiconductors but nowadays has also been used for conductive oxide semiconductors and nanocrystalline semiconductors (Jellison and Modine 1996; Chen and Shen 2005). Their formula uses a combination of the Tauc band edge and the Lorentz formulation for a collection of uncoupled atoms to determine the imaginary part of the dielectric function, given by

$$\varepsilon_2(E) = \begin{cases} \dfrac{AE_oC(E - E_g)^2}{\left[(E^2 - E_o^2)^2 + C^2E^2\right]}\dfrac{1}{E} & E > E_g \\ 0 & E \le E_g \end{cases} \tag{4.21}$$

The real part is determined using Kramers–Kronig integration (Fujiwara 2007) from zero to infinite.

$$\varepsilon_1(E) = n^2(E) - k^2(E) = \varepsilon_1(\infty) + \frac{2}{\pi}P\int_{E_g}^{\infty} \frac{\xi\varepsilon_2(E)}{\xi^2 - E^2}\mathrm{d}\xi \tag{4.22}$$

where P stands for the Cauchy principal part of the integral and an additional parameter $\varepsilon_1(\infty)$ is also included. Normally, $\varepsilon_1(\infty) = 1$, but it can be greater than 1 if there are significant optical transitions at energies greater than those sampled by the ellipsometer.

- Other models have also been defined for specific classes of materials.

For all of them, the film thickness and the $n(E)$ and $k(E)$ relationships in the spectral range are obtained, as shown in Fig. 4.21 for the same Bi_2O_3 film shown in Fig. 4.20.

Then, the optical gap can be calculated using some of the equations shown in Sect. 4.5. However, as shown in Fig. 4.22a–c the selection of the adequate equation requires previous knowledge about the structure (crystalline or amorphous) and the band structure (direct or indirect gap). Otherwise, any of the models can be applied and a linear region can be used to fit the plot, but the optical gap is completely different according to the model used, as shown in Fig. 4.22a–c, where the gap changes from 3.4 eV assuming a direct gap to 1.55 eV for indirect gap calculation

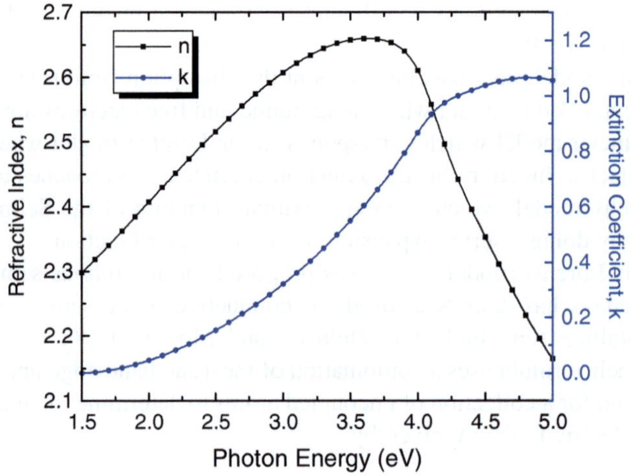

Fig. 4.21 Optical properties obtained for the Bi_2O_3 thin film using the Tauc–Lorentz model

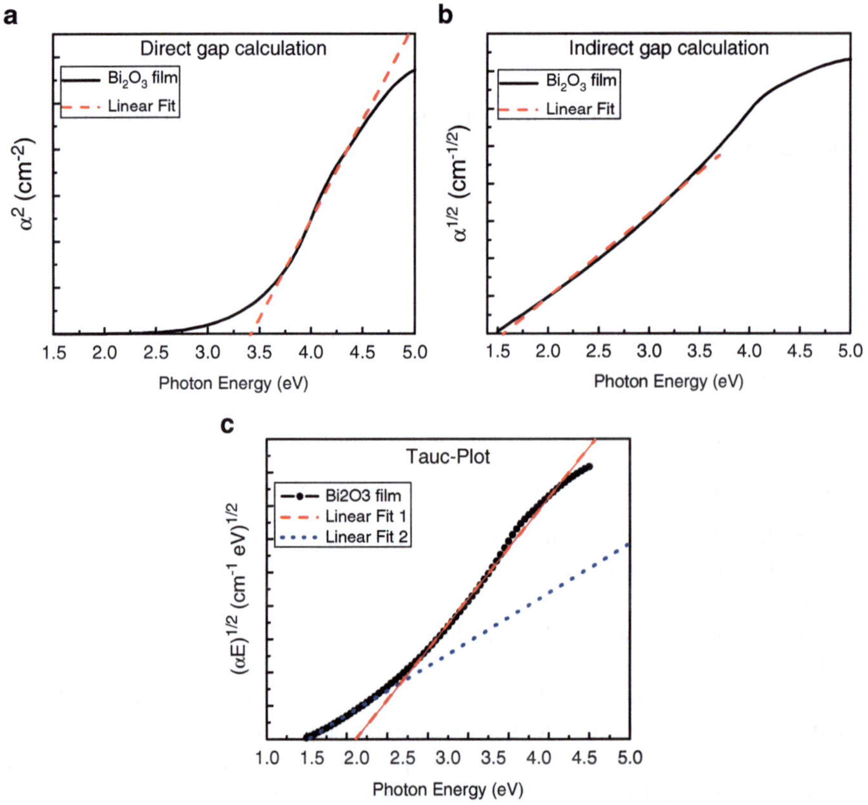

Fig. 4.22 Calculations of the optical band gap for the Bi_2O_3 thin films. (**a**) Using the direct gap model, (**b**) using the indirect gap model, and (**c**) using the Tauc-plot

and assuming amorphous semiconductor, the estimated gap depends strongly in the linear region choose to fit the spectra (2.1 or 1.5 eV), since not a good linear relationship is observed.

Spectroscopic ellipsometry can also provide other electronic and structural parameters. Using appropriate models and analyzing the fitting parameters, information about the energy for inter-/intraband transitions can be obtained, as well as parameters, such as porosity, volume fraction of a second phase, roughness, etc. There is also the capability to perform in situ analysis due to the fast acquisition instrumentation.

4.5.4 Photoluminescence (PL) Spectroscopy

Photoluminsecence spectroscopy is a complimentary technique to UV–VIS DRS in that it also gives information on the electronic structure from light excitation. Whereas UV–VIS DRS looks at the transition from the ground state to the excited state, photoluminescence (PL) spectroscopy differs by examining the transition from the excited state to the ground state. The emission of photons from this phenomenon can be measured as fluorescence. This process is relevant to the study of photocatalysts since fluorescence occurs when electrons in the conduction band recombine with holes in the valence band. Thus, this characterization technique can give direct information on electron transfer kinetics during the photocatalytic process. The PL intensity can be used to compare recombination rates for different photocatalytic systems, detect impurities and defects in the sample, and estimate the band gap through the band to band transitions (Anpo et al. 2009; Gilliand 1997), as it is shown in Fig. 4.23 for a photocatalytic ZnO film. The peak centered at 383 nm corresponds to the band gap of the material (3.23 eV), and the broad band centered at 495 nm indicates a very large number of defects and intrinsic impurities of the ZnO lattice. Moreover, surface species can have a large effect on the PL spectra since surface sites and the presence of adsorbed surface molecules can act as efficient electron traps which help to decrease PL intensity, preventing electron–hole recombination (Anpo et al. 2009; Anpo and Kamat 2010).

The band diagram corresponding to the photoluminescence process in a direct or an indirect gap material is similar to the one given in Fig. 4.15; the only difference is that the arrows point in the opposite direction. Photons are absorbed from an excitation source such as laser of lamp, and this injects electrons into the conduction band and holes into the valence band. This will be possible if the energy of the source $h\nu_L$ is greater than E_g (usually UV light is needed). The electrons can be initially created in states high up in the conduction band, but they do not remain in those states for very long. They rapidly (~10^{-13} s) lose their energy by emitting phonons to satisfy the conservation laws. After the electrons and holes have relaxed as far as they can by phonon emission, they must wait at the bottom of the bands until they can emit a photon or recombine non-radiatively. In Fig. 4.24 the PL

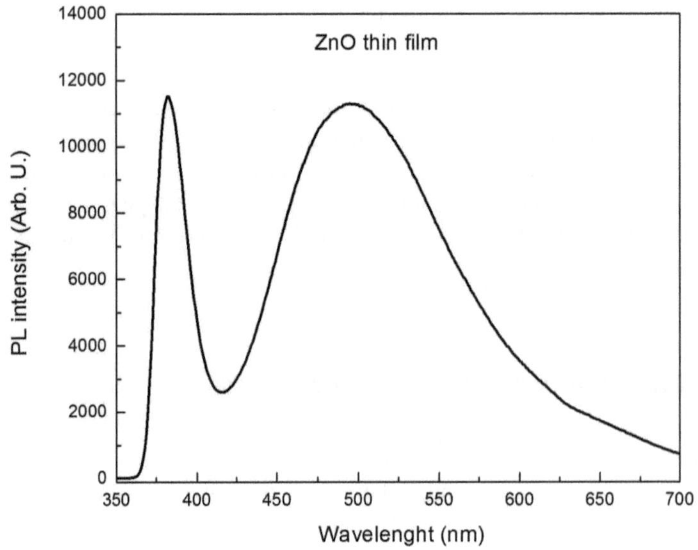

Fig. 4.23 PL spectrum of a photocatalytic ZnO thin film. The peak centered at 383 nm corresponds to the band to band transition ($E_g = 3.23$ eV)

spectra of undoped and Cu-doped photocatalytic ZnS films are shown, where a very broad band in the visible region is observed. The maximum of each band varies within 561 and 599 nm, and the peaks become wider as the Cu concentration increases. These wide bands give orange-red emissions that are a consequence of the overlapping of at least two peaks, one that corresponds to Cu centers—formed if the Zn site is replaced by a Cu^{+} ion—(Khomchenko et al. 2005) and the low energy peak that is caused by interstitial Zn and S, which are deep defects that decay in the red region of the spectrum (Chauhan et al. 2013, 2014). The systematic red shift of the band confirms the presence of Cu^{2+} ions in the ZnS matrix (Muthukumaran and Kumar 2013). The highest photocatalytic activity was achieved for the sample with 12 at.% Cu; this amount of impurities provided enough localized states to facilitate the electron transport to the conduction band for the photocatalytic process without compromising it with the radiative recombination.

Time-resolved picosecond in situ PL (TR-PL) spectroscopy is often utilized to examine the recombination lifetimes of electron/hole pairs. Whereas conventional or steady-state PL spectroscopy uses continuous excitation from a light source, TR-PL spectroscopy relies on pulsed excitation and measures photoluminescence at certain time intervals after the pulse excitation (Gilliand 1997; Matsuoka et al. 2008). The PL emission intensity over time is then fitted to an exponential decay model to determine the lifetimes of photogenerated electron/hole pairs (Selby et al. 2003; Quickenden et al. 1996). An increased lifetime of photogenerated electron/hole pairs has been shown to correlate with TiO_2 for photocatalytic splitting (Tang et al. 2008).

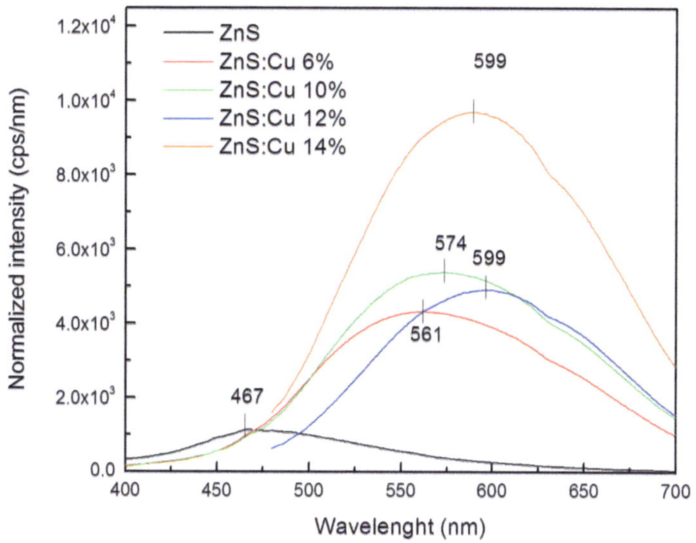

Fig. 4.24 PL spectra of undoped and Cu-doped ZnS photocatalytic films. The highest photocatalytic performance was achieved with ZnO:Cu 12 %, where the PL was not the most intense

4.5.5 Other Methods

As mentioned briefly above, the low energy region of the EEL spectra, less than 50 eV, contains information about the excitation of valence band electrons, such as interband transition (single excitation) and plasmon excitation (collective excitation), and can be used to obtain the local electronic and optical properties of materials at the nanoscale. This technique is called valence electron energy loss spectroscopy (VEELS). The other advantage of the TEM–VEELS method compared to the optical spectroscopies is the wider energy range (~0 to 50 eV). However, the disadvantage is the difficulty of implementation and data analysis that demands the proper substraction of the zero-loss signal.

Similarly, ultraviolet photoelectron spectroscopy, UPS, described qualitatively in Fig. 4.3, used ultraviolet light to ionize valence band electrons, and so the Fermi level and other electronic transition can be determined.

4.5.6 Examples

The band gap determination in a photocatalyst is of primary interest, especially if visible light absorption is pursued. Modifications in the synthesis of a photocatalyst can lead to a decrease in the band gap. The most common way to affect the band gap of semiconductors is by means of doping. This is the case of TiO_2 that has been doped with F, N, C, etc. (Maeda and Watanabe 2006; Abadias et al. 2010) resulting in smaller band gaps when the proper amount of doping element is incorporated. For thin films the transmittance spectra are usually measured, while for nanopowders or nontransparent samples, the diffuse reflectance is more adequate. Li et al. (2005) reported the effect of N-doped, F-doped, and N–F-codoped TiO_2 powders. The UV–Vis absorption spectra of the compounds were measured, and a clear difference in the absorption edge was observed, where the samples doped with N and codoped with N–F presented an absorption band in the visible range of 400–550 nm. Similarly, impurities of copper in the ZnS semiconductor show a diminution of the band gap from 3.4 to 3.2 eV.

Finally, the high sensitivity of the spectroscopy ellipsometry to detect small changes in the refractive index was investigated by Carretero-Genevrier et al. (2012) that measured the in situ photodegradation of organic species by TiO_2 nanoparticles during UV irradiation. The changes in the refractive index of a different series of layers were measured as a function of the irradiation time. Such changes were then correlated to the nanostructure of the samples, previously studied by GISAXS. Not much experimental details are given about the ellipsometric measurements, as well as concerning the models used. However, it is the proposal for an interesting field of research.

Concluding Remarks

In this chapter we presented an overview of the most important techniques used in the characterization of photocatalytic materials, going from the composition to the structure and properties of the materials. A brief description of each technique was presented in order to provide the physical aspects involved during the use of each one and the information that can be extracted from them. In the examples provided, we pointed the importance of using a combination of several techniques to obtain a better knowledge of the materials under study, especially in the case of complex materials, where independent results or isolated techniques cannot give concluding information of the samples. The proper choice of a number of characterization techniques will allow a better understanding of the properties of the photocatalyst under study.

Acknowledgements Funding from the BisNano project (European Community Seven Framework rogramme and CONACYT under grant agreements n° 263878 and 125141, respectively) and the Phoscleen IRSES project 318977.

References

Abadias G, Paumier F, Eyidi D, Guerin P, Girardeau T (2010) Structure and properties of nitrogen-doped titanium dioxide thin films produced by reactive magnetron sputtering. Surf Interface Anal 42:970–973

Ahmad M, Ahmed E, Hong ZL, Ahmed W, Elhissi A, Khalid NR (2013) Photocatalytic, sonocatalytic and sonophotocatalytic degradation of Rhodamine B using ZnO/CNTs composites photocatalysts. Ultrason Sonochem 21:761–773

Amelinckx S, Van-Dyck D, Van-Landuyt J, Van-Tendeloo G (2008) Electron microscopy: principles and fundamentals. Wiley, Weinheim. ISBN 978-3-527-61455-4

Anpo M, Kamat PV (eds) (2010) Environmentally benign photocatalysts: applications of titanium oxide-based materials. Springer, New York

Anpo M, Dzwigaj S, Che M (2009) Chapter 1: Applications of photoluminescence spectroscopy to the investigation of oxide-containing catalysts in the working state. In: Gates BC, Jentoft FC (eds) Advances in catalysis, vol 52. Elsevier, Amsterdam, pp 1–42

Balaji S, Djaoued Y, Robichaud J (2006) Phonon confinement studies in nanocrystalline anatase-TiO_2 thin films by micro Raman spectroscopy. J Raman Spectrosc 37:1416–1422

Barrett EP, Joyner LG, Halenda PP (1951) The determination of pore volume and area distributions in porous substances. I. Computations from nitrogen isotherms. J Am Chem Soc 73:373–380

Beuvier T, Richard-Plouet M, Brohan L (2009) Accurate methods for quantifying the relative ratio of anatase and TiO_2(B) nanoparticles. J Phys Chem C 113:13703–13706

Briggs D, Sheah MP (1990) Practical surface analysis by Auger and X-ray photoelectron spectroscopy. Wiley, New York

Briggs D, Sheah MP (1996) Ion and neutral spectroscopy. Wiley, New York

Brunauer S, Emmett PH, Teller E (1938) Adsorption of gases in multimolecular layers. J Am Chem Soc 60:309–319

Bunker G (2010) Introduction to XAFS: a practical guide to X-ray absorption fine structure spectroscopy. Cambridge University Press, Cambridge. ISBN 978-0-521-76775-0

Carretero-Genevrier A, Boissiere C, Nicole L, Grosso D (2012) Distance dependence of the photocatalytic efficiency of TiO_2 revealed by in situ ellipsometry. J Am Chem Soc 134:10761–10764

Catlow CRA, Guo ZX, Miskufova M, Shevlin SA, Smith AGH, Sokol AA, Walsh A, Wilson DJ, Woodley SM (2010) Advances in computational studies of energy materials. Philos Trans R Soc A 368:3379–3456

Chauhan R, Kumar A, Chaudhary RP (2013) Photocatalytic degradation of methylene blue with Fe doped ZnS nanoparticles. Spectrochim Acta A 113:250–256

Chauhan R, Kumar A, Chaudhary RP (2014) Photocatalytic degradation of methylene blue with Cu doped ZnS nanoparticles. J Lumin 145:6–13

Chen H, Shen WZ (2005) Perspectives in the characteristics and applications of Tauc-Lorentz dielectric function model. Eur Phys J B 43:503–507

Chipera SJ, Bish DL (2002) FULLPAT: a full-pattern quantitative analysis program for X-ray powder diffraction. International Union of Crystallography, Commission on Powder Diffraction Newsletter, vol 27. pp 27–28.

Chipera SJ, Bish DL (2013) Fitting full X-ray diffraction patterns for quantitative analysis: a method for readily quantifying crystalline and disordered phases. Adv Mater Phys Chem 3:47–53

Chung FH (1974) Quantitative interpretation of X-ray diffraction patterns of mixtures. I. Matrix-flushing method for quantitative multicomponent analysis. J Appl Crystallogr 7:519–525

Cody GD, Brooks BG, Abeles B (1982) Optical absorption above the optical gap of amorphous silicon hydride. Sol Energy Mater 8:231–240

Condon JB (2006) Surface area and porosity determination by physisorption: measurement and theory. Elsevier, Amsterdam. ISBN 978-0-444-51964-1

Dante RC, Martín-Ramos P, Sánchez-Arévalo FM, Huerta L, Bizarro M, Navas-Gracia LM, Martín-Gil J (2013) Synthesis of crumpled nanosheets of polymeric carbon nitride from melamine cyanurate. J Solid State Chem 201:153–163

Davis LE (1996) Handbook of Auger electron spectroscopy physical electronics. A reference book of standard data for identification and interpretation of Auger electron spectroscopy data. Physical Electronics, Eden Prairie, MN

Delgass WN, Haller G, Kellerman RK, Lunsford JH (1979) Spectroscopy in heterogeneous catalysis. Elsevier, Amsterdam. ISBN 978-0-12-210150-2

Deng Q, Duan X, Ng DHL, Tang H, Yang Y, Kong M, Wu Z, Cai W, Wang G (2012) Ag nanoparticle decorated nanoporous ZnO microrods and their enhanced photocatalytic activities. ACS Appl Mater Interfaces 4:6030–6037

Dombrowski RJ, Lastoskie CM, Hyduke DR (2001) The Horvath–Kawazoe method revisited. Colloids Surf A 187–188:23–39

Eaton P, West P (2010) Atomic force microscopy. Oxford University Press, Oxford. ISBN 978-0-19-957045-4

Egerton RF (2011) Electron energy-loss spectroscopy in the electron microscope, 3rd edn. Springer, New York. ISBN 978-1-4419-9583-4

Fagerlund G (1973) Determination of specific surface by the BET method. Matériaux et Construction 6:239–245

Ferraro J, Nakamoto K, Brown CW (2003) Introductory Raman spectroscopy. Elsevier, Amsterdam. ISBN 978-0-12-254105-6

Fox M (2010) Optical properties of solids. Oxford University Press, Oxford. ISBN 10: 0199573379

Froitzheim H (1977) Electron energy loss spectroscopy. In: Ibach H (ed) Electron spectroscopy for surface analysis, vol 4. Springer, Berlin, pp 205–250

Fujiwara H (2007) Spectroscopy ellipsometry; principle and applications. Wiley, West Sussex. ISBN 9780470016084

Gauden PA, Terzyk AP, Rychlicki G, Kowalczyk P, Cwiertnia MS, Garbacz JK (2004) Estimating the pore size distribution of activated carbons from adsorption data of different adsorbates by various methods. J Colloid Interface Sci 273:39–63

Gilliand GD (1997) Photoluminescence spectroscopy of crystalline semiconductors. Mater Sci Eng R 18:99–399

Glatter VO, Kratky O (1982) Small angle X-ray scattering. Academic, London. ISBN 0-12-286280-5

Goldstein JI, Newbury DE, Echlin P, Joy DC, Lyman CE, Lifshin E, Sawyer L, Michael JR (2003) Scanning electron microscopy and X-ray microanalysis . Kluwer-Plenum, New York. ISBN 0-306-47292-9

Goodhew PJ, Humphreys J, Beanland R (2000) Electron microscopy and analysis, 3rd edn. Taylor & Francis, London. ISBN 0-7484-0968-8

Gualtieri AF (2000) Accuracy of XRPD QPA using the combined Rietveld and RIR method. J Appl Crystallogr 33:267–278

Guillén C, Montero J, Herrero J (2014) Anatase and rutile TiO_2 thin films prepared by reactive DC sputtering at high deposition rates on glass and flexible polyimide substrates. J Mater Sci 49:5035–5042

Hafez HS (2012) Highly active ZnO rod-like nanomaterials: synthesis characterization and photocatalytic activity for dye removal. Physica E Low Dimens Syst Nanostruct 44:1522–1527

Hardcastle FD (2011) Raman spectroscopy of titania (TiO_2) nanotubular water-splitting catalysts. J Ark Acad Sci 65:43–48

Haugstad G (2012) Atomic force microscopy: understanding basic modes and advanced applications. Wiley, Hoboken, NJ. ISBN 978-0-470-63882-8

Haw JF (2002) In situ spectroscopy in heterogeneous catalysis. Wiley-VCH, Weinheim. ISBN 9783527302482

Horvath-Bordon E, Kroke E, Svoboda I, Fuess H, Riedel R, Neeraj S, Cheetham AK (2004) Alkalicyamelurates $M_3[C_6N_7O_3]xH_2O$ M = Li, Na, K, Rb, Cs: UV-luminescent and thermally very stable ionic tri-s-triazine derivatives. Dalton Trans 22:3900–3908

Huang L, Xu H, Zhang R, Cheng X, Xia J, Xu Y, Li H (2013) Synthesis and characterization of g-C_3N_4/MoO_3 photocatalyst with improved visible-light photoactivity. Appl Surf Sci 283:25–32

Jellison GE, Modine FA (1996) Parameterization of the optical functions of amorphous materials in the interband region. Appl Phys Lett 69:371–373

Johansson SAE, Campbell JL, Malqvist KG (eds) (1995) Particle induced X-ray emission spectrometry (PIXE). Wiley, Chichester. ISBN 978-0-471-58944-0

Keast VJ, Scott AJ, Brydson R, Williams DB, Bruley J (2001) Electron energy-loss near-edge structure—a tool for the investigation of electronic structure on the nanometre scale. J Microsc 203:135–175

Khomchenko V, Fedorenko L, Yusupov N, Rodionov V, Bacherikov Y, Svechnikov G, Zavyalova L, Roshchina N, Lytvyn P, Mukhlio M (2005) Laser processing and characterization of ZnS Cu thin films. Appl Surf Sci 247:434–439

Kowalczyk P, Terzyk AP, Gauden PA, Leboda R, Szmechtig-Gauden E, Rychlicki G, Ryu Z, Rong H (2003) Estimation of the pore-size distribution function from the nitrogen adsorption isotherm. Comparison of density functional theory and the method of Do and co-workers. Carbon 41:1113–1125

Kundu S, Ciston J, Senanayake SD, Arena DA, Fujita E, Stacchiola D, Barrio L, Navarro RM, Fierro JLG, Rodriguez JA (2012) Exploring the structural and electronic properties of Pt/Ceria-modified TiO_2 and its photocatalytic activity for water splitting under visible light. J Phys Chem C 116:14062–14070

Kuzmany H (2010) Solid-state spectroscopy: an introduction. Springer, Berlin. ISBN 978-3-642-01479-6

Lai Y, Meng M, Yu Y (2010) One-step synthesis, characterizations and mechanistic study of nanosheets-constructed fluffy ZnO and Ag/ZnO spheres used for Rhodamine B photodegradation. Appl Catal B Environ 100:491–501

Lamberti C, Zecchina A, Groppo E, Bordiga S (2010) Probing the surfaces of heterogeneous catalysts by in situ IR spectroscopy. Chem Soc Rev 39:4951–5001

Lastoskie CM, Gubbins KE (2000) Characterization of porous materials using density functional theory and molecular simulation. Stud Surf Sci Catal 128:41–50

Li X, He J (2013) Synthesis of raspberry-like SiO_2-TiO_2 nanoparticles toward antireflective and self-cleaning coatings. ACS Appl Mater Interfaces 5:5282–5290

Li D, Ohashi N, Hishita S, Kolodiazhnyi T, Haneda H (2005) Origin of visible-light-driven photocatalysis: a comparative study on N/F-doped and N–F-codoped TiO_2 powders by means of experimental characterizations and theoretical calculations. J Solid State Chem 178:3293–3302

Li G, Zhang D, Yu JC (2009) Thermally stable ordered mesoporous CeO_2/TiO_2 visible-light photocatalysts. Phys Chem Chem Phys 11:3775–3782

Liang Q, Shi Y, Ma W, Li Z, Yang X (2012) Enhanced photocatalytic activity and structural stability by hybridizing Ag_3PO_4 nanospheres with graphene oxide sheets. Phys Chem Chem Phys 14:15657–15665

Lin D, Wu H, Zhang R, Pan W (2009) Enhanced photocatalysis of electrospun Ag–ZnO heterostructured nanofibers. Chem Mater 21:3479–3484

Liu H, Ji S, Zheng Y, Li M, Yang H (2013a) Porous TiO_2-coated magnetic core-shell nanocomposites: preparation and enhanced photocatalytic activity. Chin J Chem Eng 21:569–576

Liu L, Bai H, Liu J, Sun DD (2013b) Multifunctional graphene oxide-TiO_2-Ag nanocomposites for high performance water disinfection and decontamination under solar irradiation. J Hazard Mater 261C:214–223

Lowell S, Shields JE (1984) Powder surface area and porosity, Powder Technology Series. Springer, Berlin. ISBN 978-94-010-8953-1

Lowell S, Shields JE, Thomas MA, Thommes M (2004) Characterization of porous solids and powders: surface area, pore size and density. Springer, Berlin. ISBN 1402023022

Maeda M, Watanabe T (2006) Visible light photocatalysis of nitrogen-doped titanium oxide films prepared by plasma-enhanced chemical vapor deposition. J Electrochem Soc 153:C186–C189

Matsuoka M, Kamegawa T, Anpo M (2008) Photoluminescence spectroscopy an its application to the characterization of active sites and reaction dynamics in catalysis. In: Ertl G, Knézinger H, Schiith F, Weitkamp J (eds) Handbook of heterogeneous catalysis. Wiley, Weinheim, pp 1065–1073

McDermott EJ, Wirnhier E, Schnick W, Virdi KS, Scheu C, Kauffmann Y, Kaplan WD, Kurmaev EZ, Moewes A (2013) Band gap tuning in poly(triazine imide), a nonmetallic photocatalyst. J Phys Chem C 117:8806–8812

Moria S, Wiesendanger R, Meyer E (2002) Noncontact atomic force microscopy. Springer, New York

Moser EM, Chappuis S, Olleros J (2013) Production of photocatalytically active titania layers: a comparison of plasma processes and coating properties. Surf Coat Technol 227:2–9

Murphy AB (2006) Modified Kubelka–Munk model for calculation of the reflectance of coatings with optically-rough surfaces. J Phys D Appl Phys 39:3571–3581

Murphy AB (2007a) Band gap determination from diffuse reflectance measurements of semiconductor films, and application to photoelectrochemical water-splitting. Sol Energy Mater Sol Cells 91:1326–1337

Murphy AB (2007b) Optical properties of an optically rough coating from inversion of diffuse reflectance measurements. Appl Opt 46:3133–3143

Muthukumaran S, Kumar MA (2013) Structural, FTIR and photoluminescence properties of ZnS: Cu thin films by chemical bath deposition method. Mater Lett 93:223–225

Qi B, Yu Y, He X, Wu L, Duan X, Zhi J (2012) Series of transition metal-doped TiO_2 transparent aqueous sols with visible-light response. Mater Chem Phys 135:549–553

Quickenden TI, Green TA, Lennon D (1996) Luminescence from UV-Irradiated Amorphous H2O Ice. J Phys Chem 100:16801

Reimer L, Kohl H (2008) Transmission electron microscopy. Springer, Berlin. ISBN 978-0-387-40093-8

Rengifo-Herrera JA, Mielczarski E, Mielczarski J, Castillo NC, Kiwi J, Pulgarin C (2008) Escherichia coli inactivation by N, S co-doped commercial TiO_2 powders under UV and visible light. Appl Catal Environ 84:448–456

Rietveld HM (1969) A profile refinement method for nuclear and magnetic structures. J Appl Crystallogr 2(2):65–71

Rouquerol F, Rouquerol J, Sing KSW, Llewellyn P, Maurin G (2014) Adsorption by powders and porous solids. Principles, methodology and applications. Academic, Oxford. ISBN 978-0-08-097035-6

Schartl W (2007) Light scattering from polymer solutions and nanoparticle dispersions. Springer, New York. ISBN 978-3-540-71951-9

Schiettekatte F, Chicoine M, Forster J.S, Geiger J.S, Gujrathi S, Kolarova R, Paradis A, Roorda S, Wei P (2004) Nuclear Instruments and Methods in Physics Research B ERD,15 N external beam for NRRA in air, HIRBS: ion beam analysis developments on the HVEC EN1 Tandem 219–220:430–434

Selby BJ, Quickenden TI, Freeman CG (2003) The fitting of luminescence rises and decays. Kinet Catal 44:5–15

Sério S, Melo Jorge ME, Nunes Y, Barradas NP, Alves E, Munnik F (2012) Incorporation of N in TiO_2 films grown by DC-reactive magnetron sputtering. Nucl Instrum Methods Phys Res B 273:109–112

Siketic Z, Radovic IB, Jaksic M, Skukan N (2010) Time of flight elastic recoil detection analysis with a position sensitive detector. Rev Sci Instrum 81:033305

Sing KSW (1985) Reporting physisorption data for gas/solid systems with special reference to the determination of surface area and porosity. Pure Appl Chem 54:2201–2218

Sridharan K, Jang E, Park TJ (2013) Novel visible light active graphitic C_3N_4–TiO_2 composite photocatalyst: synergistic synthesis, growth and photocatalytic treatment of hazardous pollutants. Appl Catal B Environ 142–143:718–728

Stohr J (2003) NEXAFS spectroscopy. Springer, Berlin. ISBN 3-540-54422-4

Suryanarayana C, Norton MG (1998) X-ray diffraction: a practical approach. Springer, New York. ISBN 978-1-4899-0148-4

Swamy V, Kuznetsov A, Dubrovinsky LS, Caruso RA, Shchukin DG, Muddle BC (2005) Finite-size and pressure effects on the Raman spectrum of nanocrystalline anatase TiO_2. Phys Rev B 71:184302

Tang J, Durrant JR, Klug DR (2008) J Am Chem Soc 130:13885

Tauc J, Grigorovici R, Vancu A (1966) Optical properties and electronic structure of amorphous germanium. Phys Status Solidi B 15:627–637

Tavares CJ, Marques SM, Viseu T, Teixeira V, Carneiro JO, Alves E, Barradas NP, Munnik F, Girardeau T, Rivière J-P (2009) Enhancement in the photocatalytic nature of nitrogen-doped PVD-grown titanium dioxide thin films. J Appl Phys 106:113535

Tesmer JR, Nastasi MA (1995) Handbook of modern ion beam analysis. Materials Research Society, Pittsburgh. http://www.quantachrome.com/technotes.html

Thommes M. Pore size analysis by gas adsorption. Part I: Aspects of the application of Density Functional Theory (DFT) and Monte Carlo simulation (MC) for micro/mesopore size analysis. Quantachrome Instruments, Technical & Application Notes N. 31, pp. 1–7

Tompkins HG, Irene EA (2005) Handbook of ellipsometry. William Andrew, Norwich, NY. ISBN 0-8155-1499-9

Ulyanenkov A (1998) Grazing-incidence X-ray diffraction from multilayers taking into account diffuse scattering from rough interfaces. Appl Phys A 66:193–199

Wang S, Liu S, Guo J, Guo Q (2013) Surface electronic structure and morphology of silver on iron oxide films. Surf Sci 607:124–129

Waseda Y, Matsubara E, Shinoda K (2011) X-ray diffraction crystallography: introduction, examples and solved problems. Springer, Berlin. ISBN 978-3-642-16635-8

Watts JF, Wolstenholme J (2003) An introduction to surface analysis by XPS and AES. Wiley, New York. ISBN 978-0-470-84713-8

Weckhuysen BM, Vand Der Voort P, Catana G (2000) Spectroscopy of transition metal ions on surfaces. Leuven University Press, Leuven. ISBN 90-5867-025-2

Williams DDB, Carter CB (1996) Transmission electron microscopy: a textbook for materials science. Plenum Press, New York

Xie J, Jiang D, Chen M, Li D, Zhu J, Lü X, Yan C (2010) Preparation and characterization of monodisperse Ce-doped TiO_2 microspheres with visible light photocatalytic activity. Colloids Surf A 372:107–114

Xu M, Gao Y, Moreno EM, Kunst M, Muhler M, Wang Y, Idriss H, Wöll C (2011) Photocatalytic activity of bulk TiO_2 anatase and rutile single crystals using infrared absorption spectroscopy. Phys Rev Lett 106:138302

Yan SC, Li ZS, Zou ZG (2009) Photodegradation performance of g-C_3N_4 fabricated by directly heating melamine. Langmuir 25:10397–10401

Zhu M, Li Z, Xiao B, Lu Y, Du Y, Yang P, Wang X (2013) Surfactant assistance in improvement of photocatalytic hydrogen production with the porphyrin noncovalently functionalized graphene nanocomposite. ACS Appl Mater Interfaces 5:1732–1740

Chapter 5
Electrochemical Characterization of Photocatalytic Materials

Erika Bustos, Juan Manríquez, Juan Manuel Peralta-Hernández, and Edgar J. Ruiz-Ruiz

Abstract The semiconductor–electrolyte interface have interesting similarities and differences with their semiconductor–metal (or metal oxide) and metal–electrolyte counterparts. Thus, approaches to garnering a fundamental understanding of these interfaces have stemmed from both electrochemistry and solid-state physics perspectives and have proven to be equally fruitful. Electron transfer theories were also rapidly evolving during this period, starting from homogeneous systems to heterogeneous metal–electrolyte interfaces leading, in turn, to semiconductor–electrolyte junctions.

To facilitate a self-contained description, this chapter will start with well-established aspects related to the thermodynamic properties as semiconductor energy band model and the electrostatics at semiconductor–electrolyte interfaces in the dark. Additionally, this chapter examines the kinetic properties in the processes of light absorption, electron–hole generation, and charge separation at these interfaces. The steady state and dynamic aspects of charge transfer are then briefly considered. Nanocrystalline semiconductor films and size quantization are then discussed as are issues related to electron transfer across chemically modified semiconductor–electrolyte interfaces to determine the photocatalytic efficiency of semiconductor materials.

E. Bustos (✉) • J. Manríquez
Centro de Investigación y Desarrollo Tecnológico en Electroquímica, S. C., Parque Tecnológico Querétaro s/n, Sanfandila, Pedro Escobedo, 76703 Querétaro, Qro., México
e-mail: ebustos@cideteq.mx

J.M. Peralta-Hernández
Centro de Innovación Aplicada a Tecnologías Competitivas, A. C., Calle Omega No. 201 - Fracc. Ind. Delta, 37545 León, Gto., México

E.J. Ruiz-Ruiz
Universidad Autónoma de Nuevo León, Facultad de Ciencias Químicas, Av. Pedro de Alba, Ciudad Universitaria, San Nicolás de los Garza, Nuevo León, México

A. Hernández-Ramírez, I. Medina-Ramírez (eds.), *Photocatalytic Semiconductors*,
DOI 10.1007/978-3-319-10999-2_5

Table 5.1 Common semiconductors used as photoanodes

Semiconductor	Conductivity type (s)	Optical band gap energy (eV)
Si	*n,p*	1.11
InP	*n,p*	1.35
GaAs	*n,p*	1.42
CdTe	*n,p*	1.50
CdSe	*n*	1.70
α-Fe_2O_3	*n*	2.20
GaP	*n,p*	2.26
$BiVO_4$	*n*	2.40
CdS	*n*	2.5
WO_3	*n*	2.80
TiO_2	*n*	3.00 (rutile)
		3.20 (anatase)
$SrTiO_3$	*n*	3.2
ZnO	*n*	3.35
SnO_2	*n*	3.8

5.1 Characterization of Thermodynamic Properties in the Semiconductor–Electrolyte Interface Using Electrochemical Techniques

5.1.1 The Double Layer at Semiconductor

Electrodes prepared with a photocatalytic semiconductor can be used in a photoelectrochemical cell to measure its properties including the band gap energy, flat-band potential, and kinetics of hole and electron transfer.

The semiconductor must be supported on an electronic conductor substrate and put it into an electrochemical cell in contact with electrolytic solution. Since in a photoelectrode the electron transfer occurs necessarily at the interface between the semiconducting photoanode and the electrolyte, first, an explanation about the semiconductor–electrolyte interface is important. The band bending is a result of interface phenomena and the characteristic charge transfer reactions across the solid electrolyte that represents the basis of an electrode. Table 5.1 shows the most common semiconductors used as photoanodes (Rajeshwar 2002, Souza et al. 2009, Serpone and Pellizeti 1989).

The interface of semiconductor electrodes is described as either in the state of band edge level pinning or in the state of Fermi level pinning (Sato 1998). After the contact of the semiconductor surface with the electrolyte, the thermodynamic equilibrium on both sides of the interface must be established (Beranek 2011).

The charge distribution at semiconductor–electrolyte interface is summarized in Fig. 5.1. Three distinct double layers can be distinguished at the interface.

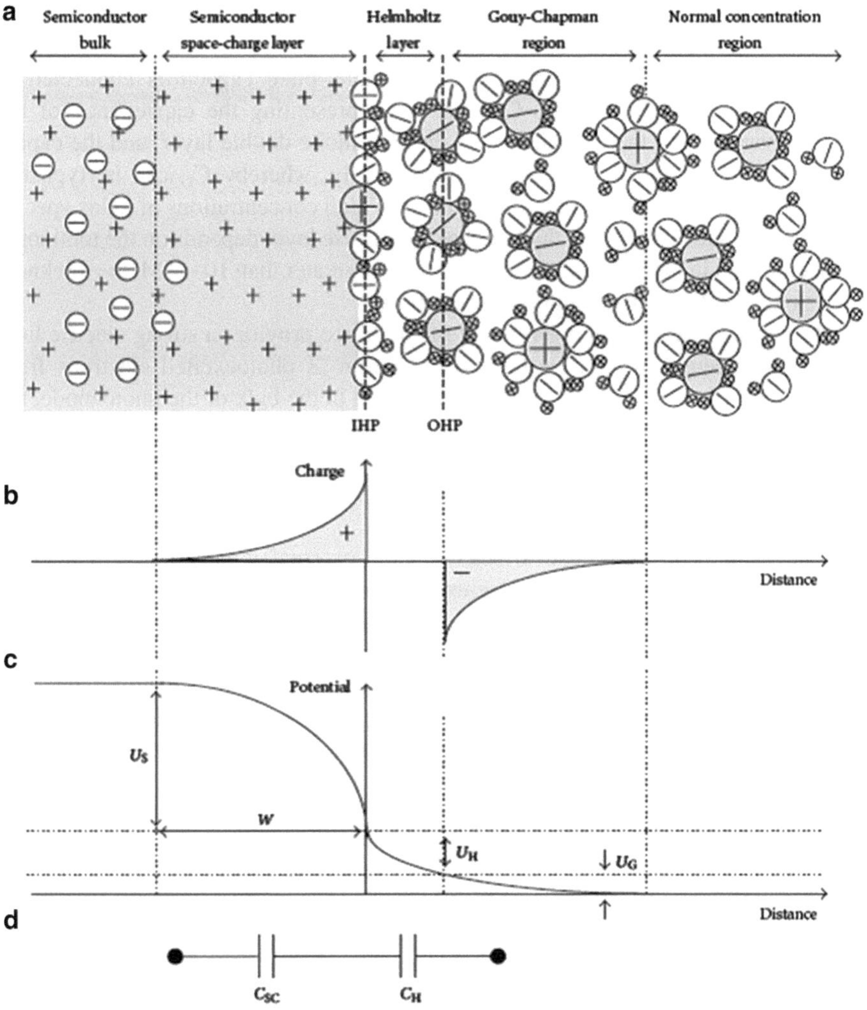

Fig. 5.1 Schematic view of the electric double layers at the n-type semiconductor–aqueous electrolyte interface (**a**) with corresponding charge (**b**) and potential (**b**) distributions. U_S is the potential drop across the space-charge layer, U_H is the potential drop in the Helmholtz layer, and U_G represents the drop in the Gouy–Chapman layer; (**d**) the equivalent circuit for the interface assuming that U_G can be neglected. From reference Beranek (2011)

First, is the semiconductor space-charge layer with positive charges in the form of ionized donors and the counter negative charge located at the surface. The second one is the Helmholtz double layer consisting of the inner Helmholtz plane (IHP) located at the semiconductor surface, and the charge is in the surface states or at the location of specifically adsorbed ions, whereas the latter denotes the position of the closest approach of hydrated mobile ions and the outer Helmholtz plane (OHP).

The third double layer is the Gouy–Chapman layer which is an extended region with an excess of free ions of one sign.

Essentially, the double layers act as parallel-plate capacitors connected in series with capacitances C_{SC}, C_H, and C_G representing the capacitance of the space-charge layer, the capacitance of the Helmholtz double layer, and the capacitance of the Gouy–Chapman layer, respectively, whereby C_G can be typically neglected for electrolytes containing relatively high concentrations of redox species (Fig. 5.1d) (Sato 1998). The thickness of the diffuse layer depends on the total ionic concentration in the solution; for concentrations greater than 10 ~ 2 M, the thickness is less than ~100 A (Bard and Faulkner 2000).

The space-charge region formed at the interface provides a strong electric field that is indispensable for an effective separation of photoexcited electrons from holes. On the other hand, when light is absorbed in the bulk of the photoanode, the photoexcited electrons and holes are created, but there is a high probability they will recombine before being used for water photolysis. Therefore, if light is absorbed in this region, a charge separation and field-assisted transport are expected (Radecka et al. 2008).

The band bending is also affected by the external voltage V_B. For a given semiconductor and electrolyte, there exists a unique potential for which the potential drop between the surface and the bulk is zero and there is no space-charge layer. This is the flat-band potential V_{Fb} or, in other words, is the applied potential (V) at which the semiconductor energy bands are "flat" leading up the solution junction. For an n-doped semiconductor, the flat-band potential is rather close to the conduction band edge, and experimentally the two concepts are sometimes treated as interchangeable (Jacobsson and Edvinsson 2012).

5.1.2 The Flat-Band Potential

There exist some electrochemical methods to determine the flat-band potential. The two most common methods are described as follows: the first one stems from the impedance spectroscopy in the dark, and the second one employs the current–voltage characteristics under a light pulse. The flat-band potential of p-type electrode is more anodic (positive) than that of n-type electrode; this difference in the flat-band potential between the two types of the same semiconductor electrode is nearly equivalent to the band gap of the semiconductor (Lee et al. 2001).

Considerations of interfacial electron transfer require knowledge of the relative positions of the participating energy levels in the two (semiconductor and solution) phases. Models for redox energy levels in solution have been exhaustively treated elsewhere. Besides the Fermi level of the redox system, the thermal fluctuation model (Kuhn-Kuhnenfeld 1972) leads to a Gaussian distribution of the energy levels for the occupied (reduced species) and the empty (oxidized species) states, respectively.

The distribution functions for the states to oxidation (D_{ox}) and reduction (D_{red}) are given by Eqs. (5.1) and (5.2), where E is the cell potential, $E_{F,redox}$ is the Fermi redox potential, λ is the solvent reorganization energy, k_B is Boltzmann's constant, and T is the temperature in Kelvin:

$$D_{ox} = \exp\left[-\frac{E - E_{F,redox} - \lambda^2}{4\, k_B T \lambda}\right] \tag{5.1}$$

$$D_{red} = \exp\left[-\frac{E - E_{F,redox} + \lambda^2}{4\, k_B T \lambda}\right] \tag{5.2}$$

Now consider the relative disposition of these solution energy levels with respect to the semiconductor band edge positions at the interface. The total potential difference across this interface is given by Eq. (5.3) (Rajeshwar 2002):

$$V_t = V_{SC} + V_H + V_G \tag{5.3}$$

In the last equation, V_t is the potential as measured between an ohmic contact on the rear surface of the semiconductor electrode and the reference electrode. The problematic factors in placing the semiconductor and solution energy levels on a common basis involve V_H and V_G, where V_H is the potential drop in the Helmholtz layer and V_G represents the drop in the Gouy–Chapman layer potential. In other words, theoretical predictions of the magnitude of the space-charge layer potential V_{SC} (and how it changes as the redox couple is varied) are hampered by the lack of knowledge on the magnitude of V_H and V_G. A degree of simplification is afforded by employing relatively concentrated electrolytes such that V_G can be ignored. As with metals, the Helmholtz layer is developed by adsorption of ions or molecules on the semiconductor surface, by oriented dipoles, or especially in the case of oxides, by the formation of surface bonds between the solid surface and species in solution. Recourse to band edge placement can be sought through differential capacitance measurements on the semiconductor–redox electrolyte interface (Takahashi 1979, Tenne and Hodes 1980).

Once V_{Fb} is known (from measurements), the Fermi level of the semiconductor at the surface is defined. It is then a simple matter to place the energies corresponding to the conduction and V_B at the surface (E_{CB} and E_{VB}, respectively) if the relevant doping levels are known. The difference between E_{CB} and E_{VB} should approximately correspond to the semiconductor band gap energy, E_g. Alternatively, if V_{Fb} is measured for one given state of doping of the semiconductor (n or p doped), the other band edge position can be fixed from knowledge of E_g. It is important to stress that the semiconductor surface band edge positions (as estimated from V_{Fb} measurements) comprises all the terms and reflects the situation in situ for a given set of conditions (solution pH, redox concentration, etc.) of the semiconductor redox electrolyte. The situation obviously becomes complex when the charge distribution and mediation at the interface changes either via surface states and illumination or both (Bard and Faulkner 2001).

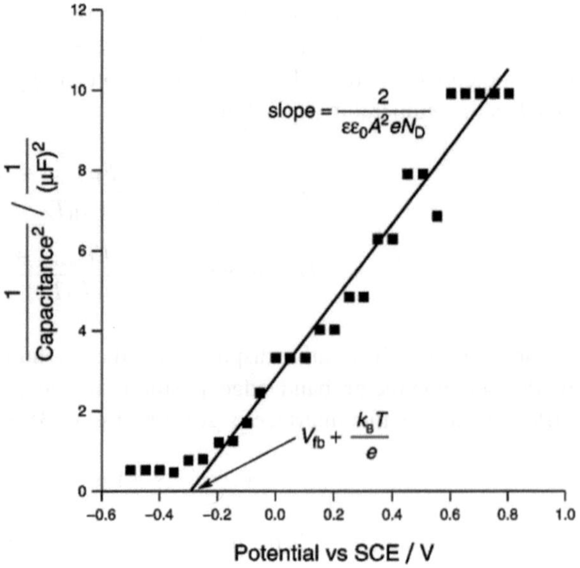

Fig. 5.2 Mott–Schottky plot of ZnO versus SCE in 7×10^{-4} M $K_3[Fe(CN)_6]$ (1 M KCl) obtained from reference Gelderman et al. (2007). Copyright 2007 American Chemical Society

In the simplest case as more fully discussed elsewhere (Takahashi 1979; Hodes et al. 1981; Tenne and Hodes 1980), one obtains the Mott–Schottky relation (for the specific instance of an n-type semiconductor) of the semiconductor depletion layer capacitance (C_{SC}). Mott–Schottky (MS) plots for layers over a surface of electrode can be obtained from electrochemical impedance spectroscopy (EIS) experiments at a constant frequency in aqueous medium. The resulting curves of the inverse square capacitance ($1/C^2$) versus cell potential (E) represent the classical Mott–Schottky equation (5.4). Generally, the *flat-band potential* (V_{Fb}) values are determined by measuring the capacitance of the electrode/electrolyte interface at different electrode potentials (V) using the Mott–Schottky equation, where C is the capacitance of the solid/electrolyte interface; ε_o and ε_s are the dielectric constants of free space and the film electrode, respectively; q is the electronic charge; V_{Fb} is the flat-band potential; T is the temperature in Kelvin; N_D is the donor density; V is the applied potential; and k_B is Boltzmann's constant. The V_{Fb} value and N_D can be determined from the extrapolation for $1/C^2$) (Cho et al. 2009; Manríquez and Godínez 2007):

$$\frac{1}{C^2} = \left[\left[\frac{2}{\varepsilon_o \varepsilon_s q N_D}\right]\left[(V_B V_{Fb}) - \left(\frac{k_B T}{q}\right)\right]\right] \tag{5.4}$$

At high concentration of multiple donor levels, an indirect tunneling of electrons through the semiconductor layer is promoted, causing a change in the slope (see Fig. 5.2). This change is determined by the distribution of relaxation times for electron emission depending on the position of the states related to the conduction band (Beranek 2011).

Figure 5.2 presents the flat-band potential of ZnO semiconductor versus SCE in 7×10^{-4} M $K_3[Fe(CN)_6]$ (1 M KCl) previously oxygen purged, using the Mott–Schottky plot. By linear extrapolation to "x" axe can be obtained the value of V_{Fb}; according with the graphic this value is about -0.316 ± 0.033 V for a single crystal of ZnO with a well-defined crystal plane, [001] synthesized by Gelderman et al. (2007). The N_D can be obtained from the slope in the Mott–Schottky plot, if the dielectric constants ε_s and charge are known. Then, from the slope, $N_D = 2 \times 10^{24}$ m^{-3}, which is comparable to previously reported values (6×10^{24} m^{-3}). Authors attribute this deviation due to the action of surface states in the polycrystalline electrode capturing and immobilizing the carriers. If the number of dopant agents per unit of volume of semiconductor (N_D) is minor to the effective number of accessible states at the conduction band (N_C), which means that $N_D < N_C$, then the photoactivity of the semiconductor is increased, becoming more intense the bending of the band; in this case, the semiconductor is nondegenerate. Nevertheless in case of $N_D > N_C$, the semiconductor exhibits a quasi-metallic behavior and is said to be degenerate.

5.1.3 Light Pulse Techniques

The light pulse technique (LPT) and the scanning light pulse technique (SLPT) are photoelectric methods used to determine electrical parameters of metal oxide semiconductor structures. The LPT method may be used to determine the V_{Fb} value of the entire semiconductor device, while the SLPT method allows determination of the distribution of local V_{Fb} values over the gate area. Most of the cases area is scanned by a small light beam (Piskorski and Przewlocki 2009).

Photocurrent vs. potential curves are typically obtained by applying a scan potential to the semiconductor–electrolyte interface in combination with an appropriated chopped light illumination in order to inhibit the effect of the electron recombination on the charge transfer kinetics (Cho et al. 2009; Gelderman et al. 2007). It requires a typical three-electrode photo-electrochemical cell (PEC) consisting of semiconductor photoanode; a cathode, made generally of Pt; and SCE as a reference electrode. The light incidence must be as a spot of controlled area.

This output photocurrent (I) signal depends on the potential V_B applied to the photoanode, and a graph as the example of Fig. 5.3 can be obtained. The magnitude of these current pulses is a function of the semiconductor surface potential S, and when $S = 0$, then current pulses disappear. Thereby, finding the dependence of the magnitude of these current peaks on the gate voltage V_B, the determination of point at which current peaks disappear is possible. This point defines directly the flat-band voltage state in semiconductor ($V_{ON} = V_{Fb}$).

Using the same three-electrode system and illumination, the photocurrent can be obtained from the transients generated under pulse visible light irradiation at a fixed bias potential. This technique was employed by Yan et al. (2011) to corroborate the

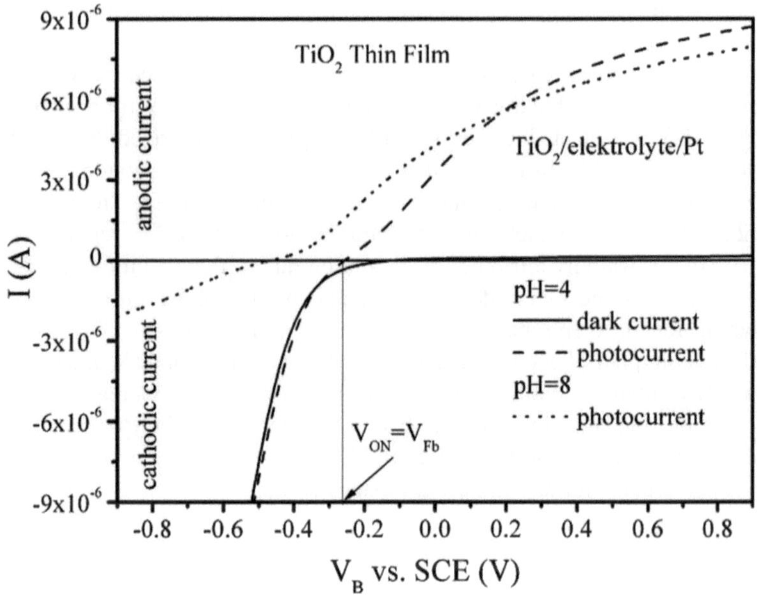

Fig. 5.3 Current–voltage I (V_B) characteristics of a TiO_2 electrode in PEC in the dark and under illumination with the white light at pH 4 and pH 8 with permission of Radecka et al. 2008

photocatalytic activity of N + S-co-doped TiO_2 nanotube arrays under visible light. They observed a constant step in photocurrent about 28 μA cm^{-2} on the photoanodes during the irradiation pulse (50s) of visible light; when light was turned off, the photocurrent returns to the initial value near zero. The results indicate that the N + S-co-doped TiO_2 nanotube arrays are sensitive to visible light and can generate a sustainable steady photocurrent under visible light irradiation.

Photocurrent intensity has been regarded as one of the most efficient methods to evaluate the photocatalytic activity of photocatalyst, and it has been recognized that a high photocurrent intensity suggests a high efficiency for electron and hole generation and separation and thus a high photocatalytic activity (Liu et al 2012).

The expected accuracy of the LPT method is typically ± 10 mV (for the most common potentiostats), while the accuracy of the V_{Fb} determination methods based on measurement of the capacitance characteristic is rarely better than ± 50 mV (Piskorski and Przewlocki 2009).

5.1.4 Electrochemical Determination of the V_{Fb} of Particles in Suspension

The abovementioned methods are appropriate for solid semiconductor electrodes or semiconductors deposited on a substrate (electronic conductor); however these techniques are not appropriate when the photocatalyst is in suspension. An electrochemical method to calculate the V_{fb} for semiconductor in suspension has been proposed by Roy et al. This method is based on the measurement of the photovoltage developed under light irradiation of a suspensions containing an electron acceptor such as methyl viologen (MV^{2+};1,1′-dimethyl-4,4′-bipyridinium dichloride). This is a very simple method for determining the E_{fb} of semiconductor particles in suspension. The experimental setup requires a work electrode of Pt (1 cm^2), a reference electrode (ECS, Ag/AgCl, or another), and a combined glass electrode for pH measurements. The pH is modified by adding HNO_3 or NaOH solutions; all electrodes must be inserted through the rubber stopper, and the suspension was stirred and sparged with N_2 before and during the measurement.

In the presence of MV^{2+}, E_{fb} (e^-) will equilibrate with the Fermi level of the redox couple in solution:

$$E_f(e^-) = E_f\left(MV^{2+}/MV^+\right) \tag{5.5}$$

Since E_f varies with the pH of the solution and E_f $(e^-) = E_{cv} = E_{fb}$, it can be rewritten as

$$E_{fb} = E_{fb}(pH = 0) - KpH \tag{5.6}$$

For any pH value,

$$E_{fb}(pH = 0) - kpH = E^0_{\left(MV^{2+}/MV^+\right)} + 0.059 \log\left(MV^{2+}/MV^+\right) \tag{5.7}$$

Since the redox potential of the MV^{2+}/MV^+ couple is pH independent, any change in photovoltage as a result of pH change must arise from changes in the energetics of electrons photogenerated at the particle surface, as E_{fb} shifts with respect to $E^0_{(MV2+)/(MV+)}$ with changes in solution pH:

$$0.059 \log\left(MV^{2+}/MV^+\right) - 0.445 = E_{fb}(pH = 0) - KpH \tag{5.8}$$

The values from the first term in Eq. (5.5) are plotted against pH to obtain E_{fb} (pH = 0) from the intercept, and the value of k is obtained from the slope; k represents the variation en E_{fb} per unit of pH.

5.1.5 The Band Gap Energy

Band gap energy E_g, i.e., the forbidden energy, is a very important parameter related to the electronic structure of a semiconducting material. This parameter has to match at least the energy difference 1.23 eV between the redox levels H_2O/H_2 and O_2/H_2O required for water splitting (Radecka et al. 2008). The absorption coefficient (α) of a crystalline material depends on the photon energy according to Eq. (5.9) (Lee et al. 2001; Rajeshwar 2002):

$$\propto = A\left(hv - E_g\right)^n / hv \tag{5.9}$$

Here, A is a proportionality constant and E_g the band gap energy. For crystalline semiconductors, n depends on the electron transition type, 1/2 for direct transition and 2 for indirect transition. However, 2 has been mostly preferred to analyze the passive film (Lee et al. 2001; Miraghaei et al. 2014). The band gap energy for a passive film can be determined from an $(i_{ph}hv)^{1/2}$ versus hv plot and is estimated at the photon energy value where the i_{ph} equals 0, provided that the photocurrent (i_{ph}) for the film is proportional to the absorption coefficient: $i_{ph} = A(hv - E_g)^n/hv$.

This method was applied to obtain the E_g of the passive film zircaloy-4 (Lee et al. 2001) using a conventional three-electrode cell of 1-l multineck flask with a quartz window as a photon inlet, using the zircaloy-4 as working electrode and SCE and Pt, respectively, as reference and counter electrodes. A 300 W Xenon (Xe) arc lamp combined with a scanning digital monochromator was used to impose a monochromatic illumination (200–800 nm) to the working electrode. The band gap energy for the inner anhydrous ZrO_2 was calculated (4.30 ± 0.15 eV) by extrapolation from $(i_{ph}hv)^{1/2}$ versus hv plot.

5.1.6 Fermi Level

Another important concept in the discussion of solid-state materials is the Fermi level. The Fermi energy, E_F, of a semiconductor is defined as the energy of the topmost filled orbital at a temperature of absolute zero (Rajeshwar 2002). For an n-type semiconductor, the Fermi level lies just below the conduction band, whereas for a p-type semiconductor, it lies just above the valence band. In addition, as with metal electrodes, the Fermi level of a semiconductor electrode varies with the applied potential; for example, moving to more negative potentials will raise the Fermi level.

The semiconductor solid-state physics community has adopted the electron energy in vacuum as reference, whereas electrochemists have traditionally used the standard hydrogen electrode (SHE) scale. While estimates vary, SHE appears to lie at −4.5 eV with respect to the vacuum level. We are now in a position to relate the redox potential E_{redox} (as defined with reference to SHE) with the Fermi level E_F, redox expressed versus the vacuum reference (Rajeshwar 2002):

$$E_{F,redox} = -4.5 \text{ eV} - eE^0_{redox} \tag{5.10}$$

Subramanian et al. (2004) have determined the Fermi level and the apparent Fermi level of the TiO_2/gold nanocomposites by attaining a Nernstian equilibrium with a known C_{60}/C_{60}^{-} redox couple.

The experiments were carried out by mixing the known concentration of deaerated TiO_2 and Au particles first, and then irradiating the composite clusters for 30 min with UV light. A known amount of deaerated C_{60} solution was injected into the preirradiated suspension, and the equilibrium concentration of C_{60}- was determined from the absorbance at 1,075 nm.

The Fermi level (E_F) of the semiconductor is directly related to the number of accumulated electrons as illustrated in the expression

$$E_F = E_{CB} + kT \ln nc/Nc \tag{5.11}$$

E_{CB} is the conduction band energy level versus NHE, nc is the density of accumulated electrons, and Nc is the charge carrier density of the semiconductor. If we accumulate more electrons in the TiO_2 or TiO_2/Au nanoparticle system, the authors would expect a negative shift in the Fermi level of the TiO_2. By shifting the Fermi level closer to the conduction band, it would therefore be possible to improve the energetics of the semiconductor system. The apparent Fermi level (E_F*) was correlated to the concentration of the redox species by using the expression

$$E_F^* \left(TiO_2(e)\right) = E_{fb} = E^0_{Ox/Red} + 0.059 \log\left([Ox]_{eq}/[Red]_{eq}\right) \tag{5.12}$$

where E_F* is the apparent Fermi level and the flat-band potential of TiO_2 (or TiO_2/Au) and $E°_{Ox/Red}$ is the standard reduction potential of the redox couple (viz., $E°(C_{60}/C_{60}^{-})$ −0.25 V versus NHE). By determining the equilibrium concentration of C_{60}^{-} in the UV-irradiated TiO_2 and TiO_2/Au suspension from the absorption at 1075 nm, the apparent Fermi level values were obtained. For the TiO_2 without gold nanoparticles was −270 mV; this represents a 270 mV more positive than the conduction band of bulk TiO_2 at neutral pH. The E_F* of the TiO_2/Au were −250, 270, and 290 using TiO_2 with gold nanoparticles of 8, 5, and 3 nm, respectively.

When the semiconductor and metal nanoparticles are in contact, the photogenerated electrons are distributed between TiO_2 and Au nanoparticles (Fermi level of Au + 0.45 V versus NHE). The transfer of electrons from the excited TiO_2 into Au continues until the two systems attain equilibration. Since the electron accumulation increases the Fermi level of Au to more negative potentials, the resultant Fermi level of the composite shifts closer to the conduction band of the semiconductor.

5.2 Characterization of Kinetic Properties in the Semiconductor–Electrolyte Interface Using Electrochemical Techniques

5.2.1 Separation of Transport, Charge Storage, and Reaction Elements in Nanostructured Oxide Semiconductor Electrodes

Electrochemical impedance spectroscopy (EIS) is an *ac* electrochemical technique very useful for characterizing, at the same time, electronic conductivity and interfacial electron transfer in nanostructured oxide electrodes permeated with electrolyte. In this sense, nanoparticulate TiO_2 films prepared on optically transparent electrodes (OTE) have been extensively studied due to their wide applications in photoelectrocatalysis.

EIS spectra are typically measured in a three-electrode cell by means of a standard potentiostat equipped with an impedance spectra analyzer. The amplitude of the *ac* signal ($|\Delta E|_{ac}$) used is about $\pm$ 10 mV, while the frequency is typically scanned between 100 kHz and 20 mHz. All the interfacial potentials must be referred against an appropriated reference electrode that controls the *dc* potential (E_{dc}) of the electrode/electrolyte interface. The counter electrodes can be Pt, Au, or W wires. The obtained spectra for films having a thickness l can be fitted to the equivalent circuit proposed by Bisquert and coworkers (Fig. 5.4) (Fabregat-Santiago et al. 2002; Terezo et al. 2001).

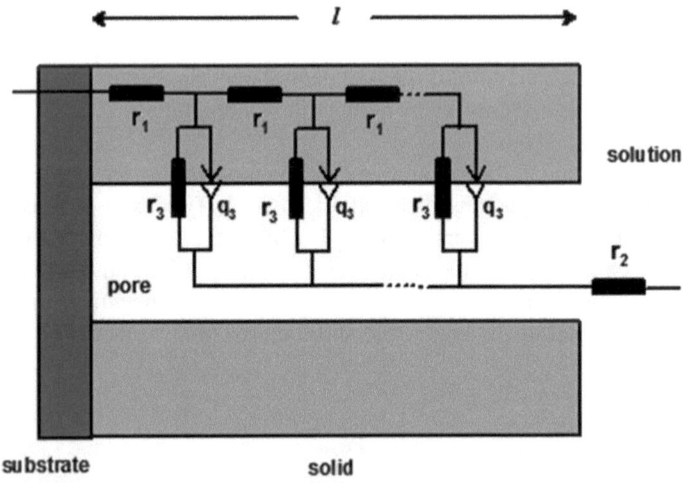

Fig. 5.4 Equivalent circuit employed to fit EIS spectra at different *dc* interfacial potentials

Electronic conductivity of the films is modeled by series resistances (r_1) representing the electron transport resistance through the electrode thickness ($R_1 = r_1 l$) and the electrolyte resistance (R_2) that separates working and reference electrodes. The electronic conductivities of the nanoporous oxide films (σ_n) are calculated from Eq. (5.13) where A and ρ represent the geometric area and porosity films, respectively:

$$\sigma_n = \frac{1}{R_1 A(1 - \rho)} \tag{5.13}$$

On the other hand, electron transfer at the oxide/electrolyte interface is modeled by constant-phase elements (CPEs, q_3) representing the charge storage into the film ($Q_3 = q_3 l$) and parallel resistances (r_3) standing for the charge transfer resistance ($R_3 = r_3/l$) which is strongly dependent from the *dc* potential.

5.2.2 Interparticle Electron Transport Through Semiconductor Nanostructured Films

Since the particle diameter is of the same order or smaller than the space-charge layer thickness in semiconductor thin films having high surface areas, the transportation of photoexcited electrons (e^-) and holes (h^+) through the films is controlled by diffusion (Kozuka et al. 2000; Manríquez and Godínez 2007).

This phenomenon is clearly plausible when the formation of necks between the particles takes place during the thermal sintering of the films. Figure 5.5 shows that the contact zone between *n-type* oxide nanoparticles is small, thus generating an energy barrier at this contact because of the space-charge layer as thick as or thicker than the contact diameter particle. In this way, photoexcited electrons cannot be easily transported through the oxide films, and the electrons photogenerated in the particles not bonded to the surface do not give photocurrent (j_{ph}) at the external circuit.

5.3 Determination of Photocatalytic Efficiency of Semiconductor–Electrolyte Interface Using Electrochemical Techniques

5.3.1 Monochromatic Quantum Efficiency

A quantum efficiency of exciton-to-charge generation is defined as the external monochromatic quantum efficiency normalized to the absorption in the active materials of the device. An upper limit of the efficiency can be determined, and

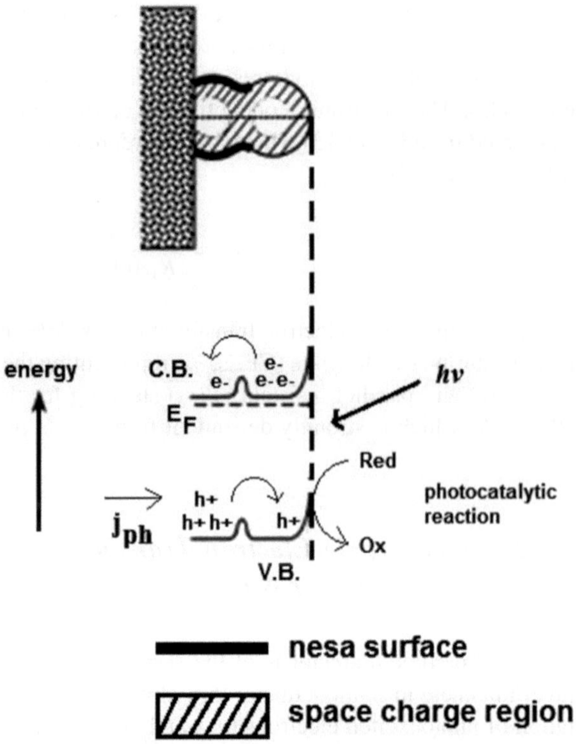

Fig. 5.5 Schematic representation of an illuminated semiconductor film containing nanometer oxide particles previously sintered, where CB and VB are the conduction and valence bands, while E_F is the Fermi level of the photogenerated electrons

results show that much of the light is absorbed in photoactive layers of the device, whereas only a fraction of the generated excitons is converted to charge carriers and can be collected as photocurrent.

Theoretical models for the action spectra of nanostructured oxide films in photoelectrochemical cells have been derived by Lindquist and coworkers (Södergren et al. 1994). These derivations where obtained by assuming that the charge carrier transport through the films is carried out by diffusion with a diffusion length constant, while recombination processes are considered to be of first order.

In this context, in Eq. (5.14) where electron flux $J_n(x,\lambda)$ is expressed as a function of electron mobility (μ_n) and density of photogenerated electrons $n(x)$, the photo-induced gradient of the electron Fermi level ($E_{F,n}$) is considered the driving force for electron transport in nanoporous photoanodes in the direction of the substrate (Vanmaekelbergh and de Jongh 1999):

$$J_n(x,\lambda) = \mu_n n(x)\left[-\frac{1}{e_0}\frac{dE_{F,n}(x)}{dx}\right]_\lambda \quad (5.14)$$

According to the nature of the electron transport activation, Lindquist and coworkers reported that the electron concentration profile in nanoporous semiconductor oxide films strongly depends from the way they are illuminated (Södergren et al. 1994). In this way, monochromatic quantum efficiencies $\Phi(\lambda)$ can be defined for semiconductor films illuminated through the electrolyte (EE-illumination) or through the substrate (SE-illumination) by means of Eqs. (5.15) and (5.16), respectively, which are expressed as a function of the photocurrent density (J^0), the light intensity (I_0), the minority carriers diffusion length (L), and the reciprocal absorption length (α):

$$\Phi_{\mathrm{SE}}(\lambda) = \frac{J^0_{\mathrm{SE}}}{e_0 I_0} = \frac{\left[-L\alpha_\lambda \cos h(l/L) + \mathrm{sen\,h}(l/L) + L\alpha_\lambda e^{-\alpha_\lambda l}\right] L\alpha}{\left(1 - L^2\alpha_\lambda^2\right) \cos h(l/L)} \tag{5.15}$$

$$\Phi_{\mathrm{EE}}(\lambda) = \frac{J^0_{\mathrm{EE}}}{e_0 I_0} = \frac{\left[L\alpha_\lambda \cos h(l/L) + \mathrm{sen\,h}(l/L) - L\alpha_\lambda e^{\alpha_\lambda l}\right] L\alpha_\lambda e^{-\alpha_\lambda l}}{\left(1 - L^2\alpha^2\right) \cos h(l/L)} \tag{5.16}$$

5.3.2 *Photochemical Thermodynamic Efficiency Factor (PTEF)*

The photochemical thermodynamic efficiency factor (PTEF) is an energy ratio equating the energy used to achieve the photocatalytic conversion of organic molecules over the energy absorbed by the photocatalyst and evaluates the performance of photocatalytic reactors on a thermodynamic basis (Serrano and De Lasa 1997), with the equation of the reactor's efficiency:

$$\mathrm{PTEF} = \eta = \frac{Q_{\mathrm{used}}}{Q_a} \tag{5.17}$$

Q_a represents the irradiation energy absorbed and Q_{used} the irradiation energy used for the desired formation of $^{\bullet}OH$ radicals which then interact with adsorbed species. The PTEF is a generally applicable parameter as it is not restricted to either a homogeneous or a heterogeneous photoconversion chemical process.

More specifically, the Q_{used} in the photoconversion process can be represented via the $r_{\bullet\mathrm{OH}}\Delta H_{\bullet\mathrm{OH}}W_{\mathrm{irr}}$ group with the PTEF being represented by

$$\mathrm{PTEF} = \eta = \frac{r_{\bullet\mathrm{OH}}\Delta H_{\bullet\mathrm{OH}}W_{\mathrm{irr}}}{Q_a} \tag{5.18}$$

where $r_{\bullet\mathrm{OH}}$ represents the rate of formation of $^{\bullet}OH$ radical groups per unit weight of irradiated catalyst, $\Delta H_{\bullet\mathrm{OH}}$ is the enthalpy of formation of an $^{\bullet}OH$ group, and W_{irr} is the total amount of irradiated catalyst. Alternatively, a PTEF definition can be introduced based on A_{irr}, the area of irradiated catalyst,

$$\mathrm{PTEF} = \eta = \frac{r'_{\bullet\mathrm{OH}}\Delta H_{\bullet\mathrm{OH}}W_{\mathrm{irr}}}{Q_a} \tag{5.19}$$

While photocatalytic reactions are frequently considered to be pseudo-homogeneous reactions with a rate based on either the unit volume of irradiated catalyst or the total reactor volume, definitions of the PTEF can be given as follows:

$$\text{PTEF} = \eta = \frac{r''_{\bullet\text{OH}} \Delta H_{\bullet\text{OH}} W_{\text{irr}}}{Q_a} \tag{5.20}$$

$$\text{PTEF} = \eta = \frac{r'''_{\bullet\text{OH}} \Delta H_{\bullet\text{OH}} W_{\text{irr}}}{Q_a} \tag{5.21}$$

where $r_{\bullet\text{OH}} W_{\text{irr}} = r'_{\bullet\text{OH}} A_{\text{irr}} = r''_{\bullet\text{OH}} V_{\text{irr}} = r'''_{\bullet\text{OH}} V$, with the rate of formation of $^{\bullet}$OH radicals and the enthalpies of $^{\bullet}$OH radical formation. Regarding the rate of $^{\bullet}$OH radical formation, it can be considered to be the sum of two terms, the rate of $^{\bullet}$OH consumption and the rate of $^{\bullet}$OH accumulation with $r_{\bullet\text{OH,c}}$ having a negative sign (consumption of $^{\bullet}$OH radicals):

$$r_{\bullet\text{OH}} = -r_{\bullet\text{OH,c}} + r_{\bullet\text{OH,acc}} \tag{5.22}$$

Evaluation of $^{\bullet}$OH radical ($r_{\bullet\text{OH}}$) formation presents inherent problems. The $^{\bullet}$OH radicals react with both the adsorbed model pollutant and the adsorbed intermediates (Pelizzetti et al. 1992; Turchi and Ollis 1990). Furthermore, the evaluation of the rate of $^{\bullet}$OH radicals involves stoichiometric coefficients such as

$$r_{\bullet\text{OH,c}} = \nu \sum_{\text{P}} \frac{r_{\text{P}}}{\nu_{\text{P}}} \tag{5.23}$$

The stoichiometric coefficient for the consumption of $^{\bullet}$OH groups is ν, and r_{p} and ν_{P} the rate and the stoichiometric number for the consumption of organic chemical species (including model pollutant and intermediate species) respectively. As a result, $r_{\bullet\text{OH}}$ can be expressed as follows:

$$r_{\bullet\text{OH}} = -\nu \sum_{\text{P}} \frac{r_{\text{P}}}{\nu_{\text{P}}} + r_{\bullet\text{OH,acc}} \tag{5.24}$$

At the beginning of the photoconversion, when the surface concentration of all chemical species is equal to the surface concentration of the model compound, the model compound is the only $^{\bullet}$OH group scavenger. Therefore

$$\sum_{\text{P}} \frac{r_{\text{P}}}{\nu_{\text{P}}} = \frac{\nu r_{1,\text{in}}}{\nu_1} \tag{5.25}$$

with $r_{1,\text{in}}$ representing the rate of consumption of the model pollutant and ν_1 the stoichiometric coefficient for the consumption of the model pollutant. Alternatively,

$$r_{\bullet\mathrm{OH}} = -\frac{\nu r_{1,\mathrm{in}}}{\nu_1} + r_{\bullet\mathrm{OH,acc}} \tag{5.26}$$

At initial conditions, the model pollutant concentration is very high, in large excess with respect to the other species, and it is very likely that the model pollutant is going to consume all the $^{\bullet}$OH radicals with no accumulation of the OH groups ($r_{\bullet\mathrm{OH,acc}} = 0$). As a result, the following is achieved:

$$r_{\bullet\mathrm{OH}} = -\frac{\nu r_{1,\mathrm{in}}}{\nu_1} \tag{5.27}$$

Under the situation described, the PTEF can be evaluated as

$$\mathrm{PTEF} = \eta = \frac{-\frac{\nu}{\nu_1} r_{1,\mathrm{in}} \Delta H_{\bullet\mathrm{OH}} W_{\mathrm{irr}}}{Q_a} \tag{5.28}$$

Or alternatively,

$$\mathrm{PTEF} = \eta = \frac{-\frac{\nu}{\nu_1} r'_{1,\mathrm{in}} \Delta H_{\bullet\mathrm{OH}} W_{\mathrm{irr}}}{Q_a} \tag{5.29}$$

$$\mathrm{PTEF} = \eta = \frac{-\frac{\nu}{\nu_1} r''_{1,\mathrm{in}} \Delta H_{\bullet\mathrm{OH}} W_{\mathrm{irr}}}{Q_a} \tag{5.30}$$

$$\mathrm{PTEF} = \eta = \frac{-\frac{\nu}{\nu_1} r'''_{1,\mathrm{in}} \Delta H_{\bullet\mathrm{OH}} W_{\mathrm{irr}}}{Q_a} \tag{5.31}$$

Frequently, it is observed that photocatalytic reactions follow the Langmuir–Hinshelwood model

$$r_{1,\mathrm{in}} = \frac{-k_1 C_1}{1 + K_1 C_1} \tag{5.32}$$

with k_1 representing the apparent intrinsic constant for the pollutant photoconversion, C_1 the model pollutant volumetric concentration, and K_1 the model pollutant adsorption constant. At large pollutant concentrations, $1 << K_1 C_1$ can be expected, and consequently the rate of photoconversion of a model pollutant shows a maximum value. Under these conditions, the PTEF reaches as well an upper limit:

$$\mathrm{PTEF}_{\max} = \eta_{\max} = \frac{-\frac{\nu}{\nu_1} (r_{1,\mathrm{in}})_{\max} \Delta H_{\bullet\mathrm{OH}} W_{\mathrm{irr}}}{Q_a} \geq \eta \tag{5.33}$$

Or

$$\mathrm{PTEF}_{\max} = \eta_{\max} = \frac{-\frac{\nu}{\nu_1}\left(r'_{1,\mathrm{in}}\right)_{\max}\Delta H_{\bullet\mathrm{OH}} W_{\mathrm{irr}}}{Q_a} \geq \eta \tag{5.34}$$

$$\mathrm{PTEF}_{\max} = \eta_{\max} = \frac{-\frac{\nu}{\nu_1}\left(r''_{1,\mathrm{in}}\right)_{\max}\Delta H_{\bullet\mathrm{OH}} W_{\mathrm{irr}}}{Q_a} \geq \eta \tag{5.35}$$

$$\mathrm{PTEF}_{\max} = \eta_{\max} = \frac{-\frac{\nu}{\nu_1}\left(r'''_{1,\mathrm{in}}\right)_{\max}\Delta H_{\bullet\mathrm{OH}} W_{\mathrm{irr}}}{Q_a} \geq \eta \tag{5.36}$$

To calculate the $\mathrm{PTEF}_{\max}$ value for a specific photocatalytic system, the initial concentration of the model pollutant has to be increased progressively until the PTEF approaches a constant value, the $\mathrm{PTEF}_{\max}$ (Serrano and De Lasa 1997):

$$\mathrm{PTEF}_{\max} = \eta_{\max} \geq \mathrm{PTEF} = \eta \tag{5.37}$$

This upper value for the PTEF is an intrinsic characteristic of a photocatalytic reactor as well as of the pollutant being photoconverted. This efficiency factor includes various reactor characteristics such as the absorbed irradiation, the rate of photoconversion, and the enthalpy of $^{\bullet}OH$ group formation. A relationship can also be established between the quantum yield and the PTEF given that the $\mathrm{PTEF}_{\max}$ can be expressed as

$$\mathrm{PTEF}_{\max} = \eta_{\max} = \frac{-\frac{\nu}{\nu_1}\left(r_{1,\mathrm{in}}\right)_{\max} W_{\mathrm{irr}}\alpha\left[\frac{N_A hc}{\lambda}\right]}{Q_a}\frac{\Delta H_{\mathrm{OH}}}{\alpha\left[\frac{N_A hc}{\lambda}\right]} \tag{5.38}$$

with α being the number of photon needed for the formation of a $^{\bullet}OH$ group, N_A the Avogadro number (6.023×10^{23} molecules mol^{-1}), h the Planck constant (6.62×10^{34} Js photon^{-1}), c the speed of light in vacuum (2.997×10^{10} cm s^{-1}), and λ the average wave length (nm). Then,

$$\mathrm{PTEF}_{\max} = \eta_{\max} = \varphi_{\mathrm{in,max}}\eta_{\bullet\mathrm{OH}} \tag{5.39}$$

with $\varphi_{\mathrm{in,max}}$ representing the maximum quantum yield (defined at initial conditions), or the maximum fraction of photons absorbed in the photocatalyst that results in the formation of $^{\bullet}OH$ radicals, and $\eta_{\bullet OH}$ being the fraction of the photon energy used in the formation of $^{\bullet}OH$ radicals. The last equation shows that energy efficiency evaluations using PTEF require not only a maximum quantum yield definition at initial conditions, based on the energy absorbed by the catalyst, but also $\eta_{\bullet OH}$, the fraction of the photon energy used in forming $^{\bullet}OH$ groups. The product of these two parameters provides an assessment of the energy efficiency of a photocatalytic reactor system.

The PTEF is a dimensionless quantity, as required by thermodynamic consistency. The PTEF definition can be broadly applied, covering various kinetic models and being appropriate for various photochemical reactors, either homogeneous (in solution) or heterogeneous (in interface). In practice, the calculation of $PTEF_{max}$ can be done using the corresponding equations for conditions where the initial photoconversion rates reach maximum values and with $\nu/\nu_1 = 1$ (De Lasa et al. 2005).

5.3.3 Relative Photonic Efficiency (ξ_r)

Photon efficiency (ξ) is used to facilitate the comparison of the efficacy of reactor designs (which differed in size and hence the residence time), as

$$\xi = \frac{C_o V}{I_o} \tag{5.40}$$

where C_o is the initial concentration, V is the volume flow rate, and I_o is the light intensity. In the case of relative photonic efficiency, a specific wavelength (ξ_r) is in order (Rajeshwar and Ibañez 1997).

5.3.4 Quantum Yield (Φ)

The experimental determination of the photocatalytic activity of heterogeneous systems has been an important topic in the proper characterization of different materials with catalytic photoactivity. Even with the commonly accepted view that the quantum yield (Φ) of a heterogeneous photochemical reaction is an important parameter in characterizing the activity of photocatalyst material, its practical application is seldom achieved. A major reason for this situation is the lack of a simple reliable experimental method to determine the quantum yield in any photocatalytic laboratory. According to the pioneer investigations carried out by Emeline et al. (2006) and Brandi et al. (2003), an interesting simple method for the direct experimental determination of quantum yields of photoreactions in liquid–solid heterogeneous systems with dispersed solid nano or microparticles based on the application of the concept of the black body-like reactor has been proposed.

The quantum yield of a heterogeneous photochemical reaction is defined exactly in accordance with the definition of the quantum yield in general homogeneous photochemistry, i.e., the number of molecules of a given reactant consumed or of a given product formed per photon of light absorbed by the photocatalyst at a given wavelength Eq. (5.41):

$$\Phi = \frac{N_m}{N_{h\nu}} \tag{5.41}$$

This definition can also be used (Emeline et al. 2006; Brandi et al. 2003) for heterogeneous photocatalytic processes when the system has reached the stationary state. As recommended by Emeline and Serpone (2002) and by Serpone and Salinaro (1999), for consistency, both quantities must be evaluated under otherwise identical conditions and preferably at the same time. In practice, the quantum yield is typically given in terms of "rates." For instance, the numerator represents the rate of reaction, and the denominator represents the rate of absorption of photons Eq. (5.42):

$$\Phi = \frac{N_m/dt}{N_{h\nu}/dt} \tag{5.42}$$

This equation should also be used for heterogeneous photoreactions whether the photocatalyst nature of material has been tested. In such cases, initial reaction rates are typically used to characterize the reproducible initial state of the photocatalyst material. The major experimental obstacle in measuring quantum yields of heterogeneous photoreactions is how to estimate the number of photons actually absorbed by the solid photocatalyst.

In this context, Eq. (5.43) denotes the balance between the fractions of reflected (R), transmitted (T), and absorbed (A) light in the system following the conservation law. Subsequently, determination of the fraction of absorbed light requires understanding of the fractions of reflected and transmitted light, a feat that cannot be achieved by application of traditionalist spectroscopic methods because of the diffuse scattering of light in dispersed photocatalytic systems. More sophisticated determinations are required. This has produced many works to report a so-called quantum yield evaluated on the basis of the incident light rather than the light actually absorbed by the heterogeneous system (Emeline et al. 2000):

$$A + R + T = 1 \tag{5.43}$$

Sun and Bolton (1996) and Serpone and Salinaro (1999) also suggested the application of an integrating sphere to estimate the fraction of absorbed light and established a standard protocol to approximate the quantum yield of any heterogeneous photoreaction using the photodegradation of phenol over TiO_2 (Degussa P25) as a typical photoreaction based on the concept of a relative photonic efficiency (ξ_r; Eq. 5.44) (Panida et al. 2009; Critchley et al. 2006; Cheadle et al. 2001):

$$\Phi = \xi_r \Phi_{st} \tag{5.44}$$

where Φ denotes the quantum yield of the photochemical process being examined and Φ_{st} is the quantum yield of the standard heterogeneous photochemical reaction. The application of this standard protocol should significantly simplify the procedure for determining the quantum yield of any photoreaction in liquid–solid heterogeneous systems, since with the known quantum yield of the standard reaction (used as a secondary actinometer), the experimental task of measuring the reaction rate of the photoreaction over the same photocatalyst is thereby simplified. The relative photonic efficiency is simply the ratio of the reaction rate of the photoreaction under examination to the standard photoreaction Eq. (5.45):

$$\xi_{\mathrm{r}} = \frac{\Phi}{\Phi_{\mathrm{st}}} = \frac{\mathrm{d}N/\mathrm{d}t}{\mathrm{d}N_{\mathrm{st}}/\mathrm{d}t} \tag{5.45}$$

Even with this simplicity, however, the proposed protocol does not find significant wide application in the photocatalytic community, with researchers continuing to report the so-called apparent quantum yields based on the photon flow of the incident light that has been termed "photonic efficiency." In the early 1970s, Basov and Solonitsyn (1973) proposed the concept of the black body reactor to estimate the (Φ) photoreaction in gas–solid heterogeneous systems.

Since then, this method was positively used to estimate the spectral dependences of the quantum yields of the photostimulated adsorption of oxygen, hydrogen, and methane on ZnO and TiO_2 semiconductors as well as on other metal oxides (Emeline et al. 2000; Basov et al. 1977; Cherkashin et al. 1980; Emeline et al. 1997).

A.V. Emeline et al. in 2006 measured the quantum yield in liquid–solid heterogeneous system for the phenol photodegradation over n-doped TiO_2 using a reactor based in the concept of the black body. Figure 5.6 shows an image of the modified reactor. The system involves a beaker containing the liquid phase with a dispersed photocatalyst and a cavity with a quartz wall located in the center of the beaker. The solution is mechanically stirred to avoid mass transfer limitation problems. The optical fiber (diameter, 2 mm) is inserted into the cavity to direct the light inside the cavity. Ever since the area of the fiber cross section is much smaller than the total area of the cavity wall; for this the authors take on that back reflection is negligible and that all the light (including the diffuse reflected light) is eventually directed inside the solution. Thus, in this reactor, the reflected light will always be negligible; that is, $R \to 0$. A very high optical density of the heterogeneous system can always be achieved so that $T \to 0$ by increasing the loading of the photocatalyst at a sufficient distance between the wall of the inner cavity and the wall of the beaker (2.5 cm, Emeline et al. 2006; Brandi et al. 2003).

Figure 5.7 shows a decay of the transmittance measured with the UV–Vis spectrometer and integrating sphere assembly with increasing light path length in a solution containing TiO_2 (Degussa P25) at a photocatalyst concentration of 1 g/L. As manifest from the data, the loss of transmittance of light for a 1 cm path length does not exceed 1 % in the spectral region of the fundamental absorption of the photocatalyst. In light of this result, the authors understand that all the light directed inside the cavity is eventually absorbed in the system. Thus, to measure the quantum yield of a heterogeneous photoreaction, one needs to know only the light irradiance at the outlet of the optical fiber and the rate of the heterogeneous

Fig. 5.6 Photography of the black body reactor for measurement of the quantum yield in liquid–solid heterogeneous system. From Emeline et al. 2006. Copyright 2006 American Chemical Society

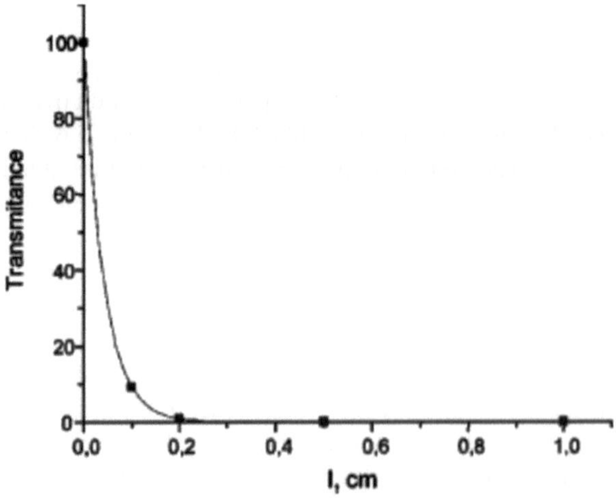

Fig. 5.7 Dependence of the transmittance ($\lambda = 365$ nm) of the solution containing TiO_2 (Degussa P25; photocatalyst loading 1 g/L) measured with an integrating sphere on the light path length

photochemical reaction determined by any conventional method (Emeline et al. 2006; Brandi et al. 2003).

With this device Emeline et al. (2006) and Brandi et al. (2003) tested the quantum yield of phenol photodegradation over N-doped TiO_2 as an early application of this method. The authors selected to measure the quantum yield of the photodegradation of organic molecule over semiconductor at two different wavelengths, 365 nm which corresponds to the fundamental absorption of TiO_2 and 436 nm corresponding to the N-doping-induced visible absorption band. In this assay the authors report the dependence of the transmittance of the heterogeneous

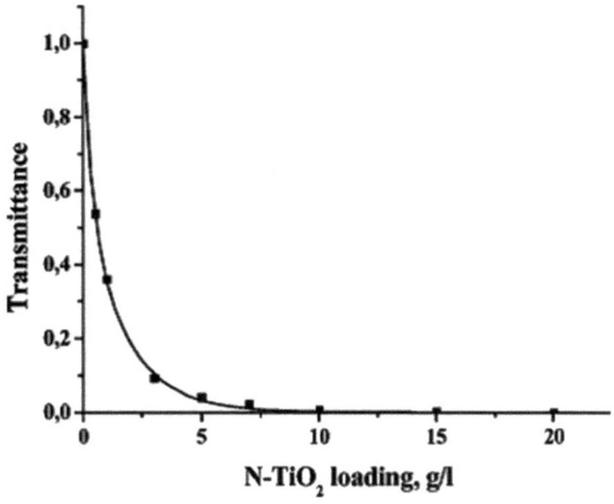

Fig. 5.8 Decay of transmittance measured with the integrating sphere with increase of the photocatalyst loading at 436 nm (light path length, 1 cm) (Emeline et al. 2006)

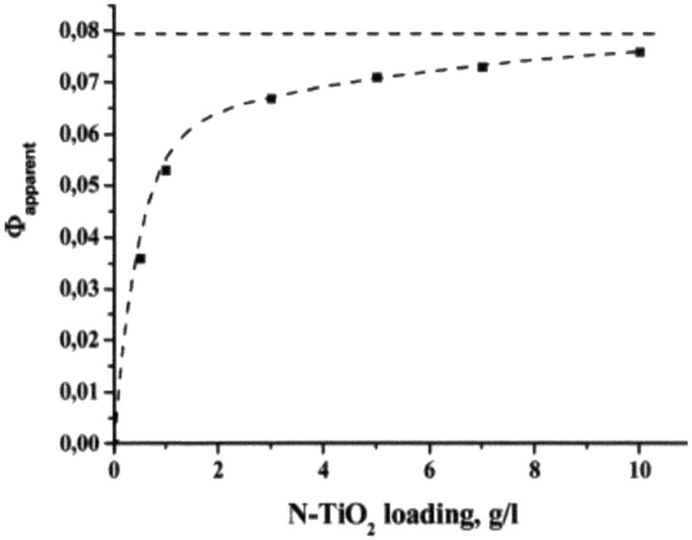

Fig. 5.9 Increase of the apparent quantum yield with increase in the photocatalyst loading (Emeline et al. 2006)

system measured with the integrating sphere assembly on photocatalyst loading at 436 nm (Fig. 5.8).

These results indicate that loss of transmittance of light for a 1 cm path length and photocatalyst loading greater than 7 g/L does not exceed 5 %, which means that most of the light does not pass outside the reactor setup. Also they measured the dependence of the apparent quantum yield on photocatalyst loading calculated with respect to the irradiance of the actinic light impinging on the reactor (Fig. 5.9).

In the light of these results, the authors used the term apparent quantum yield, instead of the otherwise more appropriate term photonic efficiency, to emphasize that regardless of the photocatalyst loading. For that reason, the system tested can also be applied to measure quantum yields in a spectral range corresponding to the extrinsic absorption caused by impurities or defects, provided that one uses a sufficiently high photocatalyst loading to satisfy the condition $T \rightarrow 0$. Under such conditions, then, the quantum yield for the photodegradation of phenol in N-doped TiO_2 aqueous dispersions under irradiation at 365 nm (intrinsic absorption of TiO_2) is $\Phi = 0.12$, whereas under irradiation in the extrinsic absorption at 436 nm induced by the N-doping Φ is 0.08 (Emeline et al. 2006; Brandi et al. 2003).

On the other hand, Thornton and Raftery in (2012) proposed the evaluation of quantum efficiency of the undoped and carbon-doped cadmium indate ($CdIn_2O_4$) powders which were prepared using a sol – gel pyrolysis method, as a function of hydrogen generation. The quantum efficiency of the optimized Pt-loaded C-doped $CdIn_2O_4$ was measured as a function of wavelength using a series of band-pass filters and irradiating an excess mass of sample (~1.0 g). The apparent quantum efficiency (Φ) was calculated using the following Eq. (5.46):

$$\mathrm{F}(\%) = (\text{no. of } \mathrm{H_2}\text{molecules} \times 2/\text{no. of incident photons}) \times 100\% \quad (5.46)$$

The number of incident photons was calculated using theme-assured power output of the lamp through each filter.

In another work, Corboz and coworkers (2000) evaluated a framework to determine the quantum efficiency η of a photoreaction in a porous layer of a photocatalyst material. This model incorporates a position-dependent source term mirroring the light intensity profile in the layer and an effective diffusion coefficient D_{eff}. It allows for a simultaneous determination of η as well as of D_{eff}. The method is applied to the photosynthesis of CH_4 from gaseous H_2O and CO_2 at the solid/gas interface of a porous layer of TiO_2 (Degussa P25).

In this research the authors describe the quantum efficiency η of the reaction as the number of product molecules divided by the total number of photons absorbed. Because of a subtle interplay between the optical penetration depth and the diffusion length of the free carriers generated, the quantum efficiency of a photochemical reaction at a semiconductor surface can even become wavelength dependent (Yu et al. 2003; Zhang and Yu 2003).

An interesting modeling of diffusion inside a porous photocatalyst was proposed where the modeling of the material flows in the porous layer is much simplified by the fact that only the flow of products must be considered. Because of the very small quantum efficiency of the reaction under consideration, the reactants in the layer are not so severely depleted during the irradiation time of 600 s that their surface coverage on the photocatalyst is expected to be altered significantly. We further assume that the product flux is proportional to the concentration gradient in accordance with the equation

$$j(x) = -D_{\text{eff}} \frac{\partial}{\partial x} c(x, t) \quad (5.47)$$

The microscopic processes involved will not be specified, and the effective diffusion coefficient D_{eff} is a phenomenological parameter determined by measurement. Mass conservation yields the diffusion equation (5.48):

$$\frac{\partial}{\partial t} C(x, t) = D_{\text{eff}} \frac{\partial^2}{\partial^2 x} C(x, t) + q(x) \quad (5.48)$$

where the source term $q(x)$ is given by Eq. (5.49):

$$q(x) = \eta K E(x) \quad (5.49)$$

Equation (5.49) implies that η is constant and independent of the light intensity E, which is not necessarily the case. Equation (5.49) states furthermore that the source strength does not depend on the concentration of the reactant, that is, that the effect of reactant consumption can be neglected.

Wang et al. in 2004 investigated the photocatalytic activity of the samples (0.1 g/l) by measuring the quantum yield, Φ_{HCHO}, of HCHO formed from aqueous methanol at pH 3.5 under different conditions. In CW photolysis of the oxygenated suspensions (300–400-nm UV light, 8×10^{-7} Einstein L^{-1} s^{-1} photon absorption rate), the platinized photocatalysts (1 wt % Pt) enhance Φ_{HCHO} by a factor of 1.5–1.7 with respect to neat colloidal TiO_2 where Φ_{HCHO} is 0.02.

Previously, Sun and Bolton (1996) studied the photocatalytic formation of formaldehyde (HCHO) from methanol in oxygenated aqueous suspensions of TiO_2 (anatase) particles and derived the quantum yield of the OH radical considered as the oxidizing intermediate (Martínez et al. 2013). Based on this knowledge authors carried out the same oxidation reaction in order to investigate the activity of different types of titanium dioxide photocatalysts (colloidal TiO_2 particles of nominally 2.4 nm diameter, Degussa P25, Sachtleben Hombikat UV 100) by the measurement of the HCHO quantum yield, Φ_{HCHO}, after continuous UV photolysis of their aqueous suspensions containing methanol (Neppolian et al. 2002; Anpo and Takeuchi 2003).

In this study the activity of the photocatalysts under different conditions of photolysis was assessed by the determination of the quantum yield, Φ, of HCHO formation by photooxidation of methanol, whereΦ is defined as

$$\Phi = R/I_{\text{a}} \quad (5.50)$$

and

$$I_{\text{a}} = I_0 F_{\text{s}} \quad (5.51)$$

R is defined as the photochemical formation rate of HCHO. I_a and I_0 are the absorbed and incident photon fluxes, respectively, in units of inset in $L^{-1}\ s^{-1}$. In CW photolysis, F_s is the integrated absorption fraction of the sample over the wavelength range used:

$$F_S = \frac{\int_{\lambda_1}^{\lambda_2} I_\lambda T_\lambda^f \mathrm{d}\lambda}{\int_{\lambda_1}^{\lambda_2} I_\lambda T_\lambda^f \mathrm{d}\lambda} \tag{5.52}$$

I_λ is the relative incident photon flux in the wavelength band $\mathrm{d}\lambda$, T_λ^f is the transmittance of the filter used in the experiment, and

$$f_\lambda^S = 1 - T_\lambda^S = 1 - 10 - A_\lambda^S \tag{5.53}$$

$f_\lambda{}^S$ is the fraction of light absorbed at wavelength λ. $T_\lambda{}^s$ and $A_\lambda{}^s$ are the transmittance and absorbance, respectively, of the sample at wavelength λ. I_0 and Fs were determined by chemical actinometry (Wang et al. 1992) and spectrophotometry. In monochromatic laser photolysis, Fs is identical with $f_\lambda{}^s$ (Eq. 5.53).

Each material was tested to evaluate the photocatalytic activity by the measurement of the quantum yield, Φ_{HCHO}, of HCHO formed by photocatalytic oxidation of methanol. Upon exposure of O_2-saturated aqueous suspensions, containing one of the photocatalysts and methanol, to 300–400 nm, HCHO was produced and identified quantitatively by HPLC analysis.

Evidently, platinization has a strong promoting effect on the formation of HCHO in TiO_2-based photocatalysis. HCHO increases by up to 50–70 % when compared with that value on neat TiO_2 particles. From Fig. 5.10 it is also seen that photochemical platinization of the TiO_2 particles yields a more efficient photocatalyst (PtTi-S1) than mixing of the colloidal components of Pt and TiO_2 (PtTi-S2).

For this case, the authors proposed the internal quantum yield (Φ_{Int}) estimation by means of the observed absorption of the film and the value of Φ_{Ext} at each wavelength, by use of $\Phi_{Int} = \Phi_{Ext}/(1-R)$, where R is the measured reflectance. For the porous electrode, Φ_{Int} was constant at a value of ~0.5 for $\lambda < 420$ nm, while the measured absorbance approached zero for $\lambda > 420$, yielding unreliable values for Φ_{Int}. In contrast, for the compact electrode, Φ_{Int} rose slowly as the illumination wavelength was decreased and only reached ~0.4 at $\lambda = 280$ nm. The porous electrode thus produced a sixfold increase in Φ_{Int} at $\lambda = 350$ nm relative to the compact film.

In the light of these sentences, internal quantum yields for porous WO_3 electrodes were proposed for use in light-driven water-splitting devices as a photoanode (Hodes et al. 1976; Tacca et al. 2012). WO_3 absorbs more of the solar spectrum than most other wide band gap metal oxides and is stable in acidic medium. WO_3 clearly does not have an optimal band gap for solar energy conversion applications, but is a useful model system to test the hypothesis of concern in this work. As shown in

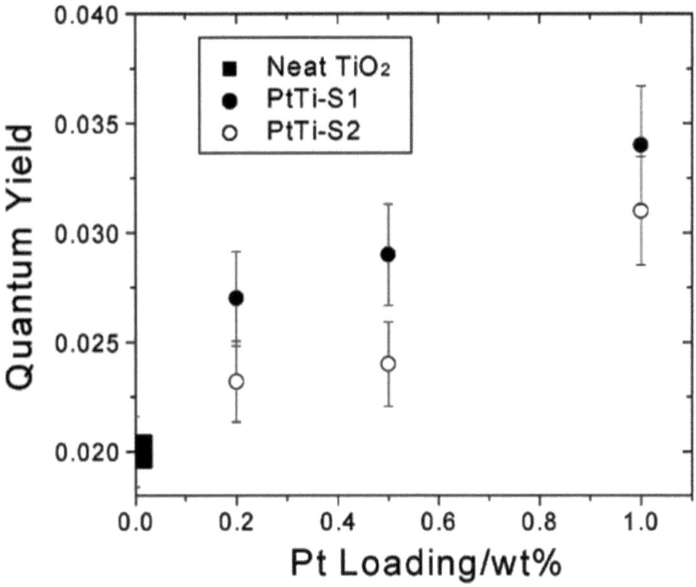

Fig. 5.10 Quantum yields of photocatalytic formaldehyde formation from methanol in the presence of PtTi-S1 and PtTi-S2 as a function of Pt loading. For comparison: neat colloidal TiO_2 (F_s) 0.24 derived from Eq. (5.13) in the wavelength range of 300–400 nm (Wang et al. 2004)

Fig. 5.10, the onset of the spectral response of both the compact and porous WO_3 electrodes occurred at $\lambda \sim 460$ nm, in agreement with the 2.6 eV band gap of WO_3 (Butler et al. 1976; Miller et al. 2006).

Although the onset of the spectral response Φ_{Ext} was principally determined by the band gap, the efficiency with which incident photons can be converted to current was dependent on the structure of the film. Even though the porous films absorbed only about half as much of the incoming light as the compact film, for $\lambda > 300$ nm, the Φ_{Ext} values of the porous WO_3 films were higher than those of the compact WO_3 films (Fig. 5.11a). Furthermore, at all wavelengths, the Φ_{Int} values of porous WO_3 films were much higher than those of the compact films (Fig. 5.11b). Notably, Φ_{Int} was not corrected for the actual optical reflection losses of the porous film, suggesting that Φ_{Int} for the porous film was higher than reported in Fig. 5.11b.

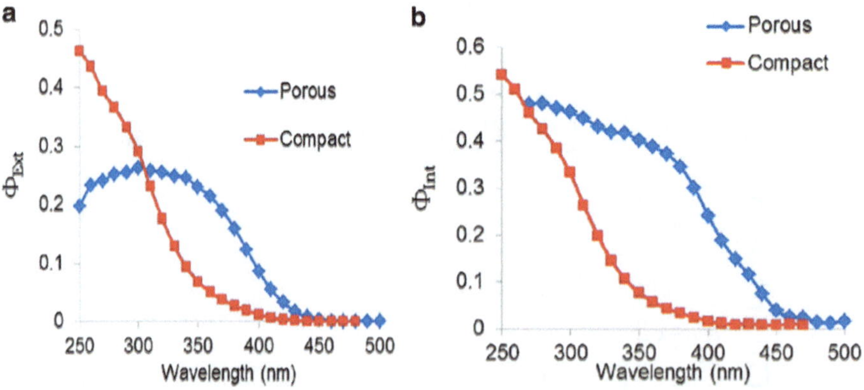

Fig. 5.11 Plot of the wavelength dependence of (**a**) the external quantum yield and (**b**) the internal quantum yield at the formal potential for water oxidation for compact (*red square*) and porous (*blue rhombohedral*) WO_3 electrodes in contact with 0.10 M HCl (aq) (pH = 1.0) (Reyes-Gil et al. 2013)

Concluding Remarks

The photoelectrochemical technique has been found to have an advantage of unique site-selective dissolution and to contribute not only to micro-tailoring of semiconductor surface but also to further characterization of grain structure and established aspects related to the semiconductor energy band model and the electrostatics at semiconductor–electrolyte interfaces in the dark, processes of light absorption, electron–hole generation, and charge separation at these interfaces. In general, the electrochemical techniques are very useful to characterize photocatalytic materials in a short time and specific thermodynamic and kinetic parameters to understand the determination of photocatalytic efficiency of semiconductor materials.

References

Anpo M, Takeuchi M (2003) The design and development of highly reactive titanium oxide photocatalysts operating under visible light irradiation. J Catal 216:505–516

Bard AJ, Faulkner LR (2000) Electrochemical methods: fundamentals and applications, 2nd edn. John Wiley& Sons, New York, p 850. ISBN ISBN: 0-471-04372-9

Bard AJ, Faulkner LR (2001) Electrochemical methods: fundamentals and applications, vol 6, 2nd edn. John Wiley & Sons, New York, pp 11–13

Basov LL, Solonitzyn YP. Patent No. 387730. Bull. InVention Committee USSR 1973, No. 28

Basov LL, Kuzmin GN, Prudnikov IM, Solonitzyn YP (1977) In: Uspekhi Fotoniki, Vilesov ThI (eds) Advances in Photonics, Leningrad State University, 6, p 82

Beranek R (2011) (Photo)electrochemical methods for the determination of the band edge positions of TiO_2-based nanomaterials. Adv Phys Chem 2011:1–20. doi:10.1155/2011/786759

Brandi RJ, Citroni MA, Alfano OM, Cassano AE (2003) Absolute quantum yields in photocatalytic slurry reactors. Chem Eng Sci 58(3–6):979–985

Butler MA, Nasby RD, Quinn RK (1976) Tungsten trioxide as an electrode for photoelectrolysis of water. Solid State Commun 19:1011–1014

Cheadle EM, Batchelder DN, Evans SD, Zhang HL, Fukushima H, Miyashita S, Graupe M, Puck A, Shmakova OE, Colorado R, Lee TR (2001) Polymerization of semi-fluorinated alkane thiol self-assembled monolayers containing diacetylene units. Langmuir 17:6616–6621

Cho IS, Kwak C, Kim DW, Lee S, Hong KS (2009) Photophysical, photoelectrochemical, and photocatalytic properties of novel $SnWO_4$ oxide semiconductors with narrow band gaps. J Phys Chem C 113(24):10647–10653

Corboz M, Alxneit I, Stoll G, Rudolf H (2000) On the determination of quantum efficiencies in heterogeneous photocatalysis. J Phys Chem B 104:10569–10577

Critchley K, Zhang LX, Fukushima H, Ishida M, Shimoda T, Bushby RJ, Evans SD (2006) Soft-UV photolithography using self-assembled monolayers. J Phys Chem B 110:17167–17174

De Lasa H, Serrano B, Salaices M (2005) Photocatalytic reaction engineering. Ed. Springer, USA, ISBN 978-0-387-27591-8

Emeline AV, Serpone N (2002) Suggested terms and definitions in photocatalysis and radiocatalysis. Int J Photoenergy 4:91–131

Emeline AV, Kuzmin GN, Purevdorj D, Shenderovich IG (1997) Spectral and temperature dependencies of simple molecules photoadsorption quantum yields for dispersed zinc oxide. Russian J Kinet Katal 38:446–450

Emeline AV, Kuzmin GN, Purevdorj D, Ryabchuk VK, Serpone N (2000) Spectral dependencies of the quantum yield of photochemical processes on the surface of wide band gap solids. 3. Gas/solid systems. J Phys Chem B 104:2989–2999

Emeline AV, Zhang X, Jin M, Murakami T, Fujishima A (2006) Application of the "black body like" reactor for the measurements of the quantum yield of photochemical reactions in heterogeneous systems. J Phys Chem B 110:7409–7413

Fabregat-Santiago F, Garcia-Belmonte G, Bisquert J, Zaban A, Salvador P (2002) Decoupling of transport, charge storage, and interfacial charge transfer in the nanocrystalline TiO_2/electrolyte system by impedance methods. J Phys Chem B 106:334–339

Gelderman K, Lee L, Donne SW (2007) Flat-band potential of a semiconductor: using the Mott–Schottky equation. J Chem Educ 84(4):685

Hodes G, Cahen D, Manassen J (1976) Tungsten trioxide as a photoanode for a photoelectrochemical cell (PEC). Nature 260:312–313

Hodes G, Manassen J, Cahen D (1981) Effect of surface etching and morphology on the stability of CdSe/Sx = photoelectrochemical cells. J Electrochem Soc 128:2325–2330

Jacobsson TJ, Edvinsson T (2012) Photoelectrochemical determination of the absolute band edge positions as a function of particle size for ZnO quantum dots. J Phys Chem C 116(29):15692–15701

Kozuka H, Takahashi Y, Zhao G, Yoko T (2000) Preparation and photoelectrochemical properties of porous thin films composed of submicron TiO_2 particles. Thin Solid Films 358:172–179

Kuhn-Kuhnenfeld F (1972) Selective photoetching of gallium arsenide. J Electrochem Soc 119:1063–1068

Lee S, Cho E, Ahn S, Kwon H (2001) Photo-electrochemical analysis of the passive film on zircaloy-4-. Electrochim Acta 46(17):2605–2611

Liu Y, Xie C, Li J, Zou T, Zeng D (2012) New insights into the relationship between photocatalytic activity and photocurrent of TiO_2/WO_3 nanocomposite. Appl Catal A 433–434:81–87

Manríquez J, Godínez LA (2007) Tuning the structural, electrical and optical properties of Ti(III)-doped nanocrystalline TiO_2 films by electrophoretic deposition time. Thin Solid Films 515:3402–3413

Martínez C, Vilariño S, Fernández MI, Faria J, Canle ML, Santaballa JA (2013) Mechanism of degradation of ketoprofen by heterogeneous photocatalysis in aqueous solution. Appl Catal B Environ 142–143:633–646

Miller EL, Marsen B, Cole B, Lum M (2006) Low temperature reactively-sputtered tungsten oxide films for solar-powered water splitting applications. Electrochem Solid-State Lett 9:G248–G250

Miraghaei S, Santamaria M, Di Quarto F (2014) Red shift in the light absorption threshold of anodic TiO_2 films induced by nitrogen incorporation. Electrochim Acta 134:150–158

Neppolian B, Choi HC, Sakthivel S, Arabindoo B, Murugesan V (2002) Solar/UV-induced photocatalytic degradation of three commercial textile dyes. J Hazard Mater 89:303–317

Panida P, Ammathnadu SA, Xiaojun H, Richard JB, Christoph W, Stephen D (2009) Improved photoreaction yields for soft ultraviolet photolithography in organothiol self-assembled monolayers. J Phys Chem C 113:21642–21647

Pelizzetti E, Minero C, Pramauro E (1992) Photocatalytic process for destruction of organic chemicals. In: de Lasa H, Dogu G, Ravella A (eds) Chemical reactor technology for environmentally safe reactors and products. Kluwer Academic Publishers, The Netherlands, pp 557–608

Piskorski K, Przewlocki HM (2009) LPT and SLPT measurement methods o flat-band voltage (VFB) in MOS devices. J Telecomm Infor Tech 1(4):76–82

Radecka M, Rekas M, Trenczek-Zajac A, Zakrzewska K (2008) Importance of the band gap energy and flat band potential for application of modified TiO_2 photoanodes in water photolysis. J Power Sources 181(1):46–55

Rajeshwar K (2002) Fundamentals of semiconductor electrochemistry and photoelectrochemistry. Encyclopedia of electrochemistry, vol 6, 1. John Wiley & Sons, New York, pp 1–52. ISBN ISBN: 978-3-527-30398-4

Rajeshwar K, Ibañez J (1997) Environmental electrochemistry: Fundamentals and applications in pollution abatement. Academic, San Diego, 723

Reyes-Gil K, Wiggenhorn C, Brunschwing BS, Lewis NS (2013) Comparison between the quantum yields of compact and porous WO_3photoanodes. J Phys Chem C 117:14947–14957

Roy AM, De GC, Sasmal N, Bhattacharyya SS (1995) Determination of the flatband potential of semiconductor particles in suspension by photovoltage measurement. Intern J Hydrogen Energy 20(8):627–630

Sato N (1998) Electrochemistry at Metal and Semiconductor Electrodes. Elsevier Science. Print Book ISBN: 9780444828064

Schiettekatte F, Chicoine M, Forster JS, Geiger JS, Gujrathi S, Kolarova R, Paradis A, Roorda S, Wei P (2004) ERD, 15N external beam for NRRA in air, HIRBS: ion beam analysis developments on the HVEC EN-1 Tandem. Nucl Instr Meth Phys Res B 219–220:430–434

Serpone N, Pelizzetti E (1989) Photocatalysis: fundamentals and applications. John Wiley & Sons, New York, pp 135–136

Serpone N, Salinaro A (1999) Terminology, relative photonic efficiencies and quantum yields in heterogeneous photocatalysis. Part I: Suggested protocol. Pure Appl Chem 71:303–320

Serrano B, De Lasa H (1997) Photocatalytic degradation of water organic pollutants. Kinetic modeling and energy efficiency. Ind Eng Chem Res 36:4705–4711

Södergren S, Hagfeldt A, Olsson J, Lindquist SE (1994) Theoretical models for the action spectrum and the current-voltage characteristics of microporous semiconductor films in photoelectrochemical cells. J Phys Chem 98:5552–5556

Souza FL, Pimenta Lopes K, Longo E, Leite ER (2009) The influence of the film thickness of nanostructured alpha-Fe_2O_3 on water photooxidation. Phys Chem Chem Phys 11(8):1215–1219

Subramanian V, Wolf EE, Kamat PV (2004) Catalysis with TiO_2/gold nanocomposites. Effect of metal particle size on the Fermi level equilibration. J Am Chem Soc 126(15):4943–4950

Sun L, Bolton JR (1996) Determination of the quantum yield for the photochemical generation of hydroxyl radicals in TiO_2 suspensions. J Phys Chem 100:4127–4134

Tacca A, Meda L, Marra G, Savoini A, Caramori S, Cristino V, Bignozzi CA, Gonzalez-Pedro V, Boix PP, Gimenez S, Bisquert J (2012) Photoanodes based on nanostructured WO_3 for water splitting. Chem Phys Chem 13:3025–3034

Takahashi K (1979) Etch pit pattern of GaAs crystals made by light irradiated electrolytic etching. Jpn J Appl Phys 18:1741–1746

Tenne R, Hodes G (1980) Improved efficiency of CdSe photoanodes by photoelectrochemical etching. Appl Phys Lett 37:428–430

Terezo AJ, Bisquert J, Pereira EC, García-Belmonte G (2001) Separation of transport, charge storage and reaction processes of porous electrocatalytic IrO_2 and IrO_2/Nb_2O_5 electrodes. J Electroanal Chem 508:59–69

Thornton JM, Raftery D (2012) Efficient photocatalytic hydrogen production by platinum-loaded carbon-doped cadmium indate nanoparticles. ACS Appl Mater Interfaces 4:2426–2431

Turchi CS, Ollis DF (1990) Photocatalytic degradation of organic water contaminants: mechanisms involving hydroxyl radical attack. J Cat 122:178–192

Vanmaekelbergh D, de Jongh PE (1999) Driving force for electron transport in porous nanostructured photoelectrodes. J Phys Chem B 103:747–750

Wang CM, Heller A, Gerischer H (1992) Palladium catalysis of O_2 reduction by electrons accumulated on TiO_2 particles during photoassisted oxidation of organic compounds. J Am Chem Soc 114:5230–5234

Wang C, Pagel R, Bahnemann DW, Dohrmann JK (2004) Quantum yields of formaldehyde formation in the presence of colloidal TiO2-based photocatalysts: effect of intermittent illumination, platinization, and deoxygenation. J Phys Chem B 108:14082–14092

Yan G, Zhang M, Houa J, Yang J (2011) Photoelectrochemical and photocatalytic properties of N + S co-doped TiO2 nanotube array films under visible light irradiation. Mater Chem Phys 129(1–2):553–557

Yu JG, Yu HG, Cheng B, Zhao XJ, Yu JC, Ho WK (2003) The effect of calcination temperature on the surface microstructure and photocatalytic activity of TiO_2 thin films prepared by liquid phase deposition. J Phys Chem B 107:13871–13879

Zhang L, Yu JC (2003) A sonochemical approach to hierarchical porous titania spheres with enhanced photocatalytic activity. Chem Commun 2003:2078–2079

[illegible]

[illegible]

[illegible] (2[illegible]) Improved efficiency of [illegible] by photo[illegible] [illegible]

[illegible] (2[illegible]) [illegible] separation [illegible] [illegible] electrode. Electrochim [illegible]

Thomas M, [illegible] (201[illegible]) Efficient photocatalytic [illegible] production by [illegible] carbon-doped [illegible] ACS Appl Mater Interfaces 6:[illegible]

[illegible] water [illegible]

Vanmaekelbergh D, de Jongh PE (1999) Driving force for electron transport in porous nanostructured photoelectrodes. J Phys Chem B 103:747–[illegible]

Wang [illegible]

[illegible]

Chapter 6
Semiconductor Materials for Photocatalytic Oxidation of Organic Pollutants in Wastewater

Laura Hinojosa-Reyes, Jorge Luis Guzmán-Mar, and Minerva Villanueva-Rodríguez

Abstract The increasing contamination of freshwater systems from a wide variety of industrial, municipal, and agricultural sources, which are released deliberately into the environment, has seriously affected water quality. Although the nature of pollution problems may vary, they are typically due to pesticides, organic dyes, and emerging contaminants (EC). Many of these chemical pollutants are able to pass through municipal wastewater treatment plants (WWTP), at least to some extent.

Recently, research activities have been centered on advanced oxidation processes (AOPs) for the destruction of pollutants resistant to conventional methods. As one of the important AOPs, the heterogeneous photocatalytic oxidation process is of special interest, especially when solar light is used.

This chapter presents a critical review of the achievements in photocatalytic semiconductors aimed at enhanced TiO_2 photocatalytic efficiency, complete mineralization of organic compounds, efficient utilization of visible and/or solar light, stability, and reuse of semiconductor materials in real wastewater treatment.

L. Hinojosa-Reyes (✉) • J.L. Guzmán-Mar • M. Villanueva-Rodríguez
Universidad Autónoma de Nuevo León, UANL, Department of Chemistry Sciences, San Nicolás de los Garza, N.L. 66451, México
e-mail: laura.hinojosary@uanl.edu.mx

A. Hernández-Ramírez, I. Medina-Ramírez (eds.), *Photocatalytic Semiconductors*,
DOI 10.1007/978-3-319-10999-2_6

6.1 Introduction

Textile dyes and other industrial dyestuffs constitute one of the largest groups of organic compounds that represent an increasing environmental danger. The dyes frequently employed on the industrial scale include azo, anthraquinone, sulfur, indigoid, triphenylmethyl (trityl), and phthalocyanine derivatives. Table 6.1 shows the chemical structure of these compounds (Konstantinou and Albanis 2004; Ahmed et al. 2010). Approximately 1–20 % of the total world production of dyes is lost during the dying process and is released into the textile effluents (Konstantinou and Albanis 2004). The release of the colored wastewaters in the environment is a considerable source of nonaesthetic pollution and eutrophication, and it can cause dangerous by-products through oxidation, hydrolysis, or other chemical reactions taking place in the wastewater (Rajeshwar et al. 2008).

Another kind of chemical that causes water contamination is the pesticides group. The worldwide pesticide usage has increased dramatically during the last two decades, coinciding with changes in farming practices and increasingly intensive agriculture. Pesticides appearing in surface water and groundwater can be classified according to their specific biological activity on a target species (e.g., herbicides, fungicides, acaricides, and insecticides) or by their chemical composition (e.g., phenylurea and phenoxy-acid pesticides, organochlorine compounds, chlorophenolic substances, triazines, carbamates, thiocarbamates, and organophosphorus insecticides) (Aragay et al. 2012). Table 6.2 provides an overview of pesticides (classified chemically) treated by heterogeneous photocatalysis, including their chemical formula and commercial group (Aragay et al. 2012).

Recently, a particular group of chemical products that has received great attention is the so-called emerging contaminants (EC), which are chemicals with little regulatory status whose middle- or long-term effects on the environment and human health caused by occurrence, frequency, or source are often unknown or are not available (Verlicchi et al. 2010). Pharmaceutical products, personal care products (PCPs), phthalates, polychlorinated biphenyl (PCBs), bisphenol A (BPA), endocrine disrupting chemicals (EDCs), flame retardants, perfluorinated compounds, and other compounds are considered as EC (Deblonde et al. 2011; Jiang et al. 2013). The long-term effect on aquatic organisms, chronic toxicity, and the possibility to reach drinking water effluents are of significant concern (Jiang et al. 2013). The structures of some pharmaceuticals and emerging pollutants treated by photocatalysis are shown in Table 6.3.

6.2 Photocatalytic Degradation of Organic Compounds in the Presence of TiO_2 Catalyst

Many results have demonstrated that Degussa P25 has been able to photodegrade various types of organic pollutants in the presence of artificial light (UV and visible range) or solar radiation. The photocatalytic degradation of dyes such as methyl orange, reactive red 195, acid orange 7, reactive yellow 81, and reactive violet

Table 6.1 Chemical classes of dyes used on an industrial scale

Compound	Structure
Anthraquinone compounds	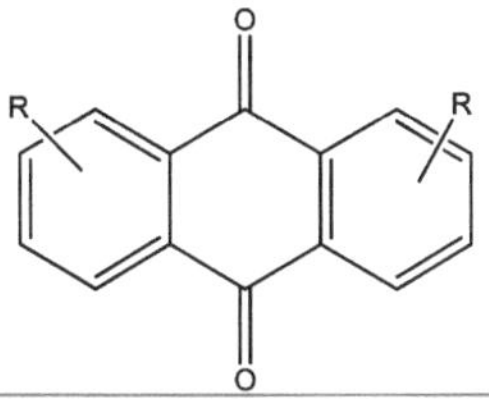
Indigoid compounds	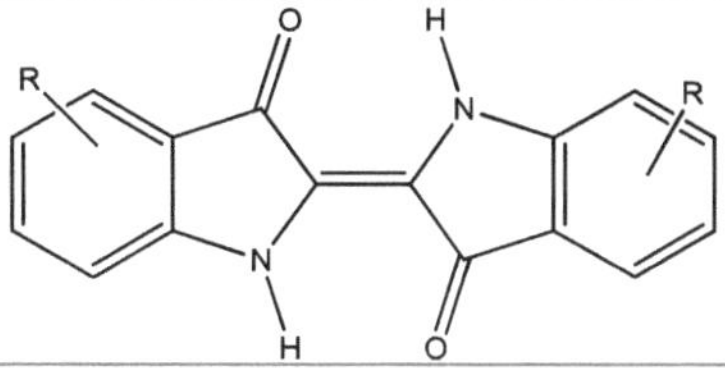
Phthalocyanine derivatives	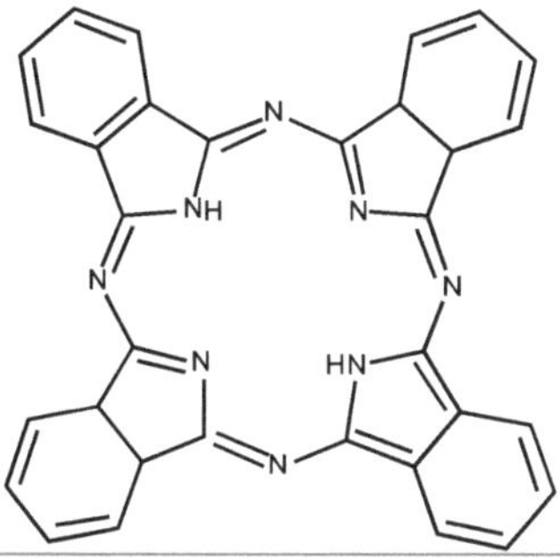
Triphenylmethyl (trityl) compounds	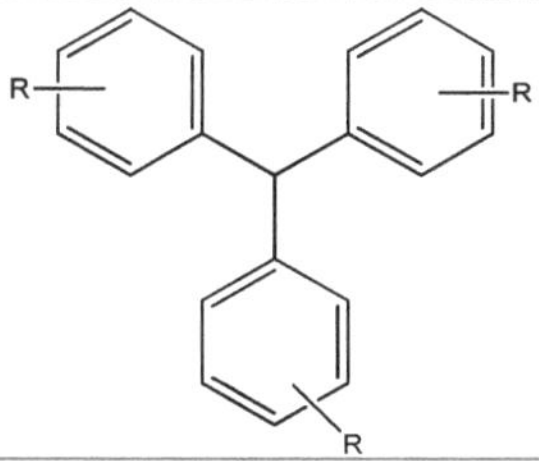
Azo compounds	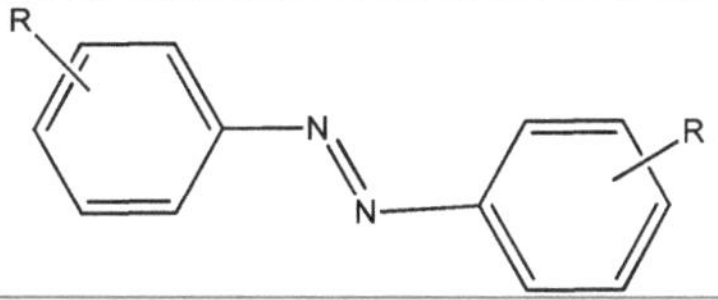

1 using Degussa P25 material has been extensively investigated (Giwa et al. 2012; Ong et al. 2013; Satuf et al. 2011; Trabelsi et al. 2013; Tzikalos et al. 2013). The benefits and drawbacks of the use of TiO_2-assisted heterogeneous photocatalysis for the decolorization of dyes have been described in extensive literature reviews (Ahmed et al. 2010; Akpan and Hameed 2009). Additionally in line with the applications and commercialization of photocatalytic devices for self-cleaning and water purification, the International Organization for Standardization (ISO)

Table 6.2 Overview of major pesticide classes

Chemical class	Chemical formula	Commercial group
Triazine derivatives	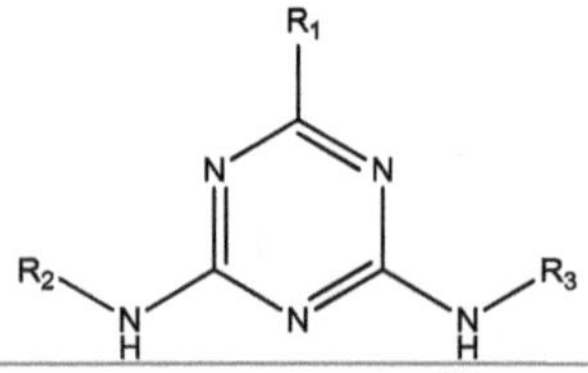	Herbicide
Phenoxyacetic acid derivatives	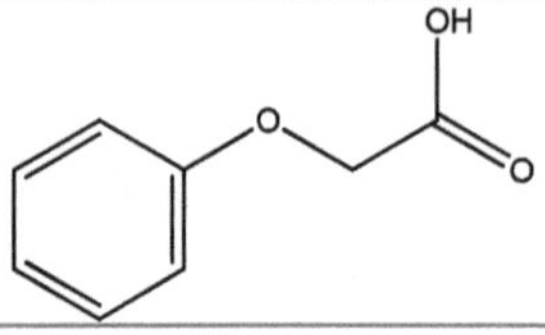	Herbicide
Organophosphorus compounds	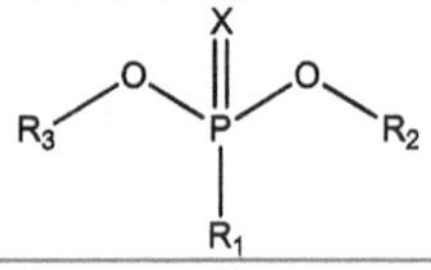	Acaricide Fungicide Rodenticide
Carbamates	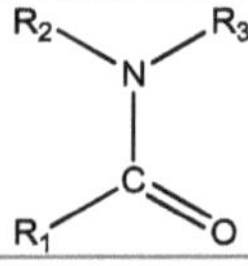	Acaricide Fungicide
Thiocarbamates	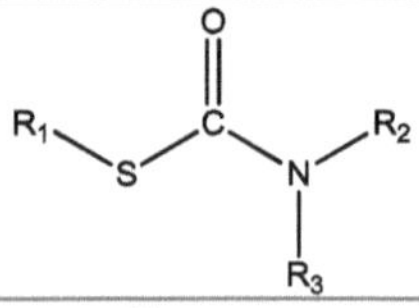	Acaricide Fungicide
Organochlorine compounds	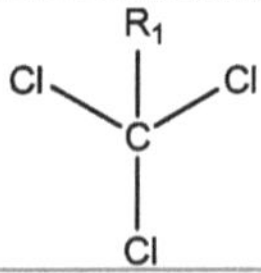	Fungicide Insecticide
Chlorophenolic compounds	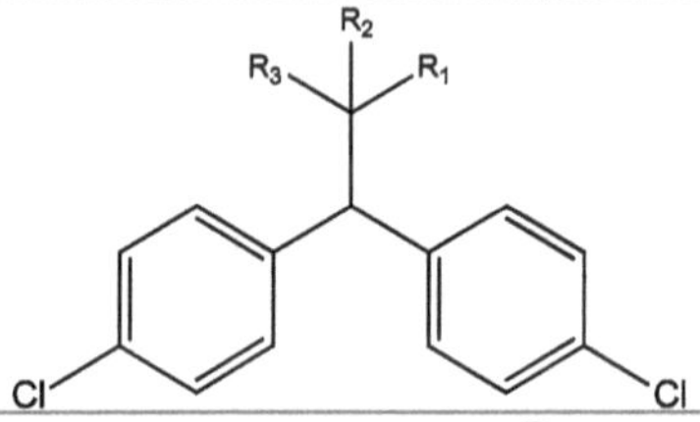	Fungicide Microbiocide Insecticide
Phenylurea compounds	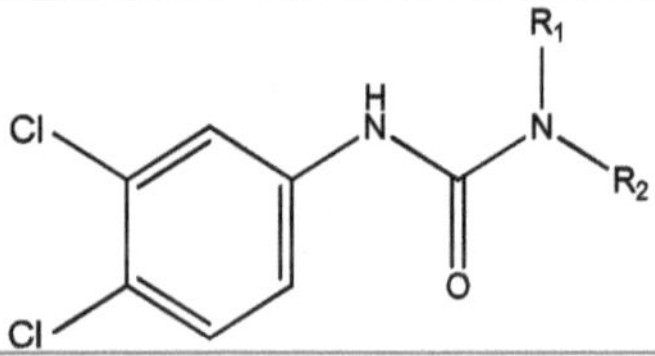	Herbicide

Table 6.3 Chemical structure of some emerging compounds treated by heterogeneous photocatalysis

Compound class	Name	Structure
Antibiotic	Tetracycline	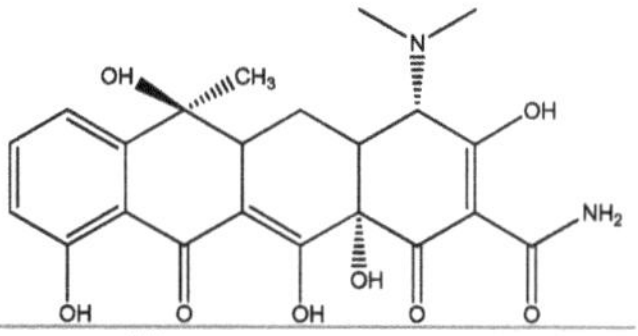
	Trimethoprim	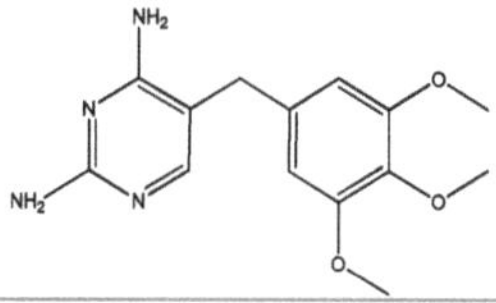
	Levofloxacin	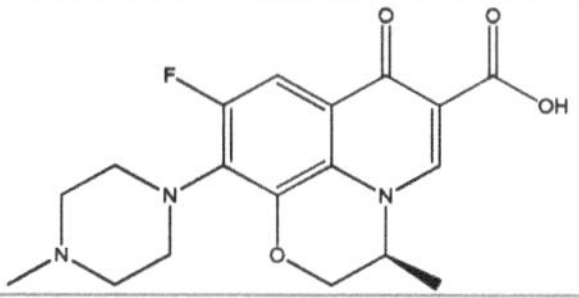
Anti-inflammatory	Diclofenac	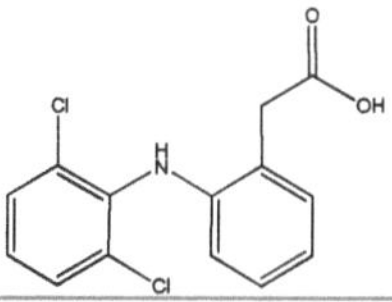
Antihypertensive	Propranolol	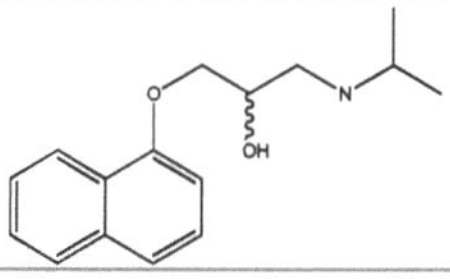
Anticonvulsant and mood stabilizing	Carbamazepine	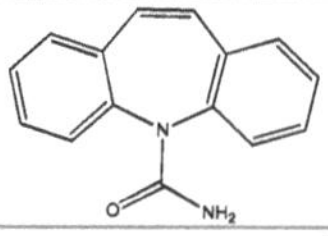
Preservative in personal care products	R-paraben	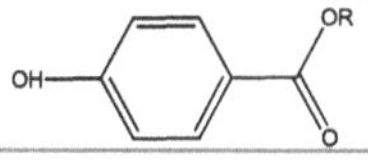
Plastic component	Bisphenol A	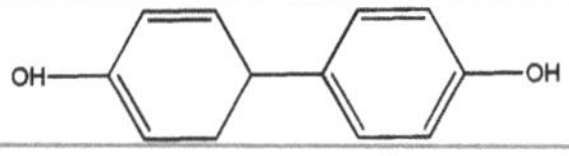
Flame retardant	Tetrabromobisphenol	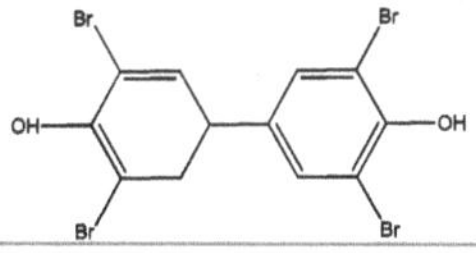
X-ray contrast media	Diatrizoate	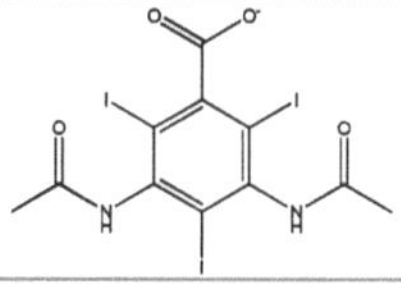

published normalized standard test: 10678:2010, namely, the "Determination of photocatalytic activity of surfaces in an aqueous medium by degradation of methylene blue," that has been clearly defined and disseminated (Mills et al. 2012; Mills 2012). However, evaluation of the mineralization by total organic carbon (TOC) analysis (Herrmann 2010) and information concerning the evolution of toxicity and biodegradability of the treated samples have been scarce. The toxicity and biodegradability of the by-products during the photocatalytic degradation of the azo dye acid orange 7 was evaluated employing artificial UV light as the source of radiation and TiO_2 Degussa P25 as catalyst. The Microtox® assay indicated that the toxicity of the pollutant was significantly reduced after the treatment. Additionally, the photocatalytic process clearly enhanced the biodegradability of the water sample after 6 h treatment. Consequently, this advanced oxidation processes (AOP) could be used as a pretreatment method to degrade bio-recalcitrant organic dyes (Satuf et al. 2011). In the case of pesticides, TiO_2 has shown better photocatalytic performance than other inorganic semiconductors studied in the catalyzed photodegradation, and TiO_2 nanomaterials have been successfully used for the degradation of several classes of pesticides (see Table 6.4), leading in some cases to complete mineralization (Konstantinou 2003; Ahmed et al. 2011). Commercially available titania sample Evonik P25 was used to treat a mixture of four commercial pesticides (oxydemeton-methyl, methidathion, carbaryl, and dimethoate) in a pilot plant reactor under natural sunlight leading the quantitative removal of pesticides in less than 300 min; this process was accompanied by a substantial reduction of acute toxicity to *Vibrio fischeri* (i.e., from an initial value of 50 to 15 %), as well as moderate mineralization, i.e., 40 % chemical oxygen demand (COD) and 25 % dissolved organic carbon (DOC) removal (Vicente et al. 2014). The photocatalyzed degradation of s-triazine-type herbicides with suspended and supported TiO_2 has been studied, and it was observed that cyanuric acid is the final photoproduct. The catalytic activity of TiO_2 was not affected when the semiconductor material was immobilized using a functionalized polymer. The durability of the supported catalyst was reduced after six runs, but it could be fully reactivated by heat treatment (Parra et al. 2004). Pharmaceutical drugs, such as lorazepam, propranolol, oxolinic acid, levofloxacin, acetaminophen, atenolol, trimethoprim, ofloxacin, enrofloxacin, clarithromycin, acetaminophen, diclofenac, caffeine, thiabendazole, carbamazepine, acetamiprid, and erythromycin, have also been treated individually or in mixture by TiO_2 photocatalytic process (Bernabeu et al. 2011; De la Cruz et al. 2013; Giraldo et al. 2010; Maroga Mboula et al. 2012; Nasuhoglu et al. 2012; Radjenović et al. 2009; Rizzo et al. 2009; Sousa et al. 2013). Commercial TiO_2 Millennium PC-500 (anatase >99 % and 320.76 $m^2\ g^{-1}$ surface area) immobilized on ceramic plates by the sol–gel method was evaluated on the photocatalytic oxidation of three pharmaceuticals compounds under UV light. The TOC (90 % removal after 16 h) and ecotoxicological results revealed that the photocatalytic process could effectively mineralize and diminish the ecotoxicity of the mixture of atenolol, metronidazole, and chlorpromazine (Khataee et al. 2013a). For the degradation of diclofenac using TiO_2 Aeroxide, toxicity of the treated water sample was determined through viability of *Daphnia magna*, *Pseudokirchneriella subcapitata*,

Table 6.4 Photocatalytic studies on the removal of organic compounds under UV/visible light using synthesized TiO_2 or ZnO

Photocatalyst	Compound degraded	Experimental conditions	Reported results	Reference
Dyes				
TiO_2 100 % anatase	Safranin T (5.0×10^{-5} M)	UV light 254 nm (6 W), H_2O_2 1.0 mM, pH 5.7, 30 °C, quartz glass photoreactor (150 mL). Catalyst loading: 12 mg in 150 mL (325 mesh)	The toxicity of the solution was inferred by measuring COD that diminish from 1,180.0 to 132.0 mg L^{-1}	Gupta et al. (2007)
TiO_2 synthesized by sol–gel	Methylene blue (MB) and Congo red (CR) (5 mg L^{-1} each dye)	Solar light irradiation, natural pH, glass photoreactor (50 mL), 150 at 900 min irradiation time. Catalyst loading: 0.5 g L^{-1}	The degradation percentage of MB and CR were 91 and 90 % in 420 and 780 min, respectively	Bin Mukhlish et al. (2013)
ZnO synthesized by sol–gel	Reactive red 15 (100 mg L^{-1})	UV light source 100–280 nm (150 W), pH 9, 2 mL L^{-1} of H_2O_2, cylindrical photoreactor (0.85 L). Catalyst loading: 0.75 g L^{-1}	90.5 % COD removal in 90 min of irradiation time	Khattab et al. (2012)
ZnO nanoparticles, by combustion method	Rhodamine B (concentration not reported)	Solar light irradiation and UV light, natural pH, 30 °C, photoreactor (100 mL). Catalyst loading: 1.5 g L^{-1}	90 % of the dye was degraded in 120 min of irradiation	Ali et al. (2013)
ZnO nanostructures synthesized by thermal evaporation method	Methyl orange (MO) and safranin O (20 mg L^{-1})	Two semicylinders with six UVA lamps 365 nm (8 W), natural pH, 25 °C. Catalyst loading: 50 mg in 125 mL	Complete degradation for MO in 20 min and 99 % in 60 min for safranin	Lucca Sánchez et al. (2013)
Pesticides				
TiO_2 (anatase-rutile 89–96/11–4) synthesized by sol–gel method. Crystalline size 57.0 nm for anatase and 86.3 nm for rutile. Surface area: 18.3 m^2 g^{-1}	2,4-Dichlorophenoxyacetic acid (2,4-D, 0.265 mM) and bentazon (0.1325 mM)	Pilot plant (CPC) using natural sunlight irradiation; total volume of aqueous solution was 5 L at natural pH. Irradiated volume was 1.8 L. Catalyst loading: 1 g L^{-1}	The toxicity on marine bacteria *Vibrio fischeri* was reduced during the photocatalytic degradation of both herbicides. TOC reduction was 74 and 81 % for 2,4-D and bentazon, respectively	Seck et al. (2013)

(continued)

Table 6.4 (continued)

Photocatalyst	Compound degraded	Experimental conditions	Reported results	Reference
TiO_2 nanocrystals (hydrolysis and peptization method). TiO_2 in anatase form. Crystalline size: 18.3 nm	Chlorpyrifos (25 mg L^{-1})	Solar photoreactor containing 100 mL of the sample at pH 4.2, temperature 30 °C. The sunlight intensity was 900–950 W m^{-2}. Catalyst loading: 1 g L^{-1}	Small crystallite size and the higher capacity to adsorb surface water hydroxyl groups make it a good catalyst. The rate constant obtained was 0.027 min^{-1}	Kanmoni et al. (2012)
TiO_2 (sol–gel technique, 100 % anatase) coated on silica gel particles (film thickness, 0–0.4 μm; TiO_2 coverage, 75 % w/w; particle size, 1.0 mm; surface area, 300 m^2 g^{-1})	Acephate (0.1 mM), dimethoate (0.1 mM), and glyphosate (0.1 mM)	Spiral glass tube reactor (200 mL) around a 6 W black light fluorescent UV lamp (365 nm, 1.4 mW cm^{-2} of light intensity), temperature 22 °C. The flow rate at the reactor was 30 mL min^{-1} bubbled with O_2 at 100 mL min^{-1}	Complete decomposition of dimethoate and glyphosate in 60 min of irradiation and 105 min for acephate. Mineralization by the oxidation process was demonstrated identifying SO_4^{2-}, NO_3^-, and PO_4^{3-}. Toxic intermediates such as methamidophos and omethoate were detected indicating the lower toxicity of the treated solution	Echavia et al. (2009)
Emerging compounds				
TiO_2 anatase, surface area: 107 m^2 g^{-1} TiO_2 rutile, surface area: 3 m^2 g^{-1} Synthesized by sol–gel	Diclofenac (8 mg L^{-1})	Hg-vapor lamp for UV ($\lambda = 254$ nm) and near UV/vis irradiation ($\lambda = 366$ nm), solution pH 6.22 Catalyst loading: 0.5 g L^{-1}	The first-order rate constant was higher under UV radiation than near UV/vis for both catalysts: 0.903 min^{-1} for anatase and 0.858 min^{-1} for rutile. 8 reaction products were identified by HPLC-MS	Martínez et al. (2011)

and *Artemia salina*. According to the results, the increase of TiO_2 catalyst loading improved the mineralization rate; however, toxicity was also enhanced during treatment; it was attributed to the formation of chlorinated and nitrogen compounds (Rizzo et al. 2009).

The photocatalytic removal of other emerging pollutants, such as endocrine disruptors, personal care products, or flame retardants (e.g., parabens and iodinated X-ray contrast media), has been conducted using TiO_2 as a catalyst. An 80 % TOC mineralization was achieved during the degradation of parabens in 360 min using this semiconductor under UV radiation, forming short-chain carboxylic acids as the final by-products (Lin et al. 2009, 2011). Diatrizoate and iopromide, used as iodinated X-ray contrast media, have been treated by heterogeneous photocatalysis using TiO_2 P25 achieving low mineralization but a significant dehalogenation (Sugihara et al. 2013).

Several studies have compared the photocatalytic activity of TiO_2 and ZnO on the removal of organic pollutants. Because ZnO has almost the same band gap energy as TiO_2, their photocatalytic capacity is anticipated to be similar. Reactive red 15 has been removed from aqueous solution under UV irradiation using TiO_2 and sol–gel ZnO. Although both materials were employed as effective photocatalysts for the elimination of color from dye wastewater, ZnO was more efficient than TiO_2. The biodegradability ratio BOD_5/COD (where BOD_5 is 5-day biochemical oxygen demand) increased from zero up to 0.52 and 0.60 within 90 min irradiation time using 1 and 0.75 g L^{-1} TiO_2 and ZnO, respectively (Khattab et al. 2012).

The photocatalytic degradation of five s-triazines (simazine, prometryn, terbutryn, atrazine, and terbuthylazine) and three chloroacetanilides (propachlor, s-metolachlor, and alachlor) in leaching water has been conducted using these two semiconductor materials at pilot plant scale under natural sunlight. The comparison of the catalysts showed that ZnO was the most efficient for the removal of the studied compounds, allowing 70 % degradation of the herbicide amount initially present in the leaching water after 240 min illumination. The photocatalytic degradation of triazine herbicides yielded a number of transient organic intermediates. Four transformation products were isolated and identified by HPLC-MS/MS analysis. The rapid transformation of the parent compounds was accompanied by the appearance of atrazine-desisopropyl, atrazine-desethyl, terbuthylazine-desethyl, and terbuthylazine-2-hydroxy compounds (Fenoll et al. 2012). The degradation percent of chloramphenicol, a pharmaceutical drug, was improved using ZnO, although the mineralization percent was lower using ZnO compared with using TiO_2 catalyst (Chatzitakis et al. 2008).

As stated in the Chap. 3, depending on the preparation method, the surface and structural semiconductor properties can be modified, affecting the photocatalytic activity of the semiconductor (Seck et al. 2013; Ruggieri et al. 2011; Kanmoni et al. 2012). In this way, Table 6.4 summarizes some examples of the application of synthesized TiO_2 and ZnO for the removal of organic pollutants.

An improved photocatalytic activity under solar light was observed using sol–gel TiO_2 compared to Degussa P25 TiO_2 on the degradation of methylene blue and congo red (Bin Mukhlish et al. 2013) and pesticides such as 2,4-dichlorophenoxyacetic acid (2,4-D) and bentazon in real water at natural pH (6.8–7.8) (Seck et al. 2012, 2013). For pesticides, the toxicity of the treated solutions was evaluated using the Microtox® test. In case of 2,4-D herbicide, the main intermediate 2,4-dichlorophenol (2,4-DCP) showed higher toxic effect than the parental herbicide that was reduced during the photocatalytic treatment by using the synthesized TiO_2 photocatalyst. Neither bentazon nor its degradation products had any significant toxic effects on *Vibrio fischeri* bacteria at the evaluated irradiation times (Seck et al. 2013).

The photocatalytic degradation of reactive red 195 from aqueous samples under UVA irradiation was evaluated using anatase/brookite TiO_2 mesoporous nanoparticles. However, despite the fact that this material showed similar decolorization efficiency compared to that of commercial titania, the synthesized TiO_2 mesoporous nanoparticles showed much lower mineralization. The composite photocatalyst was characterized by a certain degree of photochemical stability, and it may be reusable for at least three runs (Tzikalos et al. 2013). The prepared nano-sized TiO_2 via sol-gel method was evaluated on linuron degradation, a widely used phenylurea herbicide, under UV radiation. Although a complete mineralization of this herbicide was demonstrated by the TOC measurements, and the evolution of nitrogen and chlorine ions, the decrease of initial TOC and formation of ionic end products was noticeably slower (12 h of irradiation) (Ruggieri et al. 2011). Another study confirmed that nanocrystals of TiO_2 (7.5 nm) synthesized by the hydrolysis and peptization method exhibited a better photocatalytic activity under solar light than ZnO (34.3 nm) prepared by the wet chemical method on the degradation of chlorpyrifos (*O*,*O*-diethyl *O*-3,5,6-trichloro-2-pyridyl phosphorothioate) in aqueous suspensions. The photodegradation of chlorpyrifos under sunlight generates excess free radicals that are able to destroy the intermediates at a faster rate. Those results could be attributed to the TiO_2 small crystallite size and the higher capacity to adsorb surface water and hydroxyl groups (Kanmoni et al. 2012).

In case of pharmaceutical drugs, anatase nanofibers obtained by hydrothermal reaction has been evaluated in carbamazepine (10 mg L^{-1}) degradation under UVA radiation (360 nm) allowing 95 % of degradation and an increase of biodegradability which was determined through respirometric tests (Laera et al. 2011).

6.3 Degradation of Organic Pollutants in the Presence of TiO_2 or ZnO Doped with Metals

The doping of metals in semiconductors, such as ZnO and TiO_2, serves the purpose of transferring the electrons away from the catalyst surface, hence preventing them from recombining with valence band positive holes. The synthesis method, size and concentration of the dopant, and the electronegativity and electronic configuration of both the support and the dopant atoms influence the physicochemical and photocatalytic properties of the doped catalyst. Doping with suitable transition metal ions allows extending the light absorption of large band gap semiconductors, such as TiO_2 and ZnO, to the visible region. Both positive and negative results have been reported from doping with metal ions (Ahmed et al. 2011). The results are summarized in Table 6.5.

Table 6.5 Degradation of organic compounds by heterogeneous photocatalysis using metal doping catalysts

Catalyst	Pesticide	Experimental condition	Reported results	Reference
Dyes				
TiO_2/Ni (1 %) and Ru/TiO_2 (1 %) by sol–gel process and impregnation method, respectively	Amidoblack-10B (3×10^{-4} M)	High-pressure mercury lamp (365 nm, 8 W), natural pH, 420 min irradiation time, glass photoreactor (100 mL). Catalyst loading: 1.5 g L^{-1}	Complete decolorization was observed using both photocatalysts, while 65 and 50 % mineralization was achieved for Ni/TiO_2 and Ru/TiO_2, respectively	Amala Infant Joice et al. (2012)
TiO_2 nanorods-Co(II) (Co-NRs)	5,8-dihydroxy-1,4-naphthoquinone dye (99.9 μM)	Visible light irradiation Xe arc lamp (300 W), natural pH. Catalyst loading: 0.16 g^{-1}	100 % decomposition in 60 min of irradiation time was obtained	Kang et al. (2013)
TiO_2–Ag (5 mol%) by enzymatic synthesis	Black B dye (100 mg L^{-1})	Solar light irradiation, natural pH. Catalyst loading: 1 g L^{-1}	92 % of color removal in 80 min of irradiation time	Gunasekar et al. (2013)
TiO_2–Ce and TiO_2–Fe (0.8 and 1.2 mol%, respectively), both synthesized by sonochemical method	Crystal violet dye (30 mg L^{-1})	UV lamp at 365 nm (2,000 W/cm^2), pH 6.5, 35 °C. Glass photoreactor (150 mL). Catalyst loading: 0.2 g L^{-1}	The highest degradation (84 %) was achieved with Ce–TiO_2 compared to those obtained using Fe–TiO_2 (77 %) in 120 min irradiation time	Shirsath et al. (2013)
TiO_2–Cr (0.5 %) and TiO_2–V (0.5 %) by chemical vapor synthesis method	Methylene blue and acid red 27 (50 μmol L^{-1})	Solar light irradiation (200–250 W/m^2), natural pH, cylindrical Pyrex beaker (100 mL), 180 and 60 min irradiation time for MB and AR, respectively. Catalyst loading: 0.5 g L^{-1}	The effect of V on the solar photoactivity was more pronounced than that of Cr. In methylene blue, degradation kinetic rate constants were 0.0193 and 0.0041 min^{-1} and for Acid Red 27 were 0.23487 and 0.0417 min^{-1} (for TiO_2–V and TiO_2–Cr, respectively)	Khakpash et al. (2011)
ZnO–Ag (0.37 %) nanorods array (NRA) by two-step photodeposition method. Geometric area: 2 cm^2	Methylene blue (2 mg L^{-1})	Low-pressure fluorescent Hg lamp 254 nm (1.9 mW/cm^2), natural pH, glass photoreactor (15 mL)	The photodegradation was 49.3 % in 90 min of irradiation time	Ren et al. (2010)

(continued)

Table 6.5 (continued)

Catalyst	Pesticide	Experimental condition	Reported results	Reference
ZnO–Cu (0.150 wt%) and ZnO–Mn (0.150 wt %). Synthesized by crystallization	Methylene blue (5 mg L^{-1})	UV lamp (18 W) intensity 5×10^{-5} W cm^{-2}, natural pH, 20 °C, 200 min irradiation time. Catalyst loading: 0.5 g L^{-1}	The degradation percent was 100 % for ZnO–Mn in 120 min, while complete degradation was achieved in 160 min for ZnO–Cu	Donkova et al. (2010)
Pesticides				
TiO_2–Mg (1 wt %). Prepared by sol–gel method	Monocrotophos (MCP, 50 mM)	The suspension was irradiated under visible light (400–800 nm) Glass reactor containing 100 mL of standard aqueous solution, pH 3. Catalyst loading: 0.5 g L^{-1}	The degradation rate was 2.1 mg $L^{-1}min^{-1}$. The high photocatalytic activity was due to decrease in particle size that increases the surface area, as was observed by SEM image	Avasarala et al. (2011)
TiO_2–Bi (2 wt %). Prepared by sol–gel method and calcined at 500 °C TiO_2 in anatase form. Crystalline size 17.58 nm. Surface area: 102.67 m^2 g^{-1}	Isoproturon (1.14×10^{-4} M)	Open glass reactor (50 mL). The solar experiments were carried out during 10.00 A.M.–3.00 P.M. (light intensity 130,000 lx) at room temperature with volume correction (evaporation)	The high activity was due to the stronger absorption in the visible light region and lower electron–hole recombination attributed to the presence of $Bi^{3+\delta+}$ species	Reddy et al. (2011)
TiO_2–Au nanosized particles (0.83 wt %). Prepared by deposition–precipitation at pH 7 and calcination in air at 873 K TiO_2 predominantly in anatase phase	3,4-Dichlorophenylurea (3,4-DCPU) (20 mg L^{-1})	Reactor was magnetically stirred and bubbled with air at the rate of 5.7 L min^{-1} in a full-spectrum solar simulator. The temperature was maintained at 293 K. Catalyst loading: 0.5 g L^{-1}	The decrease in 3,4-dichlorophenylurea concentration follows first-order kinetics with a half-life of 36.6 ± 2.0 min	Chusaksri et al. (2011)

P25 TiO_2-Pt (0.2 wt %). Deposited onto surface by impregnation method. Crystalline size were 23.0 nm (anatase) and 32.0 nm (rutile)	Diuron (10 mg L^{-1})	The photocatalytic reactor (30 mL) was irradiated under UV radiation (300 nm). Reaction temperature was kept at 25 ± 1 °C. pH 6.5. Catalyst loading: 25 mg in 30 mL	The first-order rate degradation by Pt–TiO_2 was ca. 4 times higher than P-25 TiO_2. The mineralization degree was about 97 % after 8 h	Katsumata et al. (2009)
MoO_3 (2 at.%) particles loaded on ZnO nanorods (ZnO NRs) synthesized via hydrothermal precipitation method. The pure orthorhombic phase of α-MoO_3 nanoparticles 15–21 nm was well dispersed on ZnO NRs surface	2,4-Dichlorophenoloxyacetic acid (20 mg L^{-1})	Reactor was a 250 mL beaker. The pH was adjusted to 7. The solution was irradiated under a 55 W compact fluorescent lamp (light intensity, 14,500 lx). Catalyst loading: 1 g L^{-1}	Degradation rate was 0.0452 min^{-1}. High photocatalytic activity was attributed to the high quantum yield based on the cooperative roles between the bicomponents of MoO_3 and ZnO NRs	Lam et al. (2013)
Emerging compounds				
TiO_2 nanocrystals–Ag nanoparticles film. Surface area: 235 $m^2\ g^{-1}$. Synthesized by fast hydrolysis and colloidal chemistry route	Nalidixic acid (23 mg L^{-1})	Mercury lamp ($\lambda > 250$ nm). pH 7. Supported area: 3.6 cm^{-2}	The first-order rate constant obtained was $9.8 \times 10^{-3}\ min^{-1}$	Petronella et al. (2013)
TiO_2–Ce 0.5 % (34.1 $m^2\ g^{-1}$) TiO_2–Ce 1.0 % (56.2 $m^2\ g^{-1}$) Synthesized by sol–gel method	Propranolol (25 mg L^{-1})	Solarbox: Xe–OP lamp (290–400 nm). pH 7.5. Catalyst loading: 0.14 g L^{-1}	First-order rate constants were ~2.1 and ~1.5 ($k \times 100\ min^{-1}$) for TiO_2–Ce 0.5 % and 1.0 %, respectively. More than 30 intermediates were identified by LC/MS. Toxicity was reduced during treatment (evaluated in *P. subcapitata* and *V. fischeri*)	Santiago-Morales et al. (2013)

(continued)

Table 6.5 (continued)

Catalyst	Pesticide	Experimental condition	Reported results	Reference
TiO_2 P25–M 0.5 wt % (where M = Ag, Cu, and Ni), Surface area: 52, 53, 57 and 52 $m^2\ g^{-1}$ respectively. Synthesized by deposition–precipitation	Trimethoprim (40 mg L^{-1})	UV lamp ($\lambda = 254$ nm). pH 6. Catalyst loading: 1 g L^{-1}	The higher rate constant was obtained using TiO_2 P25–Au and TiO_2 P25–Ag (both $5.2 \times 10^{-3}\ min^{-1}$)	Oros-Ruiz et al. (2013)
0.5 % Pt–TiO_2, 0.5 % Pt (0.12 % Na)/TiO_2, 0.5 % Pt/N–TiO_2, 0.5 % Pt/P–TiO_2, 0.5 % Pt/4 % CaO–TiO_2, 0.5 % Pt (0.5 % Ag)/TiO_2, 0.5 % Pt (0.1 % K)/TiO_2, 4 % CaO–TiO_2. Surface area: 38, 39, 63, 186, 104, 42, 41, and 110 $m^2\ g^{-1}$, respectively. Synthesized by sol–gel or co-impregnation	17α-Ethynylestradiol (300 μg L^{-1})	Solar radiation simulator. pH 8. Catalyst loading: 0.5 g L^{-1}	The higher photocatalytic rate constant was obtained with 0.5 % Pt–TiO_2 and 0.5 % Pt (0.12 % Na)–TiO_2 (both with 21.2 μg^{-1} min)	Dimitroula et al. (2012)
TiO_2–Fe (0.42, 0.89, 1.33, 2.2, and 3 %). Synthesized by sol–gel method. Catalyst loading: 1 g L^{-1}	Amoxicillin (25 mg L^{-1})	Low-pressure luminescent mercury lamp ($\lambda = 365$ nm). pH 6	The higher degradation percent (25 %) was achieved using 0.89 Fe–TiO_2 in 6 h of irradiation. However, this material did not exceed the activity of TiO_2 P25 (~80 %)	Klauson et al. (2010)

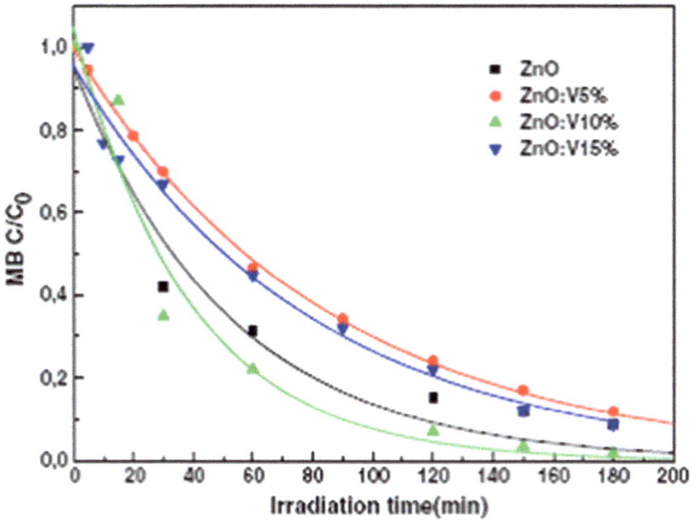

Fig. 6.1 Photocatalytic degradation of methylene blue using different percentages of vanadium-doped zinc oxide under visible irradiation (Slama et al. 2011)

Slama et al. prepared vanadium (0–15 %)-doped ZnO by the sol–gel method and evaluated the photocatalytic efficiency of the doped ZnO samples on methylene blue degradation under visible light (Fig. 6.1). The maximum degradation rate was achieved with 10 % vanadium dopant concentration, compared with undoped ZnO material (Slama et al. 2011).

The rate of degradation of monocrotophos (MCP) over Mg-doped TiO_2 synthesized by the sol–gel method was better than pure TiO_2 and Degussa P25. The maximum degradation of MCP was observed at 1.0 wt % of Mg^{2+} dopant concentration. The results could be attributed to the shift to higher wavelength, the decrease in particle size, and the increase in the surface area of the Mg-doped TiO_2 catalyst (see Fig. 6.2) (Avasarala et al. 2011).

From the studies in Table 6.5, it can be found that there is an optimum dopant concentration for the best photocatalytic performance of a particular modified TiO_2. A 2 wt % Bi-doped TiO_2 prepared by the sol–gel method was an efficient catalyst under solar light radiation for the photocatalytic degradation of isoproturon herbicide, compared to bare TiO_2. The stability/activity of the catalyst was evaluated by recycling studies showing that for the third cycle the 2 % Bi-doped TiO_2 catalyst had a slight decrease in degradation compared to first and second cycle. The decrease may be due to the accumulated organic intermediates on the surface of the catalyst, affecting the adsorption in turn reducing the activity (Reddy et al. 2011).

Certain doping agents in TiO_2 nanomaterials can provoke negative effects. Doping of Th into the TiO_2 lattice prepared by the sol–gel method was used to modify the electronic properties of titanium, a dioxide photocatalyst, enabling the absorption of solar energy irradiation. The 0.06 % Th-doped TiO_2 prepared

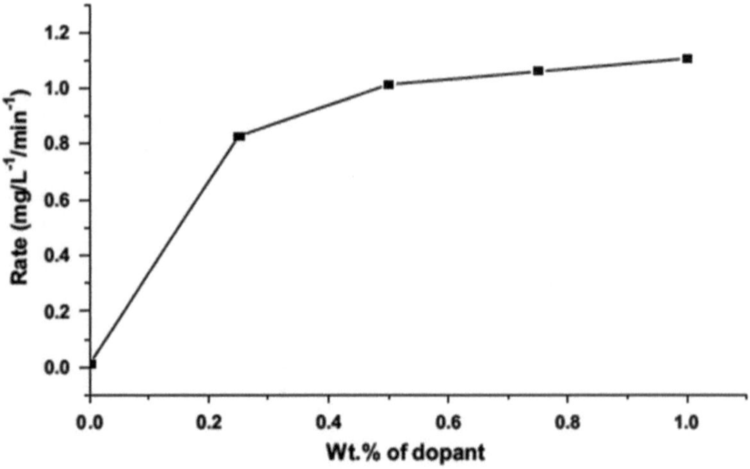

Fig. 6.2 Effect of dopant concentration on the rate of MCP degradation. Concentration of MCP = 50.0 mM (PO_4^{3-} = 515.3 mg); catalyst amount = 0.1 g; pH = 6.5 (Avasarala et al. 2011)

by the sol–gel method was a good photocatalyst for the degradation of the herbicide oryzalin under solar light irradiation. However, this modified catalyst showed lower photocatalytic efficiency than TiO_2 under UV light (Gomathi Devi and Narasimha Murthy 2008). Other doping agents evaluated in nanostructured TiO_2 include gold and platinum (Chusaksri et al. 2011; Katsumata et al. 2009). The 0.83 wt % Au–TiO_2 catalyst prepared by the deposition–precipitation method has been efficiently used to remove 3,4-dichlorophenylurea from water under simulated solar light. The process is an efficient method due to the improvement of photoinduced charge separation and enhancement of light absorption of unmodified TiO_2. The calcination temperature has an effect on the photocatalytic activity, and the best activity was achieved at 600 °C. Fifteen degradation products were identified using LC-MS/MS and ion chromatography. Degradation pathway was proposed on the basis of the observed transformation products (Chusaksri et al. 2011). Diuron, a phenylurea herbicide, was oxidized by the photocatalytic process in the presence of platinized TiO_2 photocatalyst prepared by the impregnation method. When 0.2 wt % of platinum was deposited onto the surface of TiO_2, an initial diuron concentration of 10 mg L^{-1} was completely degraded after 20 min under UV radiation. The kinetic first-order rate constant for diuron degradation by Pt–TiO_2 was approximately four times higher than for P25 TiO_2. The degree of diuron mineralization was approximately 97 % under UV irradiation after 8 h that was confirmed by the formations of the end products chloride, nitrate, and ammonium ions during the photocatalytic process. The possible degradation pathway for diuron is proposed in Fig. 6.3 (Katsumata et al. 2009).

Regarding the use of metals as dopant agents for pharmaceutical removal, low doping percent (near to 0.5 wt %) provided better results for the oxidation of these organic compounds (Santiago-Morales et al. 2013; Dimitroula et al. 2012).

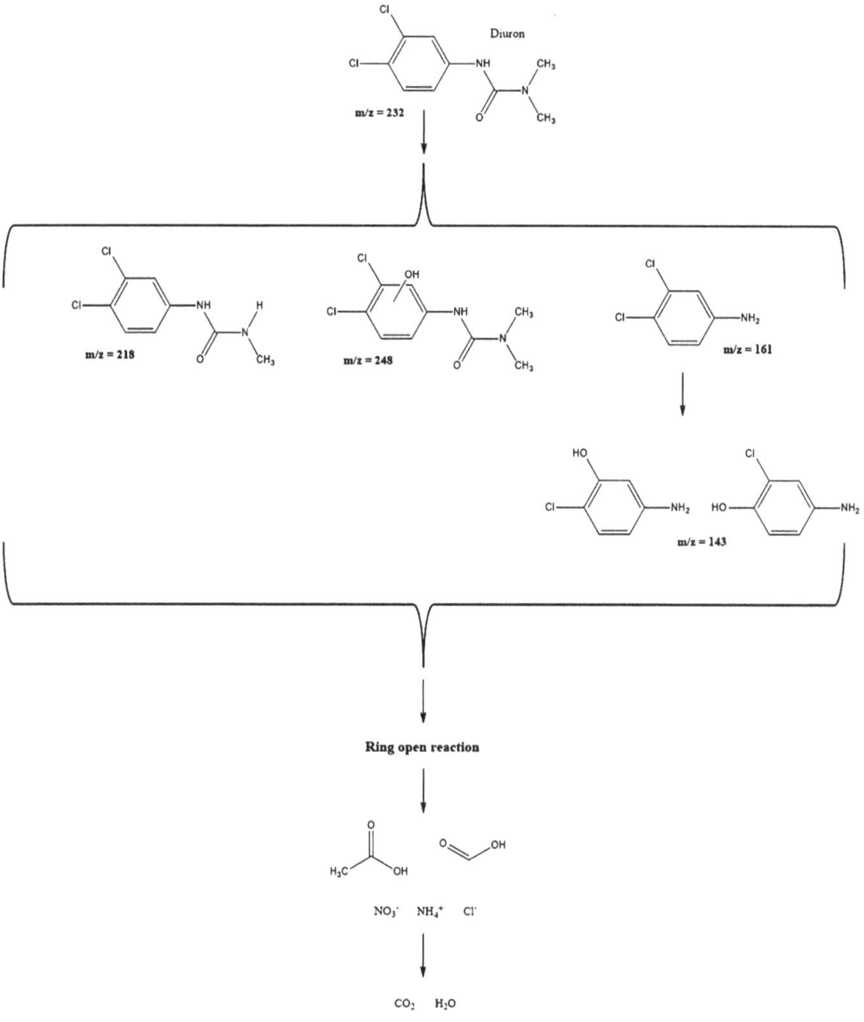

Fig. 6.3 Proposed degradation mechanism of diuron by the photocatalytic Pt–TiO_2 system (Katsumata et al. 2009)

A study conducted by Oros-Ruíz et al. in 2013 on the degradation of trimethoprim demonstrated that the better metallic dopant agents were those with high electron affinity and electronegativity in the following order: gold > silver > copper > nickel (see Fig. 6.4). The metal loading was 0.5 wt %, and the deposition of metallic particles on P25 TiO_2 produced an enhancement of the activity of the bare semiconductor. When the degradation of trimethoprim was conducted with pure TiO_2 P25, the mineralization reached only 50 % within 300 min, whereas by using metallic nanoparticles deposited on P25 TiO_2, the mineralization of organic matter increased up to 80 % (Oros-Ruiz et al. 2013).

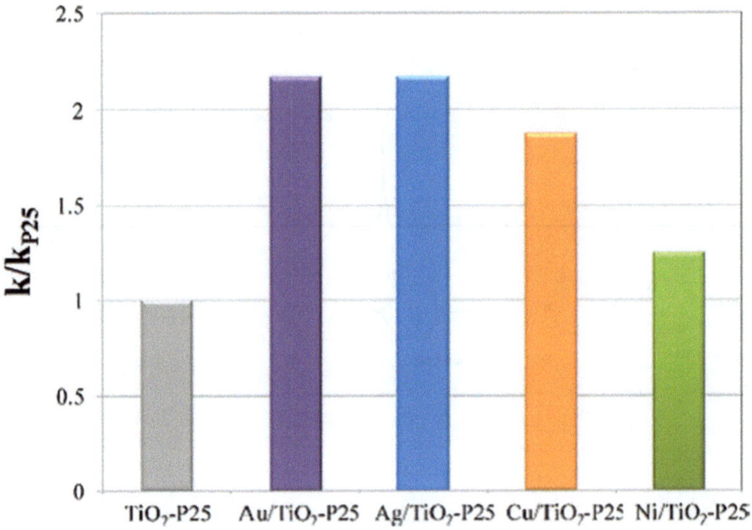

Fig. 6.4 Comparative photocatalytic enhancements for the degradation of trimethoprim, determined by TOC, of P25 TiO_2 modified with metal nanoparticles (Oros-Ruiz et al. 2013)

Those results could explain the positive effect showed when TiO_2 was doped with platinum for the oxidation of 17α-ethynylestradiol. A 0.5 % Pt–TiO_2 catalyst was highly active for the degradation of this emerging contaminant under simulated solar radiation (Dimitroula et al. 2012).

The 0.5 wt % Ce–TiO_2 using 0.14 g L^{-1} catalyst loading was evaluated on the visible photocatalytic degradation of propranolol (25 mg L^{-1}). After 2 h of reaction, the toxicity evaluated in *P. subcapitata* and *V. fischeri* was reduced; however, the accumulation of some toxic products during reaction caused growth inhibition of *V. fischeri*. More than 30 intermediates were identified by LC-MS, and the extent of mineralization reached 17.4 % after 6 h (Santiago-Morales et al. 2013).

6.4 TiO_2 Doped with Nonmetals for the Degradation of Organic Pollutants

As mentioned in Chap. 1, the doping of TiO_2 with nonmetal elements, such as C, N, and F, has improved the photocatalytic response in the visible light region (Table 6.6). These nonmetal doping agents have shown different effects on the photocatalytic activity. Kuriechen and Murugesan (2013) demonstrated that C-doped TiO_2 material synthesized by the sol–gel method showed stronger absorption in the UV-visible range, attributed to carbon substitutions at the oxygen sites in the sample and therefore exhibited better photocatalytic activity on the removal of reactive red 180 (RR180) and reactive red 141 (RR141) under visible light

Table 6.6 The applicability of TiO_2 doped with nonmetal on the removal of organic pollutants

Catalyst	Pollutant	Experimental	Results	Reference
Dyes				
TiO_2–C (5 %) prepared by sol–gel method	Reactive red 180 and reactive red 141 (50 mg L^{-1})	Tungsten-halogen lamp (250 W; $\lambda = 360$–2,000 nm), 60 min irradiation time, natural pH, and glass photoreactor (70 mL) with catalyst loading: 1.5 g L^{-1}	65 % TOC reduction was obtained with TiO_2–C (5 %) compared to 45 % TOC removal for Degussa P25 TiO_2 (within 5 h)	Kuriechen and Murugesan (2013)
Cotiox KA-100 TiO_2 powder (anatase >99 %) and nitrogen-doped TiO_2 (urea used as precursor) prepared by mechanical mixing	Basic red 46 (15 mg L^{-1})	Visible light lamp ($\lambda > 400$ nm, 100 W), 450 min irradiation time, natural pH, glass photoreactor (100 mL) with catalyst loading: 0.4 g L^{-1}	The photodegradation was 60 and 90 % for TiO_2 and N–TiO_2, respectively	Khataee et al. (2013b)
Degussa P25 TiO_2, TiO_2, and nitrogen-doped TiO_2 prepared by calcination of the hydrolysis product	Orange G (25 mg L^{-1})	Xenon arc lamp ($\lambda > 400$ nm, 500 W), pH 2, $H_2O_2 = 15.0$ mmol L^{-1} Glass photoreactor (1,000 mL) with catalyst loading: 1 g L^{-1} and 150 min irradiation time	N-doped TiO_2 exhibited the highest degradation efficiency with 96.3 %, P25 showed 42.55 %, and non-doped TiO_2 showed less than 13.0 %	Sun et al. (2008)
Pesticides				
N-doped TiO_2 prepared by modified sol–gel method (titanium isopropoxide and triethylamine in the ratio of 1:1.6 with E_g value 2.91 eV). The activity was compared with Ag^+-doped TiO_2, Fe^{3+}-doped TiO_2, Cr^{3+}-doped TiO_2, and Cr^{3+}-N co-doped	Lindane (100 μg L^{-1})	High-pressure tungsten visible lamp ($\lambda > 400$ nm, 500 W), temperature 25–30 °C, reaction time 7 h, glass photoreactor (400 mL)	100 % of degradation of lindane was observed under visible light irradiation using N-doped TiO_2 compared to 12 % achieved with P25	Senthilnathan and Philip (2010)

(continued)

Table 6.6 (continued)

Catalyst	Pollutant	Experimental	Results	Reference
N-doped TiO_2 (5 %) prepared by sol–gel and sonochemistry method (E_g 3.00 and 2.95 eV, respectively)	2,4-Dichlorophenoxyacetic acid (40 mg L^{-1})	A lamp [GaI_3 250 W (400–700 nm)], glass reactor (100 mL), catalyst loading: 0.1 g L^{-1}, irradiation time 2 h	Degradation values of 76 and 83 % were obtained for the 5 % N–TiO_2 materials synthesized by the sol–gel and sonochemical methods, respectively	Del Ángel-Sanchez et al. (2013)
Emerging pollutant				
TiO_2–C (30 and 37 wt %) synthesized by sol–gel method	Amoxicillin (25 mg L^{-1})	Low-pressure luminescent mercury lamp ($\lambda = 365$ nm), pH 6, catalyst loading: 1 g L^{-1}	Degradation percent: TiO_2 P25 = ~80 %, 30 wt % C–TiO_2 <10 % and 37 wt % C–TiO_2 = ~30 %	Klauson et al. (2010)
TiO_2–N (0.5 %, 86.7 m^2 g^{-1}, E_g 2.79 eV) synthesized by wet impregnation method	Cefazolin (45 mg L^{-1})	Black light lamp ($\lambda = 365$ nm) and solar light, pH 6.4, catalyst loading: 2 g L^{-1}	First-order rate constant (min^{-1}): Under UV radiation: 0.5 %, N–TiO_2 = 26.45 × 10^{-2} TiO_2 = 13.40 × 10^{-2} Under solar radiation: 0.5 %, N–TiO_2 = 30.18 × 10^{-2} TiO_2 = 14.28 × 10^{-2}	Gurkan et al. (2012)
Polycrystalline N-doped TiO_2 (0.5 %) thin films deposited by sol–gel dip-coating	Carbamazepine (1 mg L^{-1})	Xenon arc lamp (150 W) solar simulator, pH in the range of 5–9, Pyrex glass photoreactor (30 mL), 90 min irradiation time	Alkalinity values of 100 mg L^{-1} as $CaCO_3$ resulted in an over 40 % decrease in carbamazepine removal	Avisar et al. (2013)
CN–TiO_2 synthesized by solvothermal method	Bisphenol A (0.02 mg L^{-1})	Mercury UV lamp ($\lambda = 365$ nm) Vis lamp ($\lambda \sim 450$ nm, LED) Solar simulator (150 W xenon arc lamp) Neutral pH, catalyst loading: 0.5 g L^{-1}	Degradation in 4 h under different light source: UV lamp: 90 % White LED: 95 % Solar simulator: 90 % Toxicity was almost completely reduced by using UV, visible, and solar radiation Main reaction mechanism by $O_2^{\bullet -}$	Wang and Lim (2011)

irradiation. This material showed higher surface area, which could provide more active sites and adsorb more reactive species for photodegrading the target pollutants. Decolorization percents of 79 and 84 % were attained for RR180 and RR141 azo dyes in the presence of C–TiO_2 with 5 % carbon doping, respectively, within 60 min of visible light irradiation. Degussa P25 exhibited less than 60 % decolorization in 60 min. Although superior photocatalytic activity was observed for carbon-doped TiO_2 compared to Degussa P25, the mineralization of both dyes under visible light illumination occurred in slower rate indicating the formation of degradation intermediates. Whereas doping may be an effective strategy to decrease the band gap energy of TiO_2 and thus allow visible light photocatalysis of organic contaminants. The decrease in the band gap energy may increase the ability for recombination, resulting in unpredictability of the photocatalytic efficiency of doped TiO_2 materials. The study conducted by Klauson et al. on the photocatalytic oxidation of amoxicillin under solar radiation was favored at the highest carbon content (37 at.% C doped TiO_2), resulting in a 75 % degradation, which was moderately inferior to that of Degussa P25 (Klauson et al. 2010). The incorporation of nitrogen into TiO_2 changed the hardness, refraction index, elastic modulus, electrical conductivity, and photocatalytic reactivity toward visible light absorption. Khataee et al. prepared N-doped TiO_2 by the mechanical mixing of urea with commercial TiO_2 nanopowders in a 4:1 weight ratio and evaluated the degradation of azo dye basic red 46 (BR46) under visible light. The N-doped TiO_2 nanoparticles showed higher photocatalytic activity than undoped TiO_2 in the decolorization of BR46 solution under visible light irradiation (Khataee et al. 2013b).

The degradation of orange G (OG) on nitrogen-doped TiO_2, non-doped TiO_2, and Degussa P25 photocatalysts was investigated under visible light and sunlight irradiation by Sun et al. (2008). Figure 6.5a shows the photodegradation of OG on different catalysts under visible light irradiation. The N-doped TiO_2 exhibited the highest degradation efficiency with 96.3 % of OG being degraded in 150 min, whereas P25 showed 42.6 % conversion, and the non-doped TiO_2 showed less than

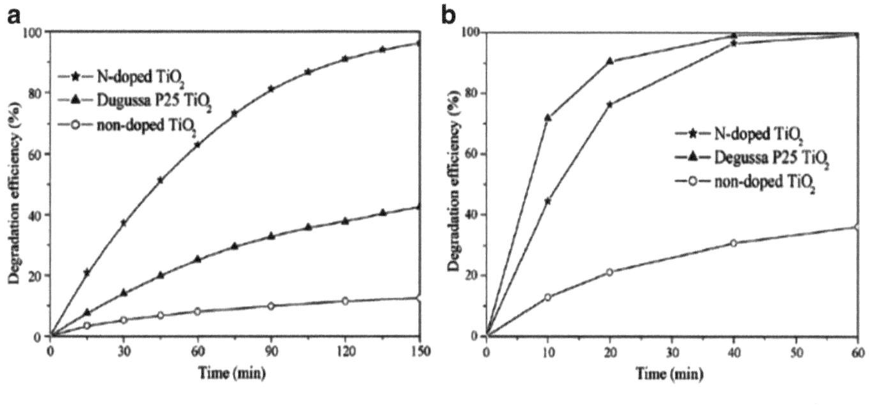

Fig. 6.5 Photodegradation of OG under (**a**) visible light (pH = 2.0, [H_2O_2] = 5.0 mmol L^{-1}) and (**b**) sunlight irradiation (pH = 2.0, [H_2O_2] = 15.0 mmol L^{-1}) (Sun et al. 2008)

13.0 % conversion in the same test period. However, the photodegradation of OG on different catalysts under sunlight irradiation (Fig. 6.5b) was superior with P25 than with N-doped TiO_2, and the non-doped TiO_2 also induced 36.24 % of OG degradation in 60 min.

N-doped TiO_2 photocatalysts for the oxidation of either a single pesticide or mixed pesticides has been extensively reported by several researchers (Del Ángel-Sanchez et al. 2013; Senthilnathan and Philip 2010, 2011). Senthilnathan and Philip reported the degradation of lindane, an organochlorine pesticide, under visible light with N-doped TiO_2 prepared by a modified sol–gel method. The synthesized material showed better photocatalytic activity compared to other metal ion-doped TiO_2 and P25 TiO_2, especially under solar light. GC-MS analysis revealed that the lindane was completely degraded and no intermediate products were observed at the end of the analysis (Senthilnathan and Philip 2010). Better photocatalytic activity under visible and solar radiation for the degradation of methyl parathion and dichlorvos pesticides was described using nitrogen (N)-doped TiO_2 from triethylamine precursor. The intermediates generated during the photocatalytic process were highly susceptible to photocatalytic degradation in the presence of N-doped TiO_2 material (Senthilnathan and Philip 2011). Del Ángel-Sanchez et al. evaluated N–TiO_2 photocatalytic material prepared by the sol–gel and sonochemical methods using urea as precursor for the visible mediated photocatalytic degradation of 2,4-dichlorophenoxyacetic acid. Both methods of synthesis showed higher absorbance when nitrogen was incorporated into the matrix of TiO_2, leading to a decrease of E_g from 3.2 to 3.0 eV, and the incorporation of nitrogen causes a decrease in the particle size. The TEM analysis shows that the nanoparticles were obtained with sizes less than 40 nm. Degradation values of 76 and 83 % were obtained for the 5 wt % N–TiO_2 materials synthesized by the sol–gel and sonochemical methods, respectively (Del Ángel-Sanchez et al. 2013). N-doped TiO_2 has been further exploited by employing a wet impregnation method based on urea nitrogen precursor for the oxidation of cefazolin, a semisynthetic antibiotic, under solar light radiation. The optimum concentration of nitrogen was 0.50 wt %, resulting in 80 % removal in 50 min (Gurkan et al. 2012). In another study, carbamazepine (CBZ, 1 mg L^{-1}), an antiepileptic drug, was removed under solar light simulation in deionized water and different real water samples (groundwater, surface water, secondary and tertiary wastewater) using 0.5 % N–TiO_2 thin films as catalyst. The photocatalytic degradation of CBZ decreased drastically from 70 % in buffered water to approximately 35 %, 21 %, and negligible values with surface water, groundwater, and either secondary or tertiary effluent, respectively. (Fig. 6.6a, b). The removal of CBZ was affected more by alkalinity than by dissolved organic matter (Avisar et al. 2013).

Co-doping TiO_2 with C and N was investigated on bisphenol A (0.02 mg L^{-1}) removal, reaching 95 % degradation compared with 7 % degradation achieved by Hombikat under visible radiation. After 4 h of treatment, toxicity evaluated by Microtox® test was almost completely reduced (Wang and Lim 2011).

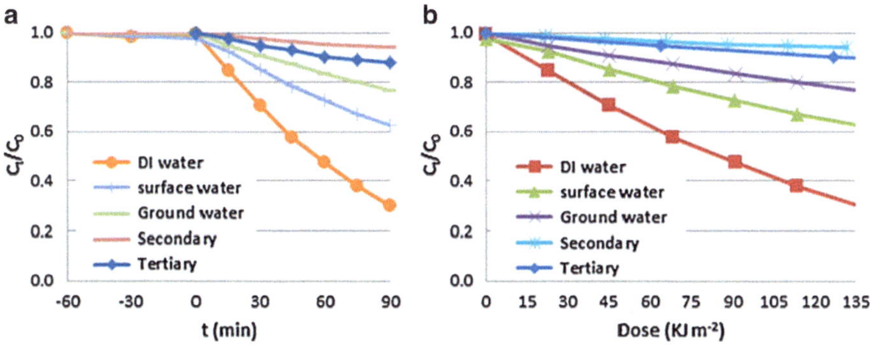

Fig. 6.6 Percent CBZ removal in different water sources as a function of time (**a**) and dose (**b**) (Avisar et al. 2013)

6.5 Coupling Two Semiconductor Systems for the Removal of Organic Pollutants

Coupling TiO_2 with a narrow band gap semiconductor makes long-term charge separation possible by decreasing the recombination and at the same time allowing the extension of the response of the photocatalyst into the visible region (Kumar and Devi 2011). The strategy has been evaluated on the degradation of organic pollutants, and some of the applications are summarized on Table 6.7. WO_3 and WO_3/TiO_2 were synthesized by the combustion synthesis method, and their photocatalytic activity was tested for the removal of orange G and methylene blue under UV and visible radiation. These two dyes were degraded in both UV and visible light conditions. A higher photocatalytic activity was observed for the WO_3/TiO_2 than for WO_3 nanoparticles for the degradation of OG and MB. A crucial advantage of WO_3/TiO_2 is that its lower optical band gap (2.5–2.8 eV) resulted in a much greater utilization of the solar spectrum for solar photocatalytic applications (Singh and Madras 2013).

In a different approach, WO_3/TiO_2 was synthesized via the sol–gel method and evaluated under solar irradiation for malathion degradation. The 2 wt % WO_3/TiO_2 material exhibited the best photocatalytic activity, achieving 76 % of TOC abatement after 300 min compared to bare TiO_2 photocatalyst, which achieved 47 % mineralization. The 2 wt % WO_3/TiO_2 catalyst exhibited the formation of smaller clusters and a higher surface area, which reduced the recombination process and resulted in better contact area between the catalyst particles and the pollutant. Additionally, this semiconductor material showed good stability against photocorrosion, and it could be recycled several times without significant loss of its activity (Fig. 6.7) (Ramos-Delgado et al. 2013).

Sensitizing ZnO with MoO_3 has also shown positive results for visible light activity in ZnO (Lam et al. 2013). The 2 % MoO_3 nanoparticles incorporated with ZnO nanorods (NRs) by hydrothermal precipitation method showed a superior

Table 6.7 Coupling two semiconductor systems for the photocatalytic degradation of organic pollutants under UV or visible light irradiation

Catalyst	Pollutant	Experimental	Results	Reference
Dyes				
WO_3 and TiO_2/WO_3 (15 wt %) synthesized by the combustion synthesis method	Orange G and methylene blue (50 mg L^{-1})	UV lamp ($\lambda = 365$ nm, 125 W), temperature = 29 °C, 120 min irradiation time. Jacketed quartz tube reactor (100 mL) and catalyst loading: 1 g L^{-1}	Degradation of orange G dye with WO_3 and TiO_2/WO_3 was 80 and 95 %, respectively. While degradation of methylene blue dye was 60 and 95 % with WO_3 and TiO_2/WO_3, respectively	Singh and Madras (2013)
Pesticides				
TiO_2–CeO_2 mixed oxides were synthesized by sol–gel method and calcined at 873 K. Crystallite size: 8.9–43.6 nm	2,4-Dichlorophenoxyacetic acid (30 mg L^{-1})	UV lamp Pen-Ray (UVP, $\lambda = 254$ nm, 2,000 μW/cm^2). Glass photoreactor (250 mL), 90 min irradiation time at room temperature with catalyst loading: 0.25 g L^{-1}	94.5 % degradation percentage compared to 53 % degradation percentage using undoped TiO_2 sample	Galindo-Hernández and Gómez (2011)
TiO_2–WO_3 (2 wt %) synthesized by the sol–gel method. The particle size of the materials was 19.4 ± 3.3 nm. WO_3 was present as a monoclinic crystalline phase with nanometric cluster sizes (1.1 ± 0.1 nm)	Malathion (12 mg L^{-1})	Reactor contained 250 mL of pesticide solution at pH 7. The solar photocatalytic experiments were done on sunny days (≈1,030 W m^{-2}) between 11:00 and 16:00 h. Catalyst loading: 1 g L^{-1}	Mineralization was 76 % in 300 min compared to the 47 % achieved with the bare TiO_2. The photocatalyst showed good stability against the photocorrosion, and it could be recycled four times	Ramos-Delgado et al. (2013)

MoO_3 (2 wt %) particles loaded on ZnO nanorods (ZnO NRs) were synthesized via hydrothermal precipitation method. The pure orthorhombic phase of α-MoO_3 nanoparticles with average sizes of 15–21 nm was well dispersed on the surface of ZnO NRs	2,4-Dichlorophenolxyacetic acid (20 mg L^{-1})	Compact fluorescent lamp (55 W; light intensity, 14,500 lx), glass cell (100 mL), 120 min irradiation time, pH 7, room temperature, catalyst loading: 1 g L^{-1}	MoO_3/ZnO NRs showed superior visible light degradation rate for herbicide (0.0452 min^{-1}) compared with the pure ZnO NRs (0.0114 min^{-1})	Lam et al. (2013)
Emerging pollutants				
Commercial anatase (9.9 $m^2\ g^{-1}$), TiO_2 (9.5 $m^2\ g^{-1}$), 9 % TiO_2–SiO_2 (234.8 $m^2\ g^{-1}$), 28 % TiO_2–SiO_2 (259.2 $m^2\ g^{-1}$), and 51 % TiO_2–SiO_2 (274.0 $m^2\ g^{-1}$) synthesized via inverse microemulsion. Crystallite size: 13 and 4 nm, for TiO_2 and SiO_2, respectively	Tetracycline (2–200 mg L^{-1})	UV 131000 lamp ($\lambda = 366$ nm with light intensity, 2.7×10^{-6} mol $photons^{-1}$), pH 4.4–9.5, 250 min irradiation time, 25 °C, glass photoreactor (8 mL), catalyst loading: 1 g L^{-1}	First-order rate constant (min^{-1}): Anatase = 16.5×10^{-3} $TiO_2 = 9.52 \times 10^{-3}$ 9 % TiO_2–$SiO_2 = 0.87 \times 10^{-3}$ 28 % TiO_2–$SiO_2 = 5.60 \times 10^{-3}$ 51 % TiO_2–$SiO_2 = 5.08 \times 10^{-3}$	Brigante and Schulz (2011)
Core–shell nanostructures: SiO_2–TiO_2 (14.6 $m^2\ g^{-1}$), SiO_2–Au (0.05 %)–TiO_2 (12.0 $m^2\ g^{-1}$), SiO_2–Au (0.1 %)–TiO_2 (10.2 $m^2\ g^{-1}$), SiO_2–Au (0.5 %)–TiO_2 (8.9 $m^2\ g^{-1}$), SiO_2–Au (1.0 %)–TiO_2 (9.4 $m^2\ g^{-1}$), and SiO_2–Fe_3O_4–Au–TiO_2 synthesized via classic Stöber method	Naproxen (2.3 mg L^{-1})	Xenon lamp with a cutoff filter ($\lambda < 420$ nm, 250 W), pH no reported, 6 h irradiation time. Batch photoreactor (50 mL) at room temperature with catalyst loading: 1 g L^{-1}	Degradation percent: SiO_2–Au (0.05 %)–TiO_2 = 55 % SiO_2–Au (0.1 %)–TiO_2 = 75 % SiO_2–Au (0.5 %)–TiO_2 = 44 % SiO_2–Fe_3O_4–Au–TiO_2 = 42 % TiO_2 P25 = 25 %	Ye et al. (2013)

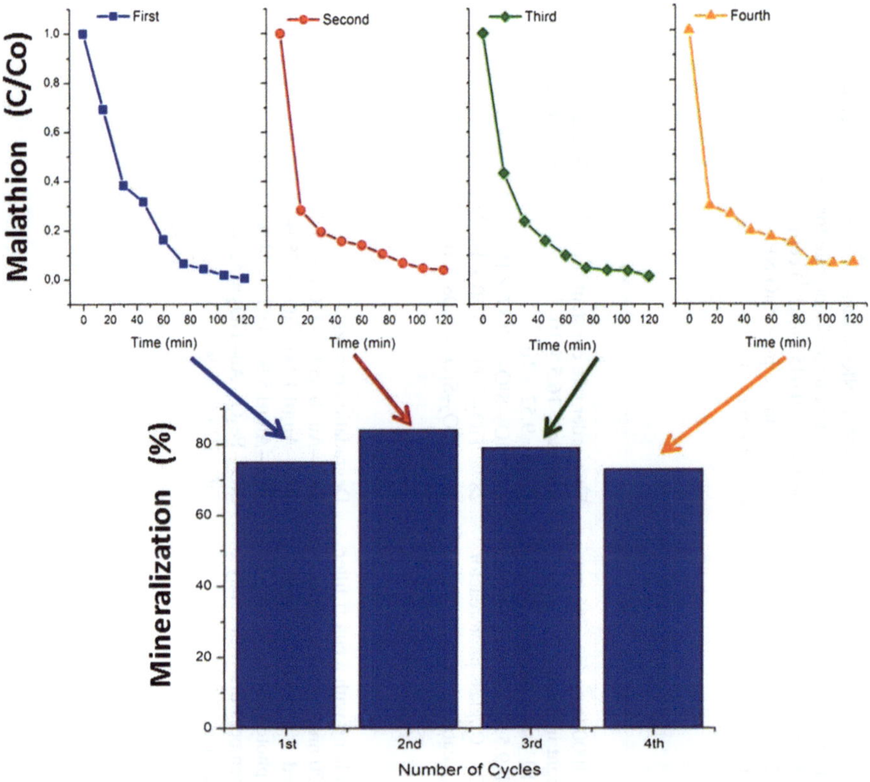

Fig. 6.7 Results of the reuse experiments of 2 wt % catalyst under solar radiation. (250 mL of malathion $C_0 = 12\ \mathrm{mg\ L^{-1}}$, 2 wt % = 1 g $\mathrm{L^{-1}}$, pH = 7). (**a**) Degradation of malathion pesticide and (**b**) mineralization efficiency (Ramos-Delgado et al. 2013)

visible light degradation rate for the herbicide 2,4-dichlorophenoxyacetic acid compared with the pure ZnO NRs due to the high quantum yield based on the cooperative roles between the bicomponents of MoO_3 and ZnO NRs (see Fig. 6.8). The catalytic efficiency of the composite was still higher than 90 % after being used for four cycles. Additionally, the authors confirmed the stability of the MoO_3/ZnO NRs composites after long irradiation time by XRD and FTIR spectroscopy.

The mixed oxide catalyst TiO_2–CeO synthesized by sol–gel method and prepared at three annealing temperatures (473, 673, and 873 K), showing E_g values in the range 2.3–2.5 eV, was evaluated under UV radiation for the oxidation of 2,4-D. The authors described that when the semiconductor strongly decreased the band gap (annealing temperature of 473 K), a fast electron–hole pair formation occurs. In addition, a fast electron–hole recombination was produced at the same time, and as a consequence, the photodegradation rate is diminished. A higher degradation percent was observed for TiO_2–CeO prepared at the annealing temperature of

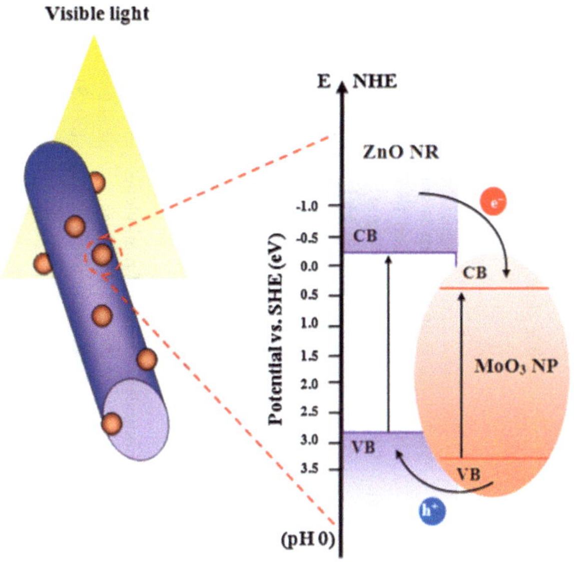

Fig. 6.8 The schematic diagram illustrating the energy band structure and occurrence of vectorial electron and hole transfer in the MoO_3/ZnO NR composites (Lam et al. 2013)

873 K, showing 94.5 % compared to the 53 % degradation using undoped TiO_2 sample (Galindo-Hernández and Gómez 2011).

Silicon oxide (SiO_2) has been evaluated as modifier to increase the surface area and improve the activity of the TiO_2 catalyst (Brigante and Schulz 2011; Ye et al. 2013). In the photocatalytic degradation of tetracycline, SiO_2 increased the adsorption of the SiO_2–TiO_2 catalyst (9, 28, and 51 wt % SiO_2), although the photocatalytic process was not favored (Brigante and Schulz 2011). Another modification in SiO_2–TiO_2 mixed oxide was proposed by Ye et al. (2013). Those authors prepared SiO_2–Au–TiO_2 core–shell nanostructures. This composite significantly increased the degradation of naproxen under visible light, and its activity was almost three times better than TiO_2 P25 when SiO_2–Au (with a Au content of 0.1 wt %)–TiO_2 was used (Ye et al. 2013).

6.6 Photocatalytic Degradation of Organic Pollutants in the Presence of Other Semiconductors

Widespread use of TiO_2 is uneconomical for large-scale water treatment; therefore, interest has been drawn toward the search for suitable alternatives to TiO_2. Table 6.8 summarizes some examples of alternative catalysts used in the removal of organic pollutants.

The photocatalytic activity of tin sulfide (SnS_2) and cadmium sulfide (CdS) nanostructures was investigated for the decolorization of methyl orange, congo red, orange II, rhodamine B, malachite green, and methylene blue under visible

Table 6.8 Different semiconductor systems for the photocatalytic degradation of organic pollutants under UV or visible light irradiation

Catalyst	Pollutant	Experimental	Results	Reference
Dyes				
SnS_2 photocatalyst prepared by hydrothermal method and CdS was commercially available	Methyl orange (MO), Congo red (CR), orange II (OII), rhodamine B (RhB), malachite green (MG), and methylene blue (MB) (20 mg L^{-1})	Visible irradiating with xenon lamp ($\lambda > 420$ nm, 300 W), 30 °C, 120 min irradiation time, and quartz reactor (50 mL), catalyst loading: 0.2 g L^{-1}	The degradation of the dyes with SnS_2 and CdS: MO 99 and 16 %, CR 90 and 84 %, OII 44 and 26 %, RhB 10 and 31 %, MG 15 and 95 %, MB 30 and 51 %, respectively	Li et al. (2012a)
SnO_2 synthesized by microwave solvothermal method, in different times: 30, 60, and 90 min	Rhodamine B (5 mg L^{-1})	UV medium-pressure Hg lamp (300 W), 25 °C, 180 min irradiation time, glass photoreactor (250 mL), catalyst loading: 0.25 g L^{-1}	Photocatalytic degradation was 88, 95, and 100 % for 90, 30, and 60 min synthesized time, respectively	He and Zhou (2013)
MoS_2 and WS_2 synthesized by thermal decomposition	Methylene blue (6 mg L^{-1})	Light source was a 300 W ozone-free xenon arc bulb ($\lambda > 400$ nm), 9 h irradiation time, glass photoreactor (200 mL), catalyst loading: 0.15 g L^{-1}	Photodegradation of methylene blue by MoS_2 and WS_2 was 35 and 60 %, respectively	James and Zubkov (2013)
Fe_2O_3 nanoparticles	Congo red (25 mg L^{-1})	Natural sunlight irradiation (3,000 lx h^{-1}), glass cell (100 mL), 4 h irradiation time, pH 4, room temperature, catalyst loading: 1 mg L^{-1}	Maximum 98.02 % photocatalytic degradation was observed for congo red under natural sunlight irradiation	Bharadwaj et al. (2012)
Tetragonal starlike ZrO_2 nanostructures prepared by hydrothermal process	Methyl orange, congo red, and rhodamine B (10 mg L^{-1})	UV high-pressure mercury lamp (300 W), pH 9, 120 min irradiation time, glass photoreactor (60 mL), catalyst loading: 0.6 mg L^{-1}	For methyl orange, congo red, and rhodamine B, the degradation of the dyes was nearly 100 % in 60 min	Shu et al. (2013)

Pesticides				
WO_3 powder. Crystallite size: 379 nm	Monuron (0.025 mM)	Low-pressure mercury lamp at 350 nm, 23 °C, pH 6, 3 h irradiation time, quartz cylinder photoreactor (150 mL), catalyst loading: 0.3 mg L^{-1}	The degradation of monuron was 90 % in 3 h under UV light radiation	Chu and Rao (2012)
Emerging contaminants				
β-Bi_2O_3 nanospheres prepared via solvothermal–calcining route	Acetaminophen (10 mg L^{-1})	Xe lamp combined with a 420 nm cutoff filter (1,000 W), 180 min irradiation time, room temperature, glass photoreactor (50 mL), catalyst loading: 0.5 g L^{-1}	Apparent rate constant (min^{-1}): Bi_2O_3 (NS) = 1.387×10^{-2} Synthetic $Bi_2O_3 = 5.43 \times 10^{-3}$ TiO_2 P25 = 1.751×10^{-4}	Xiao et al. (2013a)
Sr–Bi_2O_3 prepared via solvothermal synthesis	Tetracycline (20 mg L^{-1})	Xenon lamp with a cutoff filter ($\lambda > 420$ nm), pH 7, 120 min irradiation time, 20 °C, glass photoreactor (100 mL), catalyst loading: 0.6 g L^{-1}	Degradation percent: Sr–Bi_2O_3 = 91 % TiO_2 P25 = 100 % Toxicity reduction: Sr–Bi_2O_3 = 91.2 % TiO_2 P25 = 80 %	Niu et al. (2013)
$Bi_{24}O_{31}Br_{10}$ nanoflakes, prepared via microwave heating route BiOBr	Tetracycline (40 mg L^{-1})	Xe lamp combined with a 420 nm cutoff filter (1,000 W), pH 9–11, 60 min irradiation time, room temperature, glass photoreactor (50 mL), catalyst loading: 1.5 g L^{-1}	Degradation percent: $Bi_{24}O_{31}Br_{10} > 90$ % BiOBr = 73 % Mineralization: $Bi_{24}O_{31}Br_{10} = 71$ %	Xiao et al. (2013b)
Bi_2MoO_6 prepared by hydrothermal process	Ibuprofen (80 mg L^{-1})	Blue light-emitting diode (LED) ($\lambda = 465$ nm, 3 W), 120 min irradiation time, room temperature, glass photoreactor (50 mL), catalyst loading: 0.5 g L^{-1}	Degradation percent: Bi_2MoO_6 = 70 % (2 h) Mineralization = 76.5 % (6 h)	Zhang et al. (2012)

(continued)

Table 6.8 (continued)

Catalyst	Pollutant	Experimental	Results	Reference
$FeCu/Cu_2O$ Fe	Sulfamethoxazole, oxytetracycline, paracetamol, aspirin, triclosan (5 mg L^{-1} each drug)	Metal halide lamp ($\lambda = 590$ nm), pH 4.9, 4 h irradiation time, catalyst loading: 30 g L^{-1}	Mineralization percentage: $FeCu/Cu_2O = 89$ % Fe = 73 %	An and Zhou (2012)
Commercial Ag_3PO_4 $N–TiO_2$	Bisphenol (10 mg L^{-1})	Xe lamp ($\lambda > 420$ nm, 990 W), pH no reported, 10 min irradiation time, glass photoreactor (30 mL), catalyst loading: 0.5 g L^{-1}	Degradation percent: $Ag_3PO_4 = 100$ % $N–TiO_2 = 13$ % Mineralization in 180 min: $Ag_3PO_4 = 83$ % $N–TiO_2 = 32$ %	Katsumata et al. (2013))
BiOBr ($E_g = 2.61$ eV) prepared by solvothermal method TiO_2 P25	Tetrabromobisphenol (1 mg L^{-1})	Simulated sunlight irradiation (800 W xenon lamp), cylindrical quartz cold trap (200 mL), 60 min irradiation time, catalyst loading: 1 g L^{-1}	Apparent rate constant (min^{-1}): BiOBr = 0.388 TiO_2 P25 = 0.101	Xu et al. (2011)
$Ag/Bi_5Nb_3O_{15}$ (10 %) $Pt/Bi_5Nb_3O_{15}$ (1 %) TiO_2 P25 $Ag/Bi_5Nb_3O_{15}$ (1 %) $Bi_5Nb_3O_{15}$ Prepared by a hydrothermal method combined with photodeposition	Tetrabromobisphenol (40 mg L^{-1})	Solar simulating Xe lamp irradiation (320–680 nm, 150 mW cm^{-2}), pH 9, glass photoreactor (100 mL), catalyst loading: 0.15 g L^{-1}	Degradation percent: $Ag/Bi_5Nb_3O_{15}$ (10 %) = 95 % $Pt/Bi_5Nb_3O_{15}$ (1 %) = 94 % TiO_2 P25 = 67 % $Ag/Bi_5Nb_3O_{15}$ (1 %) = 64 % $Bi_5Nb_3O_{15} = 52$ %	Guo et al. (2011)
In_2O_3 nanoporous nanospheres (NPNSs). Synthesized by solvothermal method. Crystallite size: 100 nm. Commercial: In_2O_3 nanocrystals (NCs)	Perfluorooctanoic acid (30 mg L^{-1})	Low-pressure mercury lamp ($\lambda = 254$ nm, 23 W), pH 3.9, tubular quartz vessel reactor (100 mL), 180 min irradiation time, catalyst loading: 0.05 g L^{-1}	Pseudo first-order constant (h^{-1}): In_2O_3 (NPNSs) = 5.89 In_2O_3 (NCs) = 0.66 TiO_2 P25 = 0.11	Li et al. (2012b)

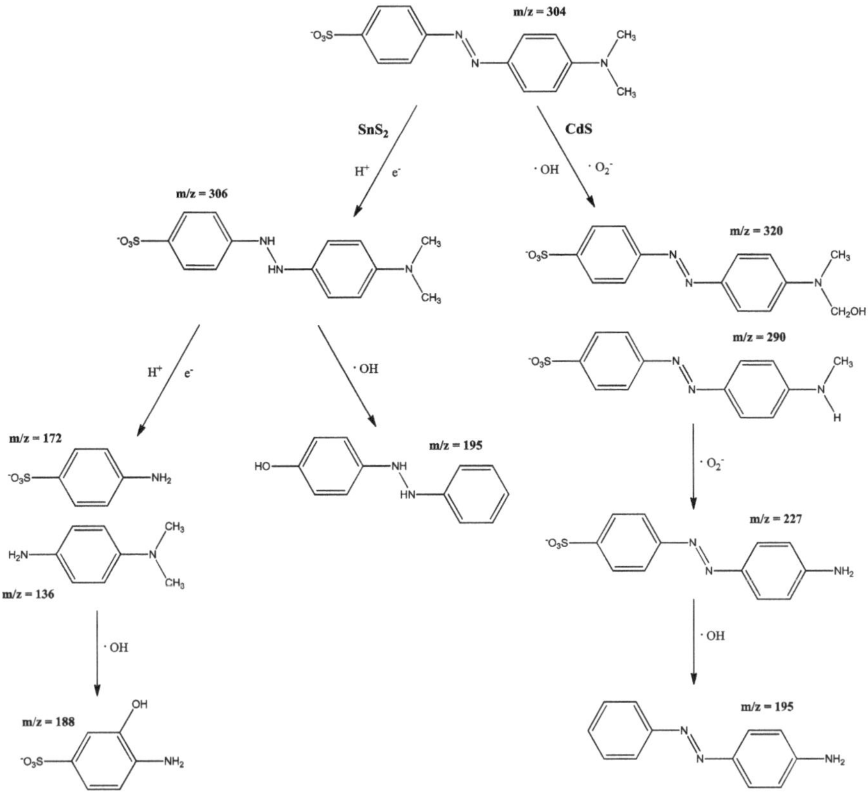

Fig. 6.9 The plausible reaction mechanisms of the methyl orange degradation on SnS_2 and CdS photocatalysts (Li et al. 2012a)

light irradiation from a xenon lamp. SnS_2 showed better results compared to CdS for the photocatalytic degradation of organic dyes containing an N=N double bond, following a reduction mechanism with photoelectrons via the Sn(IV)/Sn (II) transition, whereas the photodegradation of organic dyes without an N=N double bond on either the SnS_2 or the CdS materials followed an oxidation mechanism with $^{\bullet}O^{2-}$ and $^{\bullet}OH$ radicals. The SnS_2 exhibited much higher activity than the CdS. The reaction mechanisms of the methyl orange degradation on SnS_2 and CdS photocatalysts (Li et al. 2012a) are shown in Fig. 6.9. He and Zhou synthesized SnO_2 by a microwave solvothermal method. The photocatalytic activity of SnO_2 was evaluated through the degradation of rhodamine B (RhB) under UV irradiation. The higher activity achieved could be attributed to UV absorption, and there was a clear red shift for this material (He and Zhou 2013).

Binary metal sulfide semiconductors, such as CdS and PbS, are regarded as insufficiently stable for catalysis and are toxic. Other materials considered in recent study include MoS_2 and WS_2, which were synthesized by a thermal decomposition method. Some of these materials were reported to show some photocatalytic promise. The silica- and titania-supported MoS_2 and WS_2 nanoparticles catalyzed the photodegradation of methylene blue under visible radiation but with uncertainty over their efficiencies and spectral characteristics (James and Zubkov 2013).

Shu et al. (2013) prepared tetragonal starlike ZrO_2 nanostructures via a hydrothermal process, and the photocatalytic activity of the ZrO_2 product was evaluated for several dyes under UV irradiation. Different photodegradation activity for anionic dyes and cationic dyes was observed. For anionic dyes, such as methyl orange, congo red, and acid rhodamine B, the degradation of the dyes was up to 95 % in 40 min and nearly 100 % in 60 min. At the same photodegradation conditions, for cationic dyes, such as methyl violet, crystal violet, and rhodamine B, the degradations were 65.7, 64.0, and 86.0 %, respectively, after irradiation for 40 min. For methyl violet and crystal violet, more than 5 % of the dyes still remained even after irradiation for 120 min. The tetragonal starlike structures display excellent adsorption and photocatalytic performance for anionic dyes. However, under strong alkaline condition, the acid sites tend to adsorb a large amount of hydroxyls, and the activity sites of the catalyst are destroyed, leading to the decrease of the adsorption and photocatalytic properties for both anionic and cationic dyes.

Fe_2O_3 has been shown to be an interesting semiconductor because of its stability against photo/chemical corrosion at neutral or basic pH and a band gap energy of approximately 2.0–2.2 eV, corresponding to the absorption of 564–620 nm visible light. This material was used for the photodegradation of congo red in tap water. Agar well diffusion assay and seed germination tests were performed to assess risk upon exposure of treated and untreated CR samples to environment, and it has found satisfactory results of Fe_2O_3-mediated photocatalysis, but water still needs secondary treatment before being used for potable and irrigation purposes (Bharadwaj et al. 2012).

$FeCu/Cu_2O$ also exhibited photochemical properties to degrade pharmaceutical drugs (sulfamethoxazole, oxytetracycline, paracetamol, aspirin, and triclosan) because Cu_2O is a p-type semiconductor (2.0 eV), which can be photoactivated under visible light. This semiconductor had the following advantages: low cost, readily available material, and low toxicity (An and Zhou 2012).

Another study evaluated commercial WO_3 for the degradation of monuron, a phenylurea herbicide, under UV radiation (350 nm) (Chu and Rao 2012). Although this material is a visible light-responsive photocatalyst that absorbs light up to 480 nm, it showed low photoactivity to degrade organic pollutants due to a high recombination rate of the photogenerated electron–hole pairs (Di Paola et al. 2012). A degradation percent of 90 % was achieved for monuron with an initial concentration of 0.025 mM in 3 h under UV radiation, and the authors described that hydroxyl radicals played a major role in the decay of monuron. Although the TOC removal was insignificant, the results of the detected intermediates and the evolution of chloride, ammonium, and nitrate suggest that most aromatic compounds were ring-opened into simple aliphatic acids (Chu and Rao 2012).

Other metal oxides have been tested for the photocatalytic removal of organic pollutants. The decomposition of perfluorooctanoic acid was evaluated using

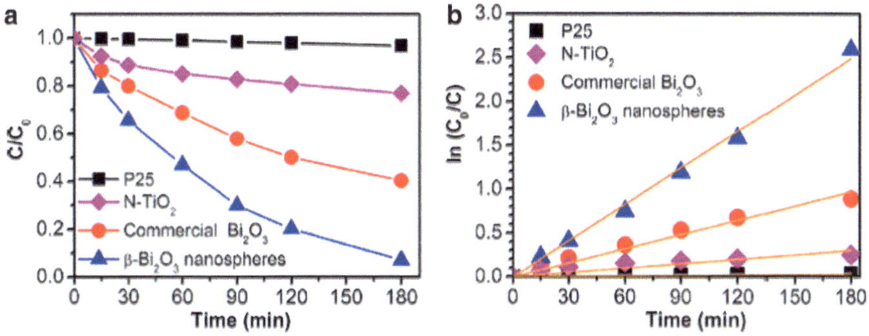

Fig. 6.10 (**a**) Photocatalytic degradation kinetics of acetaminophen over β-Bi_2O_3 nanospheres, a commercial Bi_2O_3, N-doped TiO_2, and Degussa P25 under visible light irradiation; (**b**) linear plots of ln (C_0/C_t) vs. degradation time (Xiao et al. 2013a)

synthesized In_2O_3 nanoporous nanospheres (NPNSs) by the solvothermal method as semiconductor. By means of this material, the degradation rate constant was 8 and 53 times higher than using commercial In_2O_3 and TiO_2 P25, respectively. The highest photocatalytic activity of In_2O_3 NPNSs was attributed to its high specific surface area, large number of nanopores, and the direct reaction between photogenerated holes and the pollutant (Li et al. 2012b).

Recently, bismuth oxide has been considered a very attractive photocatalytic material for the removal of organic pollutants because it can be activated under visible light (Di Paola et al. 2012). Nanospheres of Bi_2O_3 synthesized by a solvothermal-calcination process were evaluated for the degradation of acetaminophen, and the apparent kinetic constant rate obtained was approximately 79 times higher than that of Degussa TiO_2, 8 times than N–TiO_2, and 7 times than commercial Bi_2O_3 (Fig. 6.10) (Xiao et al. 2013a).

Modified bismuth oxide, such as Sr–Bi_2O_3 ($E_g = 2.21$ eV) prepared via solvothermal synthesis, was evaluated for the degradation of tetracycline under visible light. The removal ratio reached 91.2 % after 120 min of irradiation. The toxicity of photodegradation products after 120 min was evaluated by bioluminescence inhibition assay with *P. phosphoreum*. Toxicity after Sr–Bi_2O_3 photocatalysis decreased by 90.6 %, which was more substantial than with TiO_2 (80 %). Intermediate products determined by HPLC-MS indicated that the photocatalytic process using Sr–Bi_2O_3 was able to break the naphthol ring of tetracycline, which decomposes to *m*-cresol (Fig. 6.11) (Niu et al. 2013).

Another promising photocatalyst for the degradation of emerging pollutants, such as tetracycline and tetrabromobisphenol (TBBPA), a flame retardant, has been bismuth oxyhalides (Xiao et al. 2013b; Xu et al. 2011). These semiconductor materials have demonstrated notable photoactivity due to their layered structures with an internal static electric field perpendicular to each layer, which promotes the separation of photogenerated electron–holes. The use of BiOBr ($E_g = 2.88$ eV) and $Bi_{24}O_{31}Br_{10}$ ($E_g = 2.51$ eV) was applied on tetracycline degradation under visible radiation. Better results were obtained with $Bi_{24}O_{31}Br_{10}$, and this was attributed to

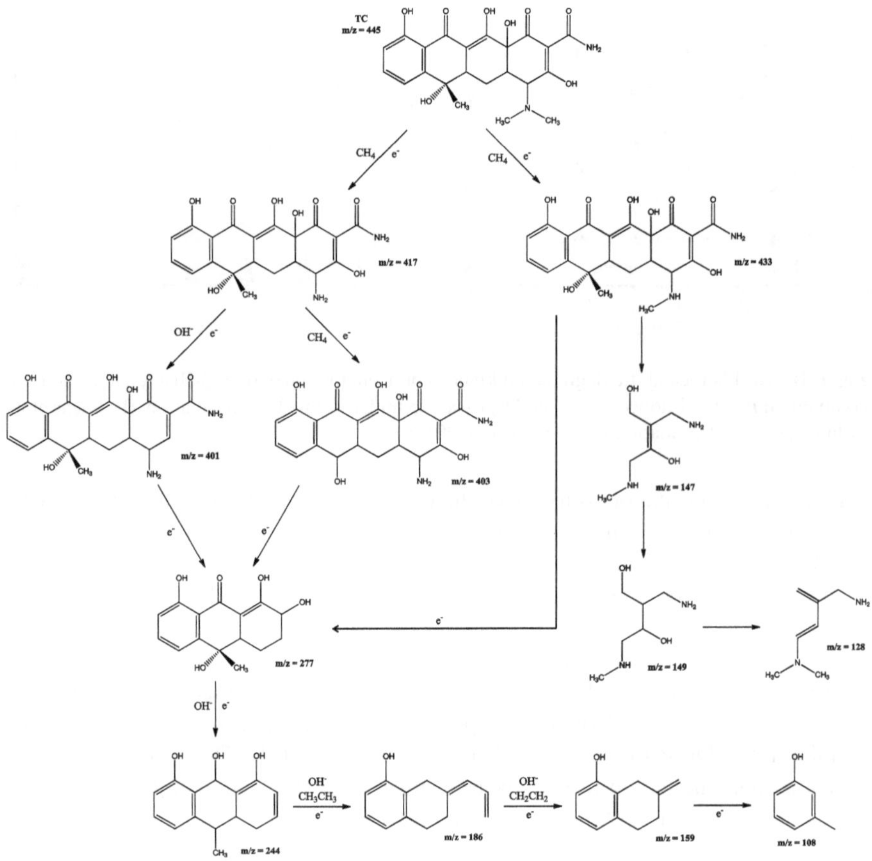

Fig. 6.11 Schematic diagram of the proposed photochemical degradation mechanisms of tetracycline by Sr–Bi_2O_3 photocatalysis. HPLC-MS was used to determine the various intermediate products (figure adapted from Niu et al. 2013)

its lower band gap and its higher surface area. The results suggested that oxidization is conducted mainly by direct photo-holes rather than hydroxyl radicals (Xiao et al. 2013b).

The photocatalytic degradation of TBBPA has been conducted using BiOBr, which presented high stability and photocatalytic performance and a short band gap, making it a suitable alternative under visible radiation (Xu et al. 2011). Another study related to the photocatalytic removal of TBBPA removal was performed using Ag/$Bi_5Nb_3O_{15}$ as catalyst. It showed better photoresponse in both UV and visible light region than $Bi_5Nb_3O_{15}$ because Ag acts as electron sink, improving the photocatalytic activity (Guo et al. 2011). In addition to $Bi_5Nb_3O_{15}$-based catalysts, other nanomaterials, such as Bi_2MoO_6, have been evaluated for the removal of pharmaceuticals under visible light by Zhang et al. (2012). The authors found that this system was effective to degrade and mineralize ibuprofen under visible radiation, reaching 70 % degradation in 2 h, and the TOC value decreased steadily reaching a value of 76.5 % after 6 h of irradiation (Zhang et al. 2012).

The applicability of the photocatalytic degradation process for the removal of bisphenol compounds has also been described by Katsumata et al. using Ag_3PO_4 as catalyst. Although Ag_3PO_4 showed low stability when recycled, the results indicated that this material presented higher catalytic activity under visible radiation to degrade and mineralize bisphenol. Additional assays conducted with some scavenger species demonstrated that the main mechanism of Ag_3PO_4 was through photogenerated holes, which react directly with organic compounds (Katsumata et al. 2013).

Concluding Remarks

Photocatalytic degradation of organic compounds is a promising technology due to the advantage of chemically transforming compounds under ambient conditions to allow total mineralization of many organic pollutants. Much of the current literature concerning the photocatalytic degradation of emerging contaminants uses undoped TiO_2 materials. The use of modified TiO_2 materials has enhanced the photodegradation of various organic contaminants, such as dyes and pesticides, under visible and solar light radiation, and the number of publications has increased exponentially in the last 5 years. However, photocatalytic processes using doped materials should be refocused to consider the following changes: (a) most of the reviewed articles described the photocatalytic applications of powdered catalysts, which lack a complete recovery of catalyst after application. Thus, new technologies are needed for complete recovery of the catalyst without any loss of activity. (b) Although a significant amount of research has been conducted on TiO_2 photocatalytic processes at laboratory scale, its application on the industrial scale has certain limitations. (c) Further research on the effect of variables is required to investigate the degradation of organic compounds in real water matrices to achieve representative results and to better understand the applicability of the process. Although the process is capable of removing a wide range of organic pollutants, the photocatalytic degradation studies are limited to some dyes, pesticides, and emerging compounds. To demonstrate the feasibility of the photocatalytic process to degrade organic pollutants from wastewater is crucial to evaluate their effectiveness after the repeated catalytic cycles and the stability of the prepared semiconductors. Moreover, in most of the reviewed articles related to doped catalysts, the efficiency of the process is limited only to the degradation percentage; however, information related to the degradation pathway and the nature of any intermediate products, the mineralization degree, and the evolution of toxicity and biodegradability of the treated samples is scarce.

References

Ahmed S, Rasul MG, Martens WN, Brown R, Hashib MA (2010) Advances in heterogeneous photocatalytic degradation of phenols and dyes in wastewater: a review. Water Air Soil Pollut 215(1–4):3–29. doi:10.1007/s11270-010-0456-3

Ahmed S, Rasul MG, Brown R, Hashib MA (2011) Influence of parameters on the heterogeneous photocatalytic degradation of pesticides and phenolic contaminants in wastewater: a short review. J Environ Manage 92(3):311–330. doi:10.1016/j.jenvman.2010.08.028

Akpan UG, Hameed BH (2009) Parameters affecting the photocatalytic degradation of dyes using TiO_2-based photocatalysts: a review. J Hazard Mater 170(2–3):520–529. doi:10.1016/j.jhazmat.2009.05.039

Ali MA, Idris MR, Quayum ME (2013) Fabrication of ZnO nanoparticles by solution-combustion method for the photocatalytic degradation of organic dye. J Nanostruct Chem 3(36):1–6. doi:10.1186/2193-8865-3-36

Amala Infant Joice J, Sivakumar T, Ramakrishnan R, Ramya G, Shanmuga Prasad KP, Arul Selvan D (2012) Visible active metal decorated titania catalysts for the photocatalytic degradation of Amidoblack-10B. Chem Eng J 210:385–397. doi:10.1016/j.cej.2012.08.103

An J, Zhou Q (2012) Degradation of some typical pharmaceuticals and personal care products with copper-plating iron doped Cu_2O under visible light irradiation. J Environ Sci 24(5):827–833. doi:10.1016/s1001-0742(11)60847-4

Aragay G, Pino F, Merkoci A (2012) Nanomaterials for sensing and destroying pesticides. Chem Rev 112(10):5317–5338. doi:10.1021/cr300020c

Avasarala BK, Tirukkovalluri SR, Bojja S (2011) Photocatalytic degradation of monocrotophos pesticide—an endocrine disruptor by magnesium doped titania. J Hazard Mater 186(2–3):1234–1240. doi:10.1016/j.jhazmat.2010.11.132

Avisar D, Horovitz I, Lozzi L, Ruggieri F, Baker M, Abel M-L, Mamane H (2013) Impact of water quality on removal of carbamazepine in natural waters by N-doped TiO_2 photo-catalytic thin film surfaces. J Hazard Mater 244–245:463–471. doi:10.1016/j.jhazmat.2012.09.058

Bernabeu A, Vercher RF, Santos-Juanes L, Simón PJ, Lardín C, Martínez MA, Vicente JA, González R, Llosá C, Arques A, Amat AM (2011) Solar photocatalysis as a tertiary treatment to remove emerging pollutants from wastewater treatment plant effluents. Catal Today 161 (1):235–240. doi:10.1016/j.cattod.2010.09.025

Bharadwaj L, Bhardwaj M, Sharma MK (2012) An analysis of Fe_2O_3 assisted photocatalytic degradation of Congo red dye. Toxicol Environ Health Sci 4(1):62–69. doi:10.1007/s13530-011-0102-5

Bin Mukhlish MZ, Najnin F, Rahman MM, Uddin MJ (2013) Photocatalytic degradation of different dyes using TiO_2 with high surface area: a kinetic study. J Sci Res 5(2):301–314. doi: 10.3329/jsr.v5i2.11641

Brigante M, Schulz PC (2011) Remotion of the antibiotic tetracycline by titania and titania–silica composed materials. J Hazard Mater 192(3):1597–1608. doi:10.1016/j.jhazmat.2011.06.082

Chatzitakis A, Berberidou C, Paspaltsis I, Kyriakou G, Sklaviadis T, Poulios I (2008) Photocatalytic degradation and drug activity reduction of chloramphenicol. Water Res 42(1–2):386–394. doi:10.1016/j.watres.2007.07.030

Chu W, Rao YF (2012) Photocatalytic oxidation of monuron in the suspension of WO_3 under the irradiation of UV-visible light. Chemosphere 86(11):1079–1086. doi:10.1016/j.chemosphere.2011.11.062

Chusaksri S, Lomda J, Saleepochn T, Sutthivaiyakit P (2011) Photocatalytic degradation of 3,4-dichlorophenylurea in aqueous gold nanoparticles-modified titanium dioxide suspension under simulated solar light. J Hazard Mater 190(1–3):930–937. doi:10.1016/j.jhazmat.2011.04.028

De la Cruz N, Dantas RF, Giménez J, Esplugas S (2013) Photolysis and TiO_2 photocatalysis of the pharmaceutical propranolol: solar and artificial light. Appl Catal Environ 130–131:249–256. doi:10.1016/j.apcatb.2012.10.003

Deblonde T, Cossu-Leguille C, Hartemann P (2011) Emerging pollutants in wastewater: a review of the literature. Int J Hyg Environ Health 214(6):442–448. doi:10.1016/j.ijheh.2011.08.002

Del Ángel-Sanchez K, Vázquez-Cuchillo O, Aguilar-Elguezabal A, Cruz-López A, Herrera-Gómez A (2013) Photocatalytic degradation of 2,4-dichlorophenoxyacetic acid under visible light: effect of synthesis route. Mater Chem Phys 139(2–3):423–430. doi:10.1016/j.matchemphys.2013.01.009

Di Paola A, García-López E, Marcì G, Palmisano L (2012) A survey of photocatalytic materials for environmental remediation. J Hazard Mater 211–212:3–29. doi:10.1016/j.jhazmat.2011.11.050

Dimitroula H, Daskalaki VM, Frontistis Z, Kondarides DI, Panagiotopoulou P, Xekoukoulotakis NP, Mantzavinos D (2012) Solar photocatalysis for the abatement of emerging micro-contaminants in wastewater: synthesis, characterization and testing of various TiO_2 samples. Appl Catal Environ 117–118:283–291. doi:10.1016/j.apcatb.2012.01.024

Donkova B, Dimitrov D, Kostadinov M, Mitkova E, Mehandjiev D (2010) Catalytic and photocatalytic activity of lightly doped catalysts M:ZnO (M = Cu, Mn). Mater Chem Phys 123(2–3):563–568. doi:10.1016/j.matchemphys.2010.05.015

Echavia GR, Matzusawa F, Negishi N (2009) Photocatalytic degradation of organophosphate and phosphonoglycine pesticides using TiO_2 immobilized on silica gel. Chemosphere 76(5):595–600. doi:10.1016/j.chemosphere.2009.04.055

Fenoll J, Hellín P, Martínez CM, Flores P, Navarro S (2012) Semiconductor-sensitized photodegradation of s-triazine and chloroacetanilide herbicides in leaching water using TiO_2 and ZnO as catalyst under natural sunlight. J Photochem Photobiol A Chem 238:81–87. doi:10.1016/j.jphotochem.2012.04.017

Galindo-Hernández F, Gómez R (2011) Degradation of the herbicide 2,4-dichlorophenoxyacetic acid over TiO_2–CeO_2 sol–gel photocatalysts: effect of the annealing temperature on the photoactivity. J Photochem Photobiol A Chem 217(2–3):383–388. doi:10.1016/j.jphotochem.2010.11.010

Giraldo AL, Peñuela GA, Torres-Palma RA, Pino NJ, Palominos RA, Mansilla HD (2010) Degradation of the antibiotic oxolinic acid by photocatalysis with TiO_2 in suspension. Water Res 44(18):5158–5167. doi:10.1016/j.watres.2010.05.011

Giwa A, Nkeonye P, Bello K, Kolawole E, Campos AO (2012) Solar photocatalytic degradation of reactive yellow 81 and reactive violet 1 in aqueous solution containing semiconductor oxides. Int J Appl Sci Technol 2(4):90

Gomathi Devi L, Narasimha Murthy B (2008) Structural characterization of Th-doped TiO_2 photocatalyst and its extension of response to solar light for photocatalytic oxidation of oryzalin pesticide: a comparative study. Cent Eur J Chem 7(1):118–129. doi:10.2478/s11532-008-0101-9

Gunasekar V, Divya B, Brinda K, Vijaykrishnan J, Ponnusami V, Rajan KS (2013) Enzyme mediated synthesis of Ag–TiO_2 photocatalyst for visible light degradation of reactive dye from aqueous solution. J Solgel Sci Technol 68(1):60–66. doi:10.1007/s10971-013-3134-2

Guo Y, Chen L, Ma F, Zhang S, Yang Y, Yuan X, Guo Y (2011) Efficient degradation of tetrabromobisphenol A by heterostructured Ag/$Bi_5Nb_3O_{15}$ material under the simulated sunlight irradiation. J Hazard Mater 189(1–2):614–618. doi:10.1016/j.jhazmat.2011.02.054

Gupta VK, Jain R, Mittal A, Mathur M, Sikarwar S (2007) Photochemical degradation of the hazardous dye safranin-T using TiO_2 catalyst. J Colloid Interface Sci 309(2):464–469. doi: http://dx.doi.org/10.1016/j.jcis.2006.12.010

Gurkan YY, Turkten N, Hatipoglu A, Cinar Z (2012) Photocatalytic degradation of cefazolin over N-doped TiO_2 under UV and sunlight irradiation: prediction of the reaction paths via conceptual DFT. Chem Eng J 184:113–124. doi:10.1016/j.cej.2012.01.011

He Z, Zhou J (2013) Synthesis, characterization, and activity of tin oxide nanoparticles: influence of solvothermal time on photocatalytic degradation of rhodamine B. Mod Res Catal 02(03):13–18. doi:10.4236/mrc.2013.23A003

Herrmann JM (2005) Heterogeneous photocatalysis: state of the art and present applications in honor of Pr. R.L. Burwell Jr. (1912–2003), Former Head of Ipatieff Laboratories, Northwestern University, Evanston (Ill). Top Catal 34(1–4):49–65. doi:10.1007/s11244-005-3788-2

Herrmann JM (2010) Photocatalysis fundamentals revisited to avoid several misconceptions. Appl Catal Environ 99(3–4):461–468. doi:10.1016/j.apcatb.2010.05.012

James D, Zubkov T (2013) Photocatalytic properties of free and oxide-supported MoS_2 and WS_2 nanoparticles synthesized without surfactants. J Photochem Photobiol A Chem 262:45–51. doi: http://dx.doi.org/10.1016/j.jphotochem.2013.04.015

Jiang J-Q, Zhou Z, Sharma VK (2013) Occurrence, transportation, monitoring and treatment of emerging micro-pollutants in waste water—a review from global views. Microchem J 110:292–300. doi:10.1016/j.microc.2013.04.014

Kang W, Spanjers CS, Rioux RM, Hoefelmeyer JD (2013) Synthesis of brookite TiO_2 nanorods with isolated Co(II) surface sites and photocatalytic degradation of 5,8-dihydroxy-1,4-naphthoquinone dye. J Mater Chem A Mater Energy Sustain 1(26):7717–7728. doi:10.1039/c3ta11038a

Kanmoni VGG, Daniel S, Raj GAG (2012) Photocatalytic degradation of chlorpyrifos in aqueous suspensions using nanocrystals of ZnO and TiO_2. React Kinet Mech Catal 106(2):325–339. doi:10.1007/s11144-012-0433-5

Katsumata H, Sada M, Nakaoka Y, Kaneco S, Suzuki T, Ohta K (2009) Photocatalytic degradation of diuron in aqueous solution by platinized TiO_2. J Hazard Mater 171(1–3):1081–1087. doi:10.1016/j.jhazmat.2009.06.110

Katsumata H, Taniguchi M, Kaneco S, Suzuki T (2013) Photocatalytic degradation of bisphenol A by Ag_3PO_4 under visible light. Catal Commun 34:30–34. doi:10.1016/j.catcom.2013.01.012

Khakpash N, Simchi A, Jafari T (2011) Adsorption and solar light activity of transition-metal doped TiO_2 nanoparticles as semiconductor photocatalyst. J Mater Sci Mater Electron 23 (3):659–667. doi:10.1007/s10854-011-0466-y

Khataee AR, Fathinia M, Joo SW (2013a) Simultaneous monitoring of photocatalysis of three pharmaceuticals by immobilized TiO_2 nanoparticles: chemometric assessment, intermediates identification and ecotoxicological evaluation. Spectrochim Acta A Mol Biomol Spectrosc 112:33–45. doi:10.1016/j.saa.2013.04.028

Khataee AR, Zarei M, Moradkhannejhad L, Nourie S, Vahid B (2013b) Nitrogen doping of commercial TiO_2 nanoparticles for enhanced photocatalytic degradation of dye under visible light: central composite design approach. Adv Chem Lett 1(1):24–31. doi:10.1166/acl.2013.1012

Khattab IA, Ghaly MY, Österlund L, Ali MEM, Farah JY, Zaher FM, Badawy MI (2012) Photocatalytic degradation of azo dye Reactive Red 15 over synthesized titanium and zinc oxides photocatalysts: a comparative study. Desalin Water Treat 48(1–3):120–129. doi:10.1080/19443994.2012.698803

Khin MM, Nair AS, Babu VJ, Murugan R, Ramakrishna S (2012) A review on nanomaterials for environmental remediation. Energy Environ Sci 5(8):8075. doi:10.1039/c2ee21818f

Klauson D, Babkina J, Stepanova K, Krichevskaya M, Preis S (2010) Aqueous photocatalytic oxidation of amoxicillin. Catal Today 151(1–2):39–45. doi:10.1016/j.cattod.2010.01.015

Konstantinou I (2003) Photocatalytic transformation of pesticides in aqueous titanium dioxide suspensions using artificial and solar light: intermediates and degradation pathways. Appl Catal Environ 42(4):319–335. doi:10.1016/s0926-3373(02)00266-7

Konstantinou IK, Albanis TA (2004) TiO_2-assisted photocatalytic degradation of azo dyes in aqueous solution: kinetic and mechanistic investigations: a review. Appl Catal B 49(1):1–14. doi:http://dx.doi.org/10.1016/j.apcatb.2003.11.010

Kumar SG, Devi LG (2011) Review on modified TiO_2 photocatalysis under UV/visible light: selected results and related mechanisms on interfacial charge carrier transfer dynamics. J Phys Chem A 115(46):13211–13241. doi:10.1021/jp204364a

Kuriechen SK, Murugesan S (2013) Carbon-doped titanium dioxide nanoparticles mediated photocatalytic degradation of azo dyes under visible light. Water Air Soil Pollut 224 (9):1671. doi:10.1007/s11270-013-1671-5

Laera G, Chong MN, Jin B, Lopez A (2011) An integrated MBR–TiO2 photocatalysis process for the removal of Carbamazepine from simulated pharmaceutical industrial effluent. Bioresour Technol 102(13):7012–7015. doi:10.1016/j.biortech.2011.04.056

Lam S-M, Sin J-C, Abdullah AZ, Mohamed AR (2013) Investigation on visible-light photocatalytic degradation of 2,4-dichlorophenoxyacetic acid in the presence of MoO_3/ZnO nanorod composites. J Mol Catal A Chem 370:123–131. doi:10.1016/j.molcata.2013.01.005

Li X, Zhu J, Li H (2012a) Comparative study on the mechanism in photocatalytic degradation of different-type organic dyes on SnS_2 and CdS. Appl Catal B 123–124:174–181. doi:http://dx.doi.org/10.1016/j.apcatb.2012.04.009

Li Z, Zhang P, Shao T, Li X (2012b) In_2O_3 nanoporous nanosphere: a highly efficient photocatalyst for decomposition of perfluorooctanoic acid. Appl Catal Environ 125:350–357. doi:10.1016/j.apcatb.2012.06.017

Lin Y, Ferronato C, Deng N, Wu F, Chovelon J-M (2009) Photocatalytic degradation of methylparaben by TiO_2: multivariable experimental design and mechanism. Appl Catal Environ 88(1–2):32–41. doi:10.1016/j.apcatb.2008.09.026

Lin Y, Ferronato C, Deng N, Chovelon J-M (2011) Study of benzylparaben photocatalytic degradation by TiO_2. Appl Catal Environ 104(3–4):353–360. doi:10.1016/j.apcatb.2011.03.006

Lucca Sánchez FA, Takimi AS, Severo Rodembusch F, Pérez Bergmann C (2013) Photocatalytic activity of nanoneedles, nanospheres, and polyhedral shaped ZnO powders in organic dye degradation processes. J Alloys Compd 572:68–73. doi:10.1016/j.jallcom.2013.03.258

Malato S, Fernández-Ibáñez P, Maldonado MI, Blanco J, Gernjak W (2009) Decontamination and disinfection of water by solar photocatalysis: recent overview and trends. Catal Today 147 (1):1–59. doi:http://dx.doi.org/10.1016/j.cattod.2009.06.018

Maroga Mboula V, Héquet V, Gru Y, Colin R, Andrès Y (2012) Assessment of the efficiency of photocatalysis on tetracycline biodegradation. J Hazard Mater 209–210:355–364. doi:10.1016/j.jhazmat.2012.01.032

Martínez C, Canle LM, Fernández MI, Santaballa JA, Faria J (2011) Aqueous degradation of diclofenac by heterogeneous photocatalysis using nanostructured materials. Appl Catal Environ 107(1–2):110–118. doi:10.1016/j.apcatb.2011.07.003

Mills A (2012) An overview of the methylene blue ISO test for assessing the activities of photocatalytic films. Appl Catal Environ 128:144–149. doi:10.1016/j.apcatb.2012.01.019

Mills A, Hill C, Robertson PKJ (2012) Overview of the current ISO tests for photocatalytic materials. J Photochem Photobiol A Chem 237:7–23. doi:10.1016/j.jphotochem.2012.02.024

Nasuhoglu D, Rodayan A, Berk D, Yargeau V (2012) Removal of the antibiotic levofloxacin (LEVO) in water by ozonation and TiO_2 photocatalysis. Chem Eng J 189–190:41–48. doi:10.1016/j.cej.2012.02.016

Niu J, Ding S, Zhang L, Zhao J, Feng C (2013) Visible-light-mediated Sr-Bi_2O_3 photocatalysis of tetracycline: kinetics, mechanisms and toxicity assessment. Chemosphere 93(1):1–8. doi:10.1016/j.chemosphere.2013.04.043

Ong S-A, Min O-M, Ho L-N, Wong Y-S (2013) Solar photocatalytic degradation of mono azo methyl orange and diazo reactive green 19 in single and binary dye solutions: adsorbability vs. photodegradation rate. Environ Sci Pollut Res 20(5):3405–3413. doi:10.1007/s11356-012-1286-1

Oros-Ruiz S, Zanella R, Prado B (2013) Photocatalytic degradation of trimethoprim by metallic nanoparticles supported on TiO_2-P25. J Hazard Mater. doi:10.1016/j.jhazmat.2013.04.010

Parra S, Elena Stanca S, Guasaquillo I, Ravindranathan Thampi K (2004) Photocatalytic degradation of atrazine using suspended and supported TiO_2. Appl Catal Environ 51(2):107–116. doi:10.1016/j.apcatb.2004.01.021

Pelaez M, Nolan NT, Pillai SC, Seery MK, Falaras P, Kontos AG, Dunlop PSM, Hamilton JWJ, Byrne JA, O'Shea K, Entezari MH, Dionysiou DD (2012) A review on the visible light active titanium dioxide photocatalysts for environmental applications. Appl Catal Environ 125:331–349. doi:10.1016/j.apcatb.2012.05.036

Petronella F, Diomede S, Fanizza E, Mascolo G, Sibillano T, Agostiano A, Curri ML, Comparelli R (2013) Photodegradation of nalidixic acid assisted by TiO_2 nanorods/Ag nanoparticles based catalyst. Chemosphere 91(7):941–947. doi:10.1016/j.chemosphere.2013.01.107

Radjenović J, Sirtori C, Petrović M, Barceló D, Malato S (2009) Solar photocatalytic degradation of persistent pharmaceuticals at pilot-scale: kinetics and characterization of major intermediate products. Appl Catal Environ 89(1–2):255–264. doi:10.1016/j.apcatb.2009.02.013

Rajeshwar K, Osugi ME, Chanmanee W, Chenthamarakshan CR, Zanoni MVB, Kajitvichyanukul P, Krishnan-Ayer R (2008) Heterogeneous photocatalytic treatment of organic dyes in air and aqueous media. J Photochem Photobiol C Photochem Rev 9(4):171–192. doi:http://dx.doi.org/10.1016/j.jphotochemrev.2008.09.001

Ramos-Delgado NA, Gracia-Pinilla MA, Maya-Trevino L, Hinojosa-Reyes L, Guzman-Mar JL, Hernandez-Ramirez A (2013) Solar photocatalytic activity of TiO_2 modified with WO_3 on the degradation of an organophosphorus pesticide. J Hazard Mater 263:36–44. doi:10.1016/j.jhazmat.2013.07.058

Reddy PAK, Srinivas B, Kala P, Kumari VD, Subrahmanyam M (2011) Preparation and characterization of Bi-doped TiO_2 and its solar photocatalytic activity for the degradation of isoproturon herbicide. Mater Res Bull 46(11):1766–1771. doi:10.1016/j.materresbull.2011.08.006

Ren C, Yang B, Wu M, Xu J, Fu Z, Lv Y, Guo T, Zhao Y, Zhu C (2010) Synthesis of Ag/ZnO nanorods array with enhanced photocatalytic performance. J Hazard Mater 182(1–3):123–129. doi:10.1016/j.jhazmat.2010.05.141

Rizzo L, Meric S, Kassinos D, Guida M, Russo F, Belgiorno V (2009) Degradation of diclofenac by TiO_2 photocatalysis: UV absorbance kinetics and process evaluation through a set of toxicity bioassays. Water Res 43(4):979–988. doi:10.1016/j.watres.2008.11.040

Ruggieri F, Antonio D'Archivio A, Fanelli M, Santucci S (2011) Photocatalytic degradation of linuron in aqueous suspensions of TiO_2. RSC Adv 1(4):611. doi:10.1039/c1ra00133g

Santiago-Morales J, Agüera A, Gómez MM, Fernández-Alba AR, Giménez J, Esplugas S, Rosal R (2013) Transformation products and reaction kinetics in simulated solar light photocatalytic degradation of propranolol using Ce-doped TiO_2. Appl Catal Environ 129:13–29. doi:10.1016/j.apcatb.2012.09.023

Satuf ML, Pierrestegui MJ, Rossini L, Brandi RJ, Alfano OM (2011) Kinetic modeling of azo dyes photocatalytic degradation in aqueous TiO_2 suspensions. Toxicity and biodegradability evaluation. Catal Today 161(1):121–126. doi:10.1016/j.cattod.2010.11.018

Seck EI, Doña-Rodríguez JM, Fernández-Rodríguez C, González-Díaz OM, Araña J, Pérez-Peña J (2012) Photocatalytic removal of 2,4-dichlorophenoxyacetic acid by using sol–gel synthesized nanocrystalline and commercial TiO_2: operational parameters optimization and toxicity studies. Appl Catal Environ 125:28–34. doi:10.1016/j.apcatb.2012.05.028

Seck EI, Doña-Rodríguez JM, Fernández-Rodríguez C, Portillo-Carrizo D, Hernández-Rodríguez MJ, González-Díaz OM, Pérez-Peña J (2013) Solar photocatalytic removal of herbicides from real water by using sol–gel synthesized nanocrystalline TiO_2: operational parameters optimization and toxicity studies. Sol Energy 87:150–157. doi:10.1016/j.solener.2012.10.015

Senthilnathan J, Philip L (2010) Photocatalytic degradation of lindane under UV and visible light using N-doped TiO_2. Chem Eng J 161(1–2):83–92. doi:10.1016/j.cej.2010.04.034

Senthilnathan J, Philip L (2011) Photodegradation of methyl parathion and dichlorvos from drinking water with N-doped TiO_2 under solar radiation. Chem Eng J 172(2–3):678–688. doi:10.1016/j.cej.2011.06.035

Shirsath SR, Pinjari DV, Gogate PR, Sonawane SH, Pandit AB (2013) Ultrasound assisted synthesis of doped TiO_2 nano-particles: characterization and comparison of effectiveness for photocatalytic oxidation of dyestuff effluent. Ultrason Sonochem 20(1):277–286. doi:http://dx.doi.org/10.1016/j.ultsonch.2012.05.015

Shu Z, Jiao X, Chen D (2013) Hydrothermal synthesis and selective photocatalytic properties of tetragonal star-like ZrO_2 nanostructures. CrystEngComm 15(21):4288–4294. doi:10.1039/C3CE40234G

Singh SA, Madras G (2013) Photocatalytic degradation with combustion synthesized WO_3 and WO_3/TiO_2 mixed oxides under UV and visible light. Sep Purif Technol 105:79–89. doi:http://dx.doi.org/10.1016/j.seppur.2012.12.010

Slama R, Ghribi F, Houas A, Barthou C, El Mir L (2011) Visible photocatalytic properties of vanadium doped zinc oxide aerogel nanopowder. Thin Solid Films 519(17):5792–5795. doi:http://dx.doi.org/10.1016/j.tsf.2010.12.197

Sousa MA, Lacina O, Hrádková P, Pulkrabová J, Vilar VJP, Gonçalves C, Boaventura RAR, Hajšlová J, Alpendurada MF (2013) Lorazepam photofate under photolysis and TiO_2-assisted photocatalysis: identification and evolution profiles of by-products formed during phototreatment of a WWTP effluent. Water Res 47(15):5584–5593. doi:10.1016/j.watres.2013.06.029

Sugihara MN, Moeller D, Paul T, Strathmann TJ (2013) TiO_2-photocatalyzed transformation of the recalcitrant X-ray contrast agent diatrizoate. Appl Catal Environ 129:114–122. doi:10.1016/j.apcatb.2012.09.013

Sun J, Qiao L, Sun S, Wang G (2008) Photocatalytic degradation of Orange G on nitrogen-doped TiO_2 catalysts under visible light and sunlight irradiation. J Hazard Mater 155(1–2):312–319. doi:http://dx.doi.org/10.1016/j.jhazmat.2007.11.062

Tong AYC, Braund R, Warren DS, Peake BM (2012) TiO_2-assisted photodegradation of pharmaceuticals—a review. Cent Eur J Chem 10(4):989–1027. doi:10.2478/s11532-012-0049-7

Trabelsi H, Khadhraoui M, Hentati O, Ksibi M (2013) Titanium dioxide mediated photodegradation of methyl orange by ultraviolet light. Toxicol Environ Chem 95(4):543–558. doi:10.1080/02772248.2013.802793

Tzikalos N, Belessi V, Lambropoulou D (2013) Photocatalytic degradation of Reactive Red 195 using anatase/brookite TiO_2 mesoporous nanoparticles: optimization using response surface methodology (RSM) and kinetics studies. Environ Sci Pollut Res Int 20(4):2305–2320. doi:10.1007/s11356-012-1106-7

Verlicchi P, Galletti A, Petrovic M, Barceló D (2010) Hospital effluents as a source of emerging pollutants: an overview of micropollutants and sustainable treatment options. J Hydrol 389(3–4):416–428. doi:10.1016/j.jhydrol.2010.06.005

Vicente R, Soler J, Arques A, Amat AM, Frontistis Z, Xekoukoulotakis N, Mantzavinos D (2014) Comparison of different TiO_2 samples as photocatalyst for the degradation of a mixture of four commercial pesticides. J Chem Technol Biotechnol 89:1259–1264. doi:10.1002/jctb.4382

Wang X, Lim T-T (2011) Effect of hexamethylenetetramine on the visible-light photocatalytic activity of C–N codoped TiO_2 for bisphenol A degradation: evaluation of photocatalytic mechanism and solution toxicity. Appl Catal Gen 399(1–2):233–241. doi:10.1016/j.apcata.2011.04.002

Xiao X, Hu R, Liu C, Xing C, Qian C, Zuo X, Nan J, Wang L (2013a) Facile large-scale synthesis of β-Bi_2O_3 nanospheres as a highly efficient photocatalyst for the degradation of acetaminophen under visible light irradiation. Appl Catal Environ 140–141:433–443. doi:10.1016/j.apcatb.2013.04.037

Xiao X, Hu R, Liu C, Xing C, Zuo X, Nan J, Wang L (2013b) Facile microwave synthesis of novel hierarchical $Bi_{24}O_{31}Br_{10}$ nanoflakes with excellent visible light photocatalytic performance for the degradation of tetracycline hydrochloride. Chem Eng J 225:790–797. doi:10.1016/j.cej.2013.03.103

Xu J, Meng W, Zhang Y, Li L, Guo C (2011) Photocatalytic degradation of tetrabromobisphenol A by mesoporous BiOBr: efficacy, products and pathway. Appl Catal Environ 107(3–4):355–362. doi:10.1016/j.apcatb.2011.07.036

Ye M, Zhou H, Zhang T, Zhang Y, Shao Y (2013) Preparation of SiO_2@Au@TiO_2 core–shell nanostructures and their photocatalytic activities under visible light irradiation. Chem Eng J 226:209–216. doi:10.1016/j.cej.2013.04.064

Zhang Z, Wang W, Ren J, Xu J (2012) Highly efficient photocatalyst $B_{i2}MoO_6$ induced by blue light-emitting diode. Appl Catal Environ 123–124:89–93. doi:10.1016/j.apcatb.2012.04.016

Chapter 7
Application of Semiconductor Photocatalytic Materials for the Removal of Inorganic Compounds from Wastewater

Jorge Luis Guzmán-Mar, Minerva Villanueva-Rodríguez, and Laura Hinojosa-Reyes

Abstract A wide range of inorganic pollutants are sensitive to photochemical transformation on the surface of catalysts. The major inorganic wastewater pollutants treated by this process include cyanide-containing waste and heavy metal pollutants, such as arsenic species and hexavalent chromium. Heterogeneous photocatalysis has been explored as an alternative technology for inorganic ion removal offering satisfactory results. The photocatalytic removal of inorganic pollutants usually has two types of mechanisms: oxidation and reduction.

In this chapter, the photocatalytic activities of various semiconductor materials for inorganic ion removal have been compiled and reviewed. The key advancements on the preparation of semiconductor materials tested for the removal of inorganic ions at low-level concentrations from natural water evaluating the matrix effects are also highlighted and discussed. In particular, the chapter focuses on enhancing the degradation efficiency; maximizing the use of illumination wavelength in the visible light region to develop solar active photocatalysts; the ease of separation from treated water, which is always of great interest; and improving the retrieval and reuse of semiconductors.

7.1 Cyanides

Cyanide, a molecule that is highly toxic to humans and aquatic organisms, is one of the common water contaminants due to industrial activities. It is used predominantly in electroplating/metal plating and the extraction of gold and silver in mining industries, where it is produced in large volumes. It is also used to make paper, textiles, and plastics (Kuyucak and Akcil 2013). Cyanide may be present in environmental matrices and waste streams in a variety of forms, such as free, simple and complex cyanides, cyanates, and nitriles. The most toxic form of

---

J.L. Guzmán-Mar • M. Villanueva-Rodríguez • L. Hinojosa-Reyes (✉)
Universidad Autónoma de Nuevo León, UANL, Department of Chemistry Sciences, San Nicolás de los Garza, N.L. 66451, México
e-mail: laura.hinojosary@uanl.edu.mx

A. Hernández-Ramírez, I. Medina-Ramírez (eds.), *Photocatalytic Semiconductors*,
DOI 10.1007/978-3-319-10999-2_7

Table 7.1 Classification of cyanide compounds

Type of compound		Examples	Remarks
Free		HCN, CN^-	Equilibrium depends on pH ($pKa = 9.2$ at 25 °C)
Simple	Soluble	NaCN, KCN, $Ca(CN)_2$	Ionize in aqueous solution at low concentration and mostly present as HCN below pH 8
	Insoluble	$Zn(CN)_2$, $Cd(CN)_2$, AgCN	
Complex	Weak	$K_2Zn(CN)_4$, $K_2Cd(CN)_4$	Ionize easily
	Moderately strong	$K_2Cu(CN)_3$, $K_2Ni(CN)_4$	Ionize moderately
	Strong	$K_2Fe(CN)_6$, $K_3Co(CN)_6$	Do not ionize easily, very stable
Inorganic		SCN^-, CNO^-	Cyanate unstable
Organic (nitriles)	Aliphatic	Acetonitrile, acrylonitrile, propionitrile	Stable
	Aromatic	Benzonitrile	Stable

cyanide is free cyanide, which includes the cyanide anion itself and hydrogen cyanide (HCN) either in a gaseous or aqueous state. The classification of cyanide compounds and their respective stabilities are provided in Table 7.1.

A promising method of cyanide detoxification is the photocatalytic process. During the removal treatment, the toxic CN^- species are converted into cyanate species (OCN^-) at the semiconductor surface and can also be transformed to N_2 and CO_2, which are environmentally friendly compounds (Harraz et al. 2013). Table 7.2 summarizes the application of semiconductor materials for the photocatalytic removal of cyanide species. The use of TiO_2 in the slurry form for the photocatalytic degradation with artificial UV light of both free and complex cyanide compounds has been extensively investigated (López-Muñoz et al. 2009; Marugan et al. 2008; Osathaphan et al. 2013; Siboni et al. 2011b). Lopez Muñoz carried out the removal of $Ag(CN)_2^-$ in the presence of oxygen using the TiO_2 photocatalytic treatment. Both the reduction of Ag(I), which was deposited as Ag^0 on the titanium surface, and the oxidation of CN^- to cyanate species were carried out simultaneously (López-Muñoz et al. 2009). The use of solar light as an irradiation source has also been investigated in the TiO_2 photocatalytic removal of cyanide species (Karunakaran et al. 2011b; Pineda Arellano and Silva Martinez 2010). Karunakaran et al. compared the photocatalytic efficiencies of different forms of TiO_2 nanocrystals (anatase, rutile, Degussa P25, Hombikat) to detoxify cyanide in an alkaline solution under natural sunlight. The photocatalytic efficiency was dependent on the specific surface area and the average crystallite size of the oxide; rutile was an exception (Karunakaran et al. 2011b). Furthermore, the effect of organic molecules on cyanide oxidation was studied by Osathaphan et al. The organic ligand used in their study was ethylenediaminetetraacetate (EDTA), which can form a strong complex with heavy metal ions. According to the study, the presence of EDTA in the reaction mixture reduced the photocatalytic oxidation rate of cyanide (Osathaphan et al. 2008). On the other hand, binary oxides have been

Table 7.2 Applications of heterogeneous photocatalysis for the removal of cyanide species

Catalyst	Radiation source	Experimental conditions	Results	Reference
Degussa P25 TiO_2, TiO_2 anatase, TiO_2 rutile, Hombikat TiO_2 nanoparticles, and TiO_2 submicron particles	Solar light 950 ± 25 Wm^{-2} sunlight (11.00 am–2.30 pm)	250 mg L^{-1} cyanide Solution pH 12.5 26.8 mg L^{-1} dissolved O_2 Catalyst loading = 4 g L^{-1} Reaction time 150 min	The reaction rate depended on the apparent area of the catalyst. Photocatalytic efficiencies: anatase > Hombikat (69 % anatase, 31 % rutile) > P 25 Degussa (81 % anatase, 19 % rutile) > rutile > microparticulate anatase	Karunakaran et al. (2011b)
10 wt % SiO_2/TiO_2 synthesized by sol–gel method Crystalline phase anatase 6 nm crystallite size Surface area 286.7 m^2g^{-1}	UV light (365 nm)	100 mg L^{-1} KCN Solution pH 10.5 Catalyst loading = 1.66 g L^{-1} Reaction time 180 min	The higher photocatalytic activity was related to the larger specific surface area and higher crystallinity. The removal efficiency was 93.3 %	Harraz et al. (2013)
Pd-TiO_2-HAP (hydroxyapatite) with 3 wt % Pd and 25 wt % TiO_2 synthesized by impregnation method Crystalline nature of the HAP and anatase was confirmed by XRD Surface area 150 m^2g^{-1}	Blue fluorescent lamp (150 W, maximum energy at 450 nm)	100 mgL^{-1} KCN Solution pH 10.5 Catalyst loading = 5.33 gL^{-1} O_2 was bubbled through the reaction solution Reaction time 60 min	The cyanide was completely degraded using the Pd–TiO_2–HAP catalyst. The material was reused for six times achieving 98 % removal	Mohamed and Baeissa (2013)

(continued)

Table 7.2 (continued)

Catalyst	Radiation source	Experimental conditions	Results	Reference
2 % Ce-doped ZnO prepared by sonochemical wet impregnation method and calcinated at 500 °C Crystalline phase wurtzite 45 nm crystallite size. Surface area 4.32 m^2g^{-1}	Solar light 950 ± 25 Wm^{-2} sunlight (11.00 am–2.30 pm)	250 mg L^{-1} KCN Solution pH 12.5 Catalyst loading = 4 g L^{-1} Dissolved O_2 = 26.8 mg L^{-1} Reaction time 150 min	Under visible light the catalytic efficiencies were of the order: Ce-doped ZnO (2.92×10^3 m^{-2} s^{-1}) > ZnO (1.81×10^3 m^{-2} s^{-1}) > CeO (1.68×10^3 m^{-2} s^{-1})	Karunakaran et al. (2010)
Sol–gel ZnO impregnated with Co (II) phthalocyanine (CoPc) Crystalline phase wurtzite Surface area 23.72 m^2g^{-1}	Visible light metal halide lamp (25 W) which emits from 400 to 700 nm	20 mg L^{-1} KCN Solution pH 11 Catalyst loading = 1 g L^{-1} Reaction time 120 min	The ZnO-CoPc showed the highest activity on KCN degradation under visible light radiation (100 % decomposition)	Salinas-Guzmán et al. (2010)
0.3 wt % Pt/Ga_2O_3-SiO_2 Ga_2O_3-SiO_2 was synthesized by sol–gel method and Pt was incorporated by photo-assisted deposition method	Visible light (fluorescent lamp, 150 W)	100 mgL^{-1} KCN Solution pH 10.5 Catalyst loading = 1 g L^{-1} Reaction time 60 min	0.3 wt % Pt/Ga_2O_3-SiO_2 allowed a complete removal. The rate constant was 160×10^{-4} min^{-1} compared to 10×10^{-4} min^{-1} for Ga_2O_3-SiO_2 catalyst. The catalyst was reused during 11 cycles with a degradation efficiency of 95 %	Baeissa and Mohamed (2013)

applied in the removal of cyanide species because the use of these coupled semiconductors could improve the photocatalytic process through a more efficient charge separation, increased lifetime of charge carriers, and enhanced interfacial charge transfer to absorbed substrates (Kumar and Devi 2011). The coupling of TiO_2/ZnO (Karunakaran et al. 2011a), TiO_2/CeO_2 (Karunakaran and Gomathisankar 2013), and TiO_2/SiO_2 (Harraz et al. 2013) has been evaluated for the removal of cyanide species under UV or visible light irradiation. The photocatalytic activity of TiO_2/ZnO was considerably higher than that of TiO_2 for the removal of cyanide under visible light (Karunakaran et al. 2011a).

Hydroxyapatite (HAP) was also coupled with TiO_2 to modify the surface charge of the composite. Thus, Mohamed et al. (Mohamed and Baeissa 2013) prepared Pd–TiO_2–hydroxyapatite nanoparticles (Pd–TiO_2–HAP) using a template-ultrasonic-assisted method and evaluated their use for the photocatalytic removal of cyanide under visible light irradiation. The Pd–TiO_2–HAP samples showed the highest visible light-induced photocatalytic activity and the highest photochemical stability compared to all of the other catalysts with a degradation rate constant that was 2.71-fold greater than that of Pd–TiO_2 and 12.2-fold greater than that of Degussa P25. Pd–TiO_2–HAP showed increased activity due to the photocatalytic activity of both the Pd–TiO_2 and the HAP materials. The vacancies formed on the surface of the excited PO_4^{3-} group under visible illumination caused the generation of active superoxide anion radicals ($O_2^{\bullet -}$) that attack the surrounding cyanide adsorbed on HAP (Fig. 7.1).

On the other hand, ZnO (Farrokhi et al. 2013), considered a promising alternative to TiO_2 catalyst, and modified ZnO materials (Karunakaran et al. 2010; Saleh et al. 2010; Salinas-Guzmán et al. 2010) have been evaluated for the removal of cyanide species. Ce-doped ZnO (2 %) prepared by the sonochemical wet impregnation method catalyzes the oxidation of cyanide under UVA light or natural sunlight. The presence of Ce enhanced the photocatalytic activity of ZnO because it reduced the intragranular resistance and recombination of the photogenerated electron–hole pairs. It also shifted the edge of the optical absorption to the visible region (Karunakaran et al. 2010).

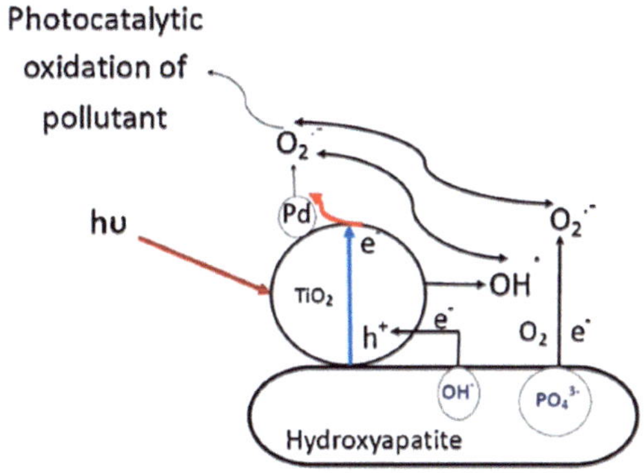

Fig. 7.1 Photocatalytic mechanism of cyanide degradation by Pd–TiO_2–HAP (Mohamed and Baeissa 2013)

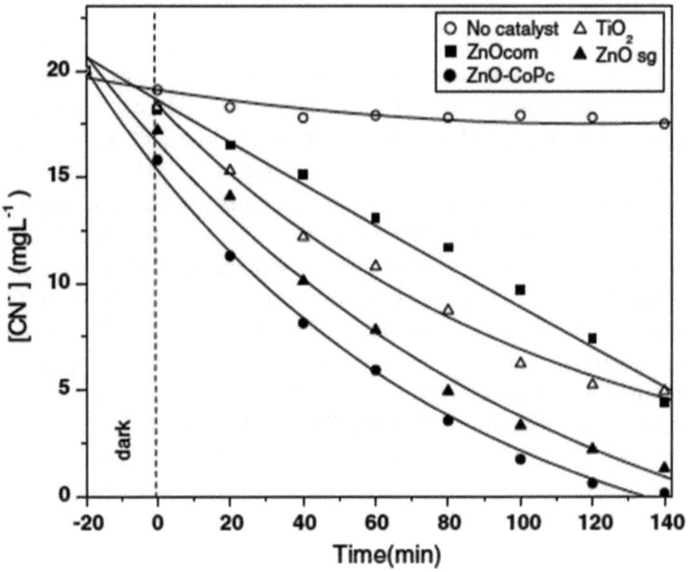

Fig. 7.2 Potassium cyanide decomposition in the presence of different semiconductor catalysts (Salinas-Guzmán et al. 2010)

Dye sensitization is a powerful approach to enable visible light harvesting for wide bandgap semiconductors, such as ZnO. Salinas-Guzmán et al. evaluated the activity of sol–gel ZnO impregnated with Co(II) phthalocyanine (CoPc) as a sensitizer for the removal of cyanide under visible light. Figure 7.2 shows the photocatalytic degradation of cyanide potassium solution using these different catalysts (Salinas-Guzmán et al. 2010). Compared with commercial ZnO and Degussa P25 TiO_2 photocatalysts, the prepared ZnO sensitized with CoPc showed the highest activities for the degradation of potassium cyanide in aqueous solution under visible light irradiation.

Alternative catalysts, such as 0.3 wt % Pt/Ga_2O_3-SiO_2, have demonstrated photocatalytic activity under visible light with the ability to completely remove cyanide. The Ga_2O_3-SiO_2 nanoparticles were prepared by the sol-gel method, and Pt was then immobilized on the surface via the photo-assisted deposition method. Pt incorporation enhanced the photocatalytic activity of this material (Baeissa and Mohamed 2013).

7.2 Removal of Heavy Metals and Metalloids

Metals are particularly problematic because they are not biodegradable and can accumulate in living tissues (Barakat 2011; Fu and Wang 2011). With the rapid development of industries such as metal plating facilities, mining operations, fertilizer industries, tanneries, batteries, paper industries, and pesticides, heavy metal wastewaters are directly or indirectly discharged into the environment, especially in developing countries (Fu and Wang 2011; Barakat 2011). Thus, the photocatalytic removal of heavy metals and metalloids is rapidly gaining importance in wastewater treatment (Litter 2009; Barakat 2011). Photocatalytic treatment processes can convert ionic species into their metallic solid forms and deposit them over the semiconductor surface or transform them into less toxic, soluble species. The redox level of the metallic couples related to the levels of the conduction and valence bands can be considered the most important parameter for predicting the feasibility of the transformation of the species. For the reduction of photocatalytic chemical species, the conduction band of the semiconductor photocatalyst must be more negative than the reduction potential of the chemical species. In contrast, photooxidation can occur only when the potential of the valence band is more positive than the oxidation potential of the chemical species (Litter 2009). The applicability of semiconductor materials for the photocatalytic removal of arsenic and chromium species from wastewater is reviewed in Sects. 7.2.1 and 7.2.2, respectively.

7.2.1 Arsenic

Arsenic contamination in natural water is a worldwide problem. Many countries, especially Taiwan, Argentina, India, Bangladesh, Mexico, Hungary, and Chile, have reported extensive arsenic contamination in their groundwater supplies. The presence of arsenic in natural water is generally associated with geochemical environments, such as volcanic deposits, inputs from geothermal sources, mining wastes, and landfills. Uncontrolled anthropogenic activities, such as smelting of metal ores and use of arsenic-based pesticides and wood preservative agents, may also release arsenic into the environment (Litter 2009; Sharma and Sohn 2009).

The inorganic forms, i.e., arsenate (As(V) and arsenite (As(III)), and the methylated forms, i.e., monomethylarsonic acid (MMA) and dimethylarsinic acid (DMA), are the main arsenic species found in natural water (Sharma and Sohn 2009). The arsenic species commonly identified in water are listed in Table 7.3 (Sharma and Sohn 2009; Aragay et al. 2012).

The relative concentrations of As(III) to As(V) vary widely depending on the redox conditions in the geological environment. As(III) is more toxic than As (V) and is difficult to remove from water by most treatment processes. Thus, As (III) is typically removed by first oxidizing it to As(V); As(V) is then removed using adsorption, precipitation, or ion exchange processes. Thus, most treatment methods used for arsenic removal exhibit higher affinity for As(V) compared to As(III) species (Sharma and Sohn 2009; Litter 2009).

TiO_2 semiconductor material has been widely used for the removal of inorganic and organic arsenic species due to their low cost and environmentally friendly nature (Choi et al. 2010; Fostier et al. 2008; Zheng et al. 2010). The final product of the oxidation of the initial arsenic species is As(V) (Choi et al. 2010).

Table 7.3 Arsenic species identified in water

Name	Abbreviation	Chemical structure	pK_a
Arsenite	As(III)	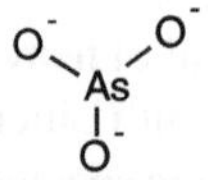	9.3
Arsenate	As(V)	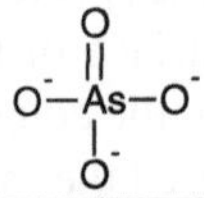	2.2 7.0 11.5
Monomethylarsonic acid	(MMA)V	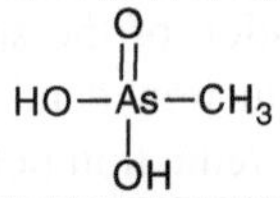	3.6 8.2
Dimethylarsinic acid	(DMA)V	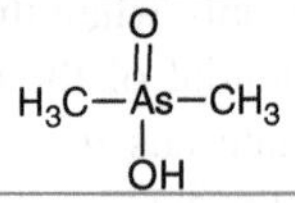	1.8 6.1

The oxidation of As(III) (Choi et al. 2010; Tsimas et al. 2009) and phenylarsonic acid species (Zheng et al. 2010) has been studied by heterogeneous photocatalysis using TiO_2 under UV light. Tsimas et al. investigated the simultaneous photocatalytic oxidation of As(III) and humic acid (HA) in aqueous Degussa P25 TiO_2 suspensions under UV light (350–400 nm). For initial As(III) concentrations in the range of 3–20 mg L^{-1}, the removal was very rapid, taking place within 10–30 min using a 50 mg L^{-1} catalyst loading. However, the oxidation of As(III) was reduced in the presence of HA (Tsimas et al. 2009). Zheng et al. demonstrated that phenylarsonic acid (PA), a compound commonly utilized in the broiler poultry industry as a feed additive, was readily degraded using Degussa P25 TiO_2. The generated mineralization product was As(V) (Zheng et al. 2010). Table 7.4 summarizes the application of photocatalytic semiconductors for the removal of arsenic species.

Palladium-modified nitrogen-doped titanium oxide (TiON/PdO) nanoparticles were evaluated for the removal of As(III) species under visible light illumination (Li et al. 2009). The As(III) concentration was reduced to less than the US Environmental Protection Agency water standard (10 μg L^{-1}). The efficient removal of As(III) was the result of the combined effects of strong adsorption and photocatalytic activity under visible light illumination by TiON/PdO, which was derived from the strong optoelectronic coupling between PdO and TiON.

Supported TiO_2 is one of the choices for field application in drinking water treatment by means of heterogeneous photocatalysis. Immobilized TiO_2 photocatalysts on porous supporting matrices, such as polymer (Fostier et al. 2008) and activated carbon (Yao et al. 2010, 2012), have been applied for the removal of As(III) species.

A 10 % TiO_2 suspension immobilized in PET (polyethylene terephthalate) plastic bottles exposed to sunlight for 4 h combined with the coprecipitation of arsenic on Fe(III) hydroxides (oxides) was evaluated for the removal of inorganic arsenic from As(III)-contaminated water (Fostier et al. 2008). A removal of over

Table 7.4 Application of photocatalytic semiconductors for the removal of arsenic species

Semiconductor	Reaction conditions	Light source	Results	Reference
Degussa P25 TiO_2	3–20 mg L^{-1} As (III) solution pH 6.5 Catalyst amount 250 mg L^{-1} Reaction time 10–30 min	UV radiation using 9 W lamp (350–400 nm)	98 % removal of As (III) was achieved in the concentration range from 3 to 20 mg L^{-1}. In the presence of humic acids, 60 % removal efficiency was reached in 30 min	Tsimas et al. (2009)
Degussa P25 TiO_2	10 mg L^{-1} phenylarsonic acid (PA) at pH 9 Catalyst amount 0.1 g L^{-1} Reaction time 30 min	UV radiation using 15 - low-pressure lamps (350 nm)	Complete mineralization of PA to As (V)	Zheng et al. (2010)
10 % Degussa P25 TiO_2 immobilized on PET bottles combined with iron salt	Standard solution and water sample spiked with 5 mg L^{-1} As(III) at pH 7 Reaction time 120 min	Sunlight radiation (1.2 mW cm^{-2})	99 % removal of total arsenic. The water matrix did not interfere in the oxidation process	Fostier et al. (2008)
Meso-TiO_2/α-Fe_2O_3 composite synthesized by impregnation method with 50 wt % iron content Surface area of 95 m^2g^{-1}	5 mg L^{-1} As (III) at pH 9 Catalyst amount 0.1 g L^{-1} Reaction time 120 min	20 W UV lamp	The complete oxidation of As (III) to As (V) was accomplished within 3 h. For single meso-TiO_2 and pure α-Fe_2O_3, oxidation was 70 and 20 %, respectively	Zhou et al. (2008)
ZnO prepared by sol–gel method Wurtzite crystalline phase Surface area: 5.3 m^2g^{-1}	Solutions of 5 mg L^{-1} As (III) at pH 7 or MMA(V) at pH 8 Catalyst loading 0.25 g L^{-1} for As(III) and 0.50 g L^{-1} for MMA(V) Reaction time 360 min	UV radiation using 15 - low-pressure lamps (350 nm)	Complete oxidation of As(III) and MMA(V). The As(V) generated as by-product was completely and strongly adsorbed onto the ZnO surface	Rivera-Reyna et al. (2013)

(continued)

Table 7.4 (continued)

Semiconductor	Reaction conditions	Light source	Results	Reference
Composite of anatase TiO_2 synthesized by sol–gel method onto activated carbon fiber (TiO_2/ACF) Fibers of 10–20 μm Surface area: 426.3 m^2g^{-1}	4 mg L^{-1} As (III) at pH 9 Catalyst loading 3 g L^{-1} Reaction time 60 min	Ultraviolet mercury lamp (125 W, 365 nm)	Complete oxidation in 60 min and the photocatalytic activity of TiO_2/ ACF remained above 90 % of its activity as-prepared after being used four times	Yao et al. (2010)

99 % of As(III) and total As was achieved, indicating that the water matrix did not interfere with the oxidation process.

Yao et al. synthesized a composite photocatalyst by loading TiO_2 onto activated carbon fiber (TiO_2/ACF) (Yao et al. 2010) and granular activated carbon (GAC–TiO_2) (Yao et al. 2012). The catalysts were evaluated for the oxidation of As(III). The photocatalytic oxidation of As(III) using carbon supporting a titanium dioxide photocatalyst took place within minutes and followed first-order kinetics. The oxidation of As(III) occurred in a wide pH range varying from 2 to 10, and the oxidation efficiency markedly increased with an increase in the pH value. The effect of anions, such as Cl^-, $SO_4{}^{2-}$, $PO_4{}^{3-}$, and $SiO_3{}^{2-}$, which are commonly present in water, on the rate of As(III) oxidation was evaluated (Fig. 7.3). Silicate and phosphate ions were found to significantly suppress As(III) oxidation due to competitive adsorption with arsenite ions on the surface of TiO_2 (Yao et al. 2012).

To increase the sorption capacity of arsenic species, some bimetal oxide adsorbents have been prepared and evaluated. Bifunctional mesoporous TiO_2 (meso-TiO_2)/α-Fe_2O_3 composites were synthesized by the impregnation of Fe^{3+} into meso-TiO_2 followed by calcination at 300 °C (Zhou et al. 2008). The composites possessed both the photocatalytic ability of meso-TiO_2 for the oxidation of As(III) to As(V) and the adsorption performance of α-Fe_2O_3 for As(V). The experimental results showed that the meso-TiO_2/-Fe_2O_3 composites could oxidize As(III) to As (V) with high efficiency at various pH values in the photocatalytic reaction. The highest efficiency of the photocatalytic oxidation of As(III) was reached when the loading amount of α-Fe_2O_3 was 50 wt.%. Meanwhile, As(V) could be effectively removed by adsorption onto the surface of the composite. Therefore, the excellent adsorption property of α-Fe_2O_3 combined with the good photocatalytic ability of meso-TiO_2 allowed the complete removal of arsenic. Moreover, As(V) was easily desorbed from the composites by heat treatment in alkali solution without destroying the structure of the composites, and the TiO_2 (meso-TiO_2)/α-Fe_2O_3 composite was very stable even after 8 recycling cycles.

Alternative catalyst semiconductors to TiO_2, such as ZnO (Rivera-Reyna et al. 2013) and $Bi_2Sn_2O_7$ (Tian et al. 2012), have been evaluated for the oxidation of arsenic species. The photocatalytic removal of As(III) and monomethylarsonic acid MMA(V) compounds using ZnO synthesized by the sol–gel technique results in

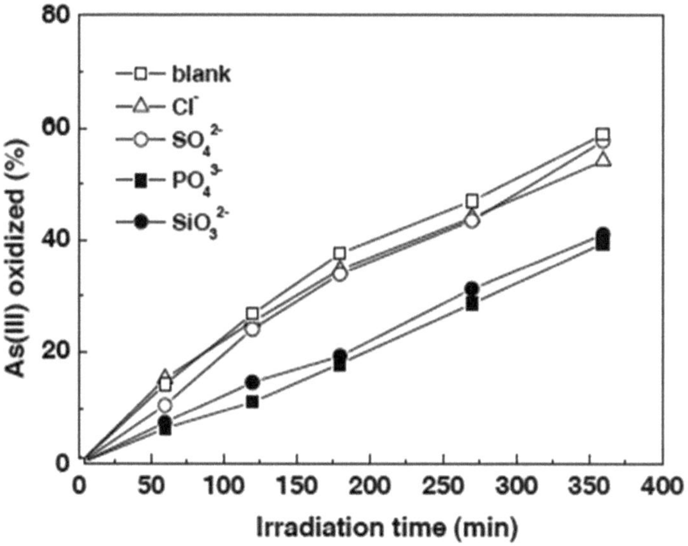

Fig. 7.3 Effect of common anions on the oxidation rate of As(III) at pH 6 using a titanium dioxide photocatalyst on a carbon support (Yao et al. 2012)

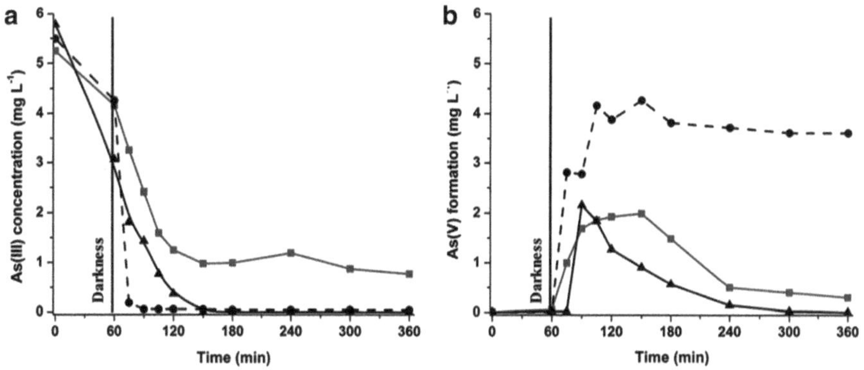

Fig. 7.4 Concentration of (**a**) As(III) and (**b**) As(V) during the photocatalytic oxidation of As(III) using (*filled square*) commercial ZnO, (*filled circle*) Degussa P25, and (*filled triangle*) sol–gel ZnO. Solution pH, 7; As(III) concentration, 5 mg L^{-1}; catalyst loading, 0.25 g L^{-1} (Rivera-Reyna et al. 2013)

their effective and rapid mineralization to less toxic inorganic arsenate As(V) in the presence of UV light (350 nm). The optimal conditions for the removal of 5 mg L^{-1} As(III) and MMA(V) aqueous solutions were observed at catalyst loadings of 0.25 and 0.50 g L^{-1} with solution pH values of 7 and 8, respectively. The photocatalytic oxidation rate of As(III) using sol–gel ZnO was faster than the MMA(V) removal process. The photocatalytic removal of As(III) in the presence of sol–gel ZnO, TiO_2 Degussa P25, and a commercial ZnO catalyst is shown in Fig. 7.4. Thus, the high adsorption capacity of the ZnO synthesized by the sol–gel method enhanced the removal of As(III) from water samples since this material could be used as a combined catalyst and adsorbent of toxic arsenic species. (Rivera-Reyna et al. 2013).

Nanocrystalline $Bi_2Sn_2O_7$ synthesized via a hydrothermal route was evaluated for the removal of As(III) from aqueous solution under visible light radiation (Tian et al. 2012). The $Bi_2Sn_2O_7$ nanoparticles exhibited high photocatalytic activity under visible light irradiation for the oxidation of As(III) (up to 96.8 %) for an initial concentration of As(III) of 2 mg L^{-1}.

7.2.2 Chromium

Chromium exists in the aquatic environment mainly in two states: hexavalent Cr (VI) and trivalent Cr(III). Cr(VI) species are known to be toxic and carcinogenic, whereas Cr(III) is less toxic and can be readily precipitated out of solution in the form of $Cr(OH)_3$ (Barrera-Díaz et al. 2012; Mukherjee et al. 2012). Chromium (VI) is a frequent contaminant in wastewater generated by industries, such as electroplating, leather tanning, and paints (Mukherjee et al. 2012). The most probable Cr(VI) species in aqueous solutions are $Cr_2O_7^{2-}$, CrO_4^{2-}, H_2CrO_4, and $HCrO_4^{-}$, although the relative distribution depends on the solution pH, the Cr (VI) concentration, and the redox potential. However, none of these Cr (VI) species form insoluble precipitates, making their separation through a direct precipitation method not feasible. In this oxidation state, the metal is extremely mobile in water and soil (Barrera-Díaz et al. 2012; Mukherjee et al. 2012). Thus, to form a chromium solid phase, which can be easily separated from the aqueous media, it is necessary to change the oxidation state. The use of heterogeneous photocatalysis for the reduction of Cr(VI) has been described as an attractive option for the removal of Cr(VI) from wastewater (Barrera-Díaz et al. 2012; Fu and Wang 2011; Litter 2009).

7.2.2.1 Photocatalytic Reduction of Cr(VI) with TiO_2-Based Semiconductors

The photocatalytic reduction of Cr(VI) over TiO_2 catalysts has been extensively investigated (Kabra et al. 2009; Wang et al. 2008, 2010; Xu et al. 2013; Yang et al. 2012a, b). Table 7.5 describes some recent studies on the removal of Cr (VI) using TiO_2-modified catalysts. The semiconductor commonly used for the photocatalytic removal of Cr(VI) has been the slurry form of Degussa P25, and the process has been based on the use of UV radiation (Yang et al. 2012a, 2012b) or solar light (Kabra et al. 2009). Very few studies have considered the reduction of Cr (VI) using immobilized TiO_2 nanoparticles. Pifferi et al. (2013) exploited an electrophoretic method to deposit titanium films (P25, Hombikat, and sol–gel

Table 7.5 TiO_2-modified catalyst used for the removal of Cr(VI)

Modifier	Light irradiation conditions	Experimental conditions	Results	Modification effect	Reference
SOA (tartaric acid)/anatase TiO_2 SOA: 0.2 mM tartaric acid Anatase TiO_2: crystallite size 15 nm and surface area of 180 m^2g^{-1}	Visible light ($\lambda > 420$ nm)	Cr(VI) concentration 10.4 mg L^{-1} Catalyst loading 1 g L^{-1} Solution pH 3 Reaction time 120 min	Cr(VI) removal of 90 % after 2 h of visible illumination	The SOA-induced photoreduction of Cr(VI) proceeded via a CTC-mediated path, being governed by chemical structures of sacrificial SOAs	Wang et al. (2010)
Surface adsorbates Polymer – Degussa P25 TiO_2	Visible light (λ 450–475 nm)	Cr(VI) concentration 15 mg L^{-1} Catalyst loading 1 g L^{-1} Solution pH 3 Reaction time 120 min	Reduction efficiency: 50 %	The PFT served not only as sensitizer but also as electron donor for Cr (VI) reduction	Qiu et al. (2012)
Dye anchoring Hydroxoaluminiumtricarboxymonoamide phthalocyanine (AlTCPc)/Degussa P25 TiO_2 7.6 μmol AlTCPc/g TiO_2	UV ($\lambda = 365$ nm) and visible light ($\lambda > 600$ nm)	Cr(VI) concentration 20.8 mg L^{-1} Catalyst loading 1 g L^{-1} Solution pH 2 Reaction time 180 min	UV degradation rates using AlTCPc)/TiO_2 was 14.5 $eistein^{-1}L^{-1}$ compared to 10.9 $eistein^{-1}L^{-1}$ obtained by TiO_2 Visible degradation rates were 6.46 and 0.18 $eistein^{-1}L^{-1}$ by AlTCPc/TiO_2 and TiO_2, respectively	The Cr(VI) removal rate depends on the photocatalyst mass. The presence of 4-CP was essential, but its concentration had no effect on the Cr (VI) decay rate	Meichtry et al. (2009)
Metal deposition 0.5 % wt Au/TiO_2 synthesized by sol–gel method TiO_2 in anatase form with surface area of 75 m^2g^{-1}	UV light ($\lambda = 365$ nm)	Cr(VI) concentration 10 mg L^{-1} Catalyst loading 1 g L^{-1} Reaction time 240 min	0.0136 min^{-1} (0.5 % wt Au/TiO_2) > 0.0092 min^{-1} (TiO_2) > 0.0049 min^{-1} (Degussa P25)	The enhanced photocatalytic performance was ascribed to the increase in light absorption intensity and range as well as the reduction of photoelectron-hole pair recombination in TiO_2 with the introduction of Au	Liu et al. (2013a)

(continued)

Table 7.5 (continued)

Modifier	Light irradiation conditions	Experimental conditions	Results	Modification effect	Reference
Doping No-metal and co-doping N/TiO_2 and Ni-F/TiO_2 synthesized by sol–gel method TiO_2 in anatase form Crystalline size: 17.7 nm	UV light (λ 300–400 nm) with irradiation intensity of 750 Wm^{-2}	Cr(VI) concentration 20.8 mg L^{-1} Catalyst loading 1 g L^{-1} Solution pH 2 Reaction time 75 min	The degradation rate (min^{-1}) was improved using N-F/TiO_2 (1:1 molar ratio): 0.039 min^{-1} (N-F/TiO_2) > 0.011 min^{-1} (TiO_2) > 0.024 min^{-1} (Degussa P25)	For the N-F/TiO_2 catalysts, the electrons in states below the conduction band of TiO_2 can reduce preferably Cr(VI)	Giannakas et al. (2013)
Hybrids with nanomaterials MWNT-COOH and TiO_2 nanoparticles in anatase form	UV light (λ = 365 nm)	Cr(VI) concentration 40 mg L^{-1} Catalyst mass 5 mg Solution pH 2.5 Reaction time 120	The photocatalytic activity of the composite was enhanced by a factor of 2.2 compared to Degussa P25	MWNT-COOH/TiO_2 composite showed higher photocatalytic activity because (1) NT originate the decrement of the recombination of photogenerated charge carriers and (2) the structure and nanoscale diameter of the NTs can promote accumulation of photoexcited electrons on the ends of the MWNT-COOH	Xu et al. (2008)
Heterogeneous composites SnS_2/TiO_2 nanocomposites synthesized by solvothermal method 44.5 (% mass) of TiO_2 in	Visible light λ (> 420 nm)	Cr(VI) concentration 50 mgL^{-1} Catalyst mass 1 g L^{-1} Solution pH not adjusted Reaction time 120 min	The degradation rate (min^{-1}) was improved using SnS_2/TiO_2 4.81×10^2 (SnS_2/TiO_2) > 0.60×10^2 (SnS_2) > 0.11×10^2 (TiO_2)	Cr(VI) was reduced to Cr (III) by SnS_2/TiO_2-mediated photocatalysis due to the tight heterojunction structure of the synthesized nanocomposite, which	Zhang et al. (2012)

anatase form Surface area 97.01 m^2g^{-1}				facilitate interfacial electron transfer and reduce the separation and self-agglomeration of two components	
Heterogeneous composites Bi_2O_3/TiO_2 TiO_2 in anatase form Crystalline size 7 nm Surface area 129 m^2g^{-1}	Visible light ($\lambda > 420$ nm)	Cr(VI) concentration 20 mg L^{-1} Catalyst mass 1 g L^{-1} Solution pH 3 Reaction time 180 min	2.0 % Bi_2O_3/TiO_2> 4.0 % Bi_2O_3/TiO_2 > 6.0%Bi_2O_3/TiO_2> 1.0%Bi_2O_3/TiO_2> 0.5%Bi_2O_3/TiO_2 > TiO_2	The photoelectrons generated by exciting Bi_2O_3 reduced Cr(VI) to Cr(III). Meanwhile, the photoholes, transferred into valence band of TiO_2 from that of Bi_2O_3, oxidized the BPA molecule	Yang et al. (2013)
Heterogeneous composites 1 % WO_3/TiO_2 NTs arrays TiO_2 in the anatase form and WO_3 in the triclinic form	Simulated solar light (λ 400–700 nm)	Cr(VI) concentration 20 mg L^{-1} Catalyst mass 1 g L^{-1} Solution pH 2 Reaction time 180 min	WO_3/TiO_2 NTs arrays showed higher photocatalytic activity compared to unmodified TiO_2. The rate constants were 110.6 and 62.5 μg L^{-1} min^{-1} cm^{-2}, respectively	The enhancement of the photocatalytic activity on WO_3/TiO_2 NTs under simulated solar light can be attributed to the increased charge-carrier separating rate and the increasing driving force with fast electron injection resulting from the shift of the TiO_2 conduction band to negative potentials due to the WO_3 doping	Yang et al. (2010)

TiO_2 calcined at 400 °C) and evaluated their photocatalytic performance for Cr (VI) reduction under UV radiation.

The removal of Cr(VI) using a TiO_2 catalyst has been favored at lower solution pH values due to an increased potential difference between the conduction band of TiO_2 and Cr(VI)/Cr(III) and the anionic-type adsorption of Cr(VI) onto the surface of TiO_2 (Yang et al. 2012a). The reaction can be accelerated by the addition of organic compounds that act as scavengers of holes or HO•, such as citric acid (Kabra et al. 2009; Wang et al. 2008), humic acid, oxalate, ethylenediaminetetraacetic acid, nitrilotriacetic acid, phenol (Yang et al. 2012b), and isopropyl alcohol (Pifferi et al. 2013).

In particular, the size and shape of the external surfaces are very important factors in determining the photocatalytic activity of the TiO_2 nanostructures. In this context, a sea-urchin-like rutile superstructure with ultrathin nanorods and exposed {110} faces exhibited excellent photocatalytic activity in the removal of Cr (VI) ions from plating wastewater exposed to sunlight (Xu et al. 2013). This special morphology exhibited a structure-enhanced catalytic activity because it facilitates the transport of photoelectrons to the rutile nanorod {110} faces and prevents them from recombining with holes. The removal effectiveness is nearly 100 % at initial Cr concentrations below 53.7 mg L^{-1} after 3 h of solar radiation.

An effective visible light active sensitization of TiO_2 has been achieved using poly(fluorene-co-thiophene) (PFT) as a surface adsorbate (Qiu et al. 2012), sensitized with hydroxoaluminiumtricarboxymonoamide phthalocyanine (AlTCPc) (Meichtry et al. 2009) and modified with small-molecular-weight organic acids (Wang et al. 2010). The sensitized TiO_2-PFT was evaluated for both Cr (VI) reduction and phenol oxidation, and the reduction efficiency of Cr (VI) reached a maximum of 50 % at pH 3 (Qiu et al. 2012). AlTCPc adsorbed on Degussa P25 TiO_2 was tested for Cr(VI) reduction in the presence of 4-chlorophenol (4-CP) as a sacrificial donor (Meichtry et al. 2009). Additionally, the photoreduction of Cr(VI) over anatase TiO_2 nanoparticles was achieved through surface modification with small-molecular-weight organic acids (SOAs), such as tartaric acid, as sacrificial organic compounds. Charge-transfer complexes (CTC) formed between TiO_2 and SOAs induced the photoreduction of Cr(VI) via a CTC-mediated path governed by the chemical structure of sacrificial SOA, allowing 95 % Cr(VI) removal in visible light-illuminated TiO_2 dispersions by the addition of 0.2 mM tartaric acid as a SOA. As depicted in Fig. 7.5, a higher energy of the highest occupied molecular orbital or lower ionization potential of the SOA was favorable for electron transfer within the TiO_2-SOA complex, thereby accelerating the photoreduction of Cr(VI) under visible light (Wang et al. 2010).

TiO_2 nanoparticles modified with Au have exhibited enhanced photocatalytic performance in the reduction of Cr(VI) under UV light irradiation because noble metals serve as both a conduction band electron trap and a co-catalyst. The TiO_2-Au photocatalyst was synthesized by the sol–gel method, and the photocatalytic performance of the TiO_2-Au composite was dependent on the proportion of Au in the composite. The composite with 0.5 wt % Au achieved the

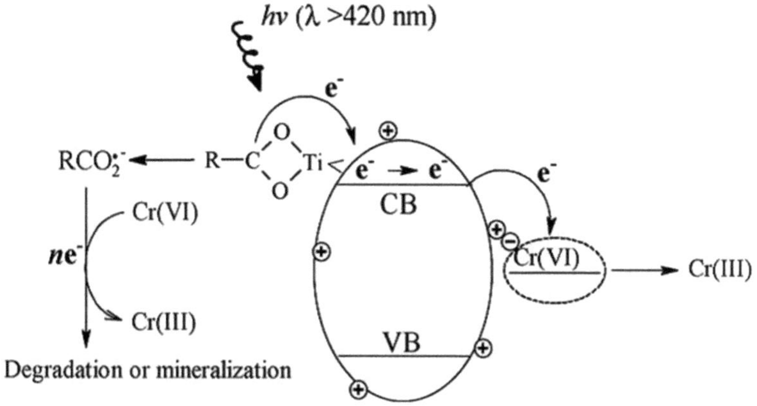

Fig. 7.5 SOA-induced photocatalytic reduction of Cr(VI) over neat TiO_2 under visible light irradiation via a charge-transfer-complex-mediated pathway(Yao et al. 2012)

highest photocatalytic reduction removal percentage (91 %) compared to the pure TiO_2 (83 %) (Liu et al. 2013a).

TiO_2 doped with one or more than two non-metallic elements has also been evaluated for the photocatalytic removal of Cr(VI) to further increase the photocatalytic activity under UV light and to make the doped TiO_2 more stable through charge compensation. The N-doped and N-F-co-doped TiO_2 catalysts prepared by the sol–gel method were tested for the photocatalytic reduction of Cr (VI) in the presence of oxalate ions. The N-F-co-doped TiO_2 material showed higher efficiency than the Degussa P25 TiO_2 catalyst for the removal of Cr (VI) (Giannakas et al. 2013).

In addition, TiO_2 particles modified with carbon nanotubes (CNTs) have been evaluated for the removal of Cr(VI) (Waldmann and Paz 2010; Xu et al. 2008). The CNTs are particularly interesting materials because they have unique electronic properties associated with their special 1D structure that facilitates charge transfer (Di Paola et al. 2012). Xu et al. (2008) evaluated composites of carboxyl-modified multi-walled carbon nanotubes (MWNT-COOH) and TiO_2 for the removal of Cr (VI). The photocatalytic activity of the nanotubes was enhanced compared with that of Degussa P25 TiO_2.

Instead of using TiO_2 alone, a coupled semiconductor configuration improves charge separation in many cases. The photocatalytic activities of various composites formed by TiO_2 and other inorganic oxides or sulfides have also been studied for Cr(VI) removal (Yang et al. 2010, 2013; Zhang et al. 2012). SnS_2/TiO_2 nanocomposites synthesized by a solvothermal method were applied for the photocatalytic reduction of Cr(VI) under visible light ($\lambda > 420$ nm) irradiation. The photocatalytic activities of SnS_2/TiO_2 nanocomposites depended on their compositions, and SnS_2/TiO_2 with a suitable content (44.5 mass %) of TiO_2 possessed the highest photocatalytic activity. This material showed good stability after the fifth reuse cycle of the catalyst on the reduction ratio of Cr(VI) (see

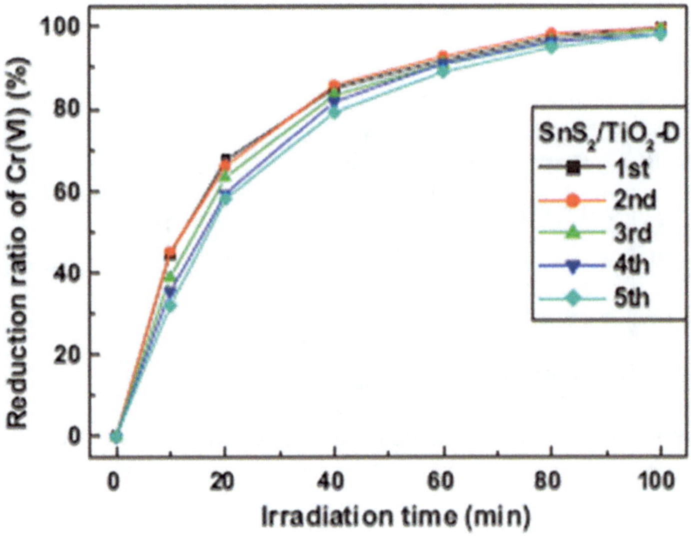

Fig. 7.6 Photocatalytic performances of SnS_2/TiO_2 for photocatalytic reduction during the first five reuse cycles (Zhang et al. 2012)

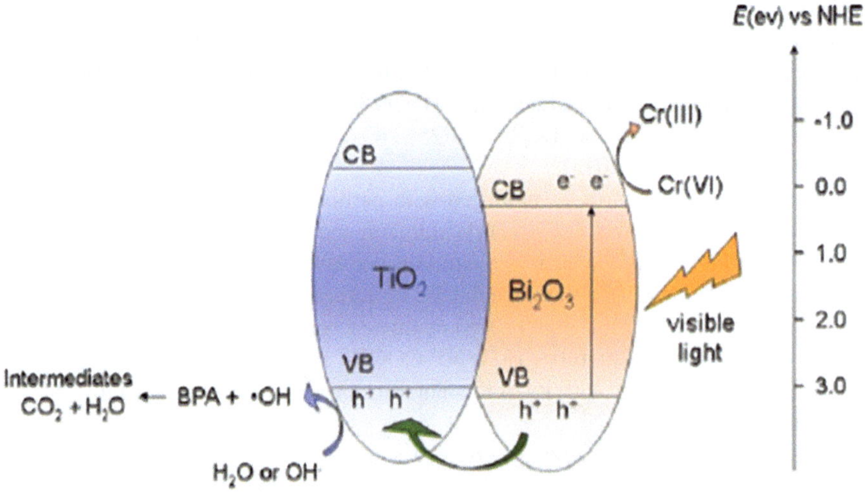

Fig. 7.7 Photocatalytic reduction of Cr(VI) and degradation of bisphenol in the Bi_2O_3/TiO_2 system under visible light irradiation (Yang et al. 2013)

Fig. 7.6), achieving a reduction of 98 % under visible light irradiation within 100 min (Zhang et al. 2012).

Additionally, Yang et al. (2013) evaluated a Bi_2O_3/TiO_2 photocatalyst under visible light on the reduction of Cr(VI) in aqueous solution and assessed its synergistic effect on the photodegradation of bisphenol (BPA) (Fig. 7.7).

Among the composite photocatalysts, 2 % (molar ratio) Bi_2O_3/TiO_2 exhibited the highest reduction rate. The removal rate of Cr(VI) in the presence of BPA was nearly threefold higher than that obtained in its absence.

The Cr(VI) photocatalytic reduction on WO_3-doped TiO_2 NT (nanotube) arrays in the presence of citric acid, which serves as both a donor scavenger and a redox reagent in the photocatalytic process, was evaluated by Yang et al. (Yang et al. 2010). The highest photoreduction efficiency of Cr(VI) was obtained with the WO_3/TiO_2 NT with a 1 % W content. The advantage of WO_3/TiO_2 NT arrays is that they can be easily removed and replaced after the photocatalytic reaction, avoiding the filtration step after the photoreaction and the immobilization process required for photocatalyst particles.

7.2.2.2 Photocatalytic Reduction of Cr(VI) with ZnO and Other Semiconductors

As mentioned previously, ZnO has long been studied as a semiconductor to replace TiO_2 for the removal of hexavalent chromium. ZnO has often been considered a valid alternative to TiO_2 due to its optoelectronic, catalytic, and photochemical properties as well as its low cost. However, the occurrence of photocorrosion and the susceptibility of ZnO to facilitate dissolution at extreme pH values have significantly limited its application in photocatalysis (Di Paola et al. 2012). Table 7.6 shows a summary of the main applications described using ZnO and modified ZnO for the removal of Cr(VI). Commercial ZnO in the slurry mode has been evaluated for the removal of Cr(VI) from synthetic solution (Chakrabarti et al. 2009) and soil leachate samples (Delgado-Balderas et al. 2012). Alternatively, Behnajady et al. immobilized commercial ZnO onto a glass plate by the heat attachment method for the removal of 5 mg L^{-1} Cr(VI) at pH 5.7 using a 254 nm UV light source (35 W m^{-2}) (Behnajady et al. 2012).

Both the activity and stability of ZnO could be affected by changing its morphology (Guo et al. 2011). Thus, ZnO synthesized by precipitation (Liu et al. 2012a) (Qamar et al. 2011; Siboni et al. 2011a) has been evaluated for Cr (VI) removal. The prepared catalysts of nanometer size were evaluated at neutral or slightly alkaline pH values using an aqueous ZnO suspension and applying UV light as the irradiation source. The photocatalytic efficiency of ZnO for the removal of Cr(VI) under UV radiation was improved by modifying the ZnO surface with reduced graphene oxide (RGO) (Liu et al. 2013b, 2012b). The ZnO-RGO composite with 1.0 wt.% RGO achieved the highest Cr(VI) removal rate of 96 % in 240 min, and the composite was stable, allowing its use after three recycling cycles (Liu et al. 2012b).

A few attempts have be made to study the photocatalytic activity of other semiconductors, such as $La_2Ti_2O_7$ (Yang et al. 2011) and Bi_2O_3-ZrO_2 (Vignesh et al. 2013), for the reduction of Cr(VI) in aqueous solution. $La_2Ti_2O_7$ was found to be a highly active photocatalyst for the reduction of 30 mg L^{-1} Cr(VI) in aqueous suspension under UV irradiation. After 3 h of irradiation, the Cr(VI) ions were reduced by 98 % in acidic solutions (pH 2). Additionally, it was found that the

Table 7.6 Application of a ZnO semiconductor for the removal of Cr(VI)

Catalyst	Light irradiation conditions	Experimental conditions	Results	Reference
Commercial ZnO Particle size 146 nm BET surface area 3.23 m^2g^{-1}	UV light ($\lambda = 254$ nm)	Cr (VI) concentration 54 mg L^{-1} 1.25 mol L^{-1} methanol as scavenger Catalyst loading 3 g L^{-1} Solution pH 4.5 Reaction time 75 min	The extent of degradation as well as its initial rate increased with the loading of the photocatalyst and intensity of UV radiation. About 90 % reduction of the substrate could be achieved over a reaction time of 75 min	Chakrabarti et al. (2009)
Commercial ZnO Particle size 30.40 nm BET surface area 3.16 m^2g^{-1}	UV light ($\lambda = 365$ nm)	Cr (VI) concentration 15 mg L^{-1} 14 mg L^{-1} humic acid as scavenger Catalyst loading 2 g L^{-1} Solution pH 7 Reaction time 360 min	An 84 % reduction of Cr(VI) for synthetic solutions while 13–99 % for soil leachates was achieved over a reaction time of 360 min. A marked synergistic effect between the photocatalytic reduction of Cr (VI) and the organic matter was observed in leachates samples and synthetic solutions with humic acids	Delgado-Balderas et al. (2012)
ZnO synthesized by precipitation method Particle size 50 nm	UV light ($\lambda = 247.3$ nm, 1,020 μWm^{-2})	Cr (VI) concentration 5 mg L^{-1} Ni (II) concentration 5 mg L^{-1} Catalyst loading 1 g L^{-1} Solution pH 7 Reaction time 120 min	After 2 h, Cr(VI) and Ni(II) removal at 1 g L^{-1} ZnO dosage was 80 % and 90 %, respectively. A photocatalytic reaction with illuminated ZnO can be effectively applied to treat wastewater contaminated with Cr(VI) or Ni(II)	Siboni et al. (2011a)
ZnO synthesized by precipitation method (microwave	UV light ($\lambda = 365$ nm)	Cr (VI concentration 10 mg L^{-1} Catalyst loading 1 g L^{-1}	Removal efficiency of 81 % was achieved under UV irradiation due to the maximal intensity of	Liu et al. (2012a)

(continued)

Table 7.6 (continued)

Catalyst	Light irradiation conditions	Experimental conditions	Results	Reference
assisted method) ZnO sheets showing wurtzite crystal structure		Solution pH 9 Reaction time 240 min	light absorption and minimal probability of electron–hole pair recombination	
ZnO synthesized by precipitation method Particle size 10–70 nm (average) ZnO sheets showing wurtzite crystal structure	UV light ($\lambda = 365$ nm, laser irradiation source)	Cr (VI) concentration 50 mg L^{-1} Catalyst loading 2 or 3 g L^{-1} Solution pH 7 Reaction time 60 min	The removal of Cr (VI) was ~95 % within short time (60 min) in aqueous suspensions of zinc oxide nanoparticles using a novel laser-induced photocatalytic process	Qamar et al. (2011)
ZnO/reduced graphene oxide (RGO)	UV light ($\lambda = 365$ nm, 500 W high pressure Hg lamp)	Cr (VI) concentration 10 mg L^{-1} Catalyst loading 1 g L^{-1} Solution pH not reported Reaction time 240 min	The ZnO-RGO 1 wt % showed 96 % removal of Cr(VI) in aqueous suspension compared with 67 % removal achieved using ZnO	Liu et al. (2012b)

photocatalytic activity of $La_2Ti_2O_7$ remained almost unchanged after nine successive cycles (Yang et al. 2011).

Vignesh et al. (2013) evaluated the activity of the mixed oxide Bi_2O_3-ZrO_2 for Cr(VI) removal under UV light irradiation. A 92.3 % removal percentage of Cr (VI) was achieved with a Bi_2O_3-ZrO_2 loading of 1.25 g L^{-1}, an initial Cr (VI) concentration of 5 mg L^{-1}, pH 2, and an irradiation time of 180 min. Moreover, the nanocomposite was highly stable with an increased surface area and possessed maximum photocatalytic activity for the reduction of Cr(VI) compared to Bi_2O_3, ZrO_2, TiO_2, and ZnO.

Concluding Remarks

Heterogeneous photocatalysis appears to have a promising potential application for the removal of inorganic pollutants by their transformation into harmless final compounds.

(continued)

(continued)

The photocatalytic process works well for all cyanide species. Most of the studies on cyanide removal have been performed using powdered suspensions. Although reports on the photocatalytic mineralization of organics under visible light are many, few focused on cyanide for the following reasons: (1) the carbon in the cyanide ion is more strongly bound to nitrogen than to nitrogen atoms in azo dye molecules (nitrogen-nitrogen double bond); (2) the photocatalytic removal of cyanide requires a stable photocatalyst in a highly alkaline medium to avoid its volatilization, because the pKa of HCN is 9.3; and (3) furthermore, in a basic solution, the semiconductor oxide surface is negatively charged due to adsorption of the hydroxide ion, which could inhibit the adsorption of the cyanide ion on the photocatalytic surface; thus effectiveness of photocatalytic treatment decrease since the substrate should be adsorbed on the photocatalyst.

In case of arsenic, the photocatalytic processes developed to remove As (III) and organic arsenic species have been based mainly on the use of TiO_2. TiO_2 photocatalysis is an effective method for oxidizing As(III)/organic arsenic to As(V). Although As(III)/organic arsenic can be oxidized to As (V) in the TiO_2/UV system, inefficient removal of As(V) is achieved using TiO_2 as an adsorbent. This shortcoming can be overcome by combining TiO_2 with other adsorbents with good adsorption properties in one system, developing bifunctional adsorbents with both great photocatalytic ability and high adsorption capacity or evaluating an alternative photocatalyst. Only few studies have been carried out to investigate the influence of coexisting ions on the TiO_2-photocatalyzed oxidation of As(III)/organic arsenic compounds.

Finally, the application of modified photocatalytic semiconductors for the removal of Cr(VI) using UV light has been extensively reviewed. Some of the evaluated modified semiconductors showed photocatalytic activity under both UV and visible light for the reduction of hexavalent chromium. The Cr(VI) photocatalytic mechanism is usually described as direct reduction by conduction band electrons, which is very slow in the absence of electron donors but can be accelerated by organic electron donors.

References

Aragay G, Pino F, Merkoci A (2012) Nanomaterials for sensing and destroying pesticides. Chem Rev 112(10):5317–5338. doi:10.1021/cr300020c

Baeissa ES, Mohamed RM (2013) Enhancement of photocatalytic properties of Ga_2O_3-SiO_2 nanoparticles by Pt deposition. Chin J Catal 34(6):1167–1172. doi:10.1016/s1872-2067(12)60570-1

Barakat MA (2011) New trends in removing heavy metals from industrial wastewater. Arab J Chem 4(4):361–377. doi:10.1016/j.arabjc.2010.07.019

Barrera-Díaz CE, Lugo-Lugo V, Bilyeu B (2012) A review of chemical, electrochemical and biological methods for aqueous Cr(VI) reduction. J Hazard Mater 223–224:1–12. doi:10.1016/j.jhazmat.2012.04.054

Behnajady MA, Mansoriieh N, Modirshahla N, Shokri M (2012) Influence of operational parameters and kinetics analysis on the photocatalytic reduction of Cr(VI) by immobilized ZnO. Environ Technol 33(3):265–271. doi:10.1080/09593330.2011.569957

Chakrabarti S, Chaudhuri B, Bhattacharjee S, Ray AK, Dutta BK (2009) Photo-reduction of hexavalent chromium in aqueous solution in the presence of zinc oxide as semiconductor catalyst. Chem Eng J 153(1–3):86–93. doi:10.1016/j.cej.2009.06.021

Choi W, Yeo J, Ryu J, Tachikawa T, Majima T (2010) Photocatalytic oxidation mechanism of As (III) on TiO_2: unique role of As(III) as a charge recombinant species. Environ Sci Technol 44 (23):9099–9104. doi:10.1021/es102507u

Delgado-Balderas R, Hinojosa-Reyes L, Guzman-Mar JL, Garza-Gonzalez MT, Lopez-Chuken UJ, Hernandez-Ramirez A (2012) Photocatalytic reduction of Cr(VI) from agricultural soil column leachates using zinc oxide under UV light irradiation. Environ Technol 33 (23):2673–2680. doi:10.1080/09593330.2012.676070

Di Paola A, García-López E, Marcì G, Palmisano L (2012) A survey of photocatalytic materials for environmental remediation. J Hazard Mater 211–212:3–29. doi:10.1016/j.jhazmat.2011.11.050

Farrokhi M, Yang JK, Lee SM, Shirzad-Siboni M (2013) Effect of organic matter on cyanide removal by illuminated titanium dioxide or zinc oxide nanoparticles. J Environ Health Sci Eng 11(1):23. doi:10.1186/2052-336x-11-23

Fostier AH, Pereira MSS, Rath S, Guimarães JR (2008) Arsenic removal from water employing heterogeneous photocatalysis with TiO_2 immobilized in PET bottles. Chemosphere 72 (2):319–324. doi:10.1016/j.chemosphere.2008.01.067

Fu F, Wang Q (2011) Removal of heavy metal ions from wastewaters: a review. J Environ Manag 92(3):407–418. doi:10.1016/j.jenvman.2010.11.011

Giannakas AE, Seristatidou E, Deligiannakis Y, Konstantinou I (2013) Photocatalytic activity of N-doped and N-F co-doped TiO_2 and reduction of chromium(VI) in aqueous solution: an EPR study. Appl Catal B-Environ 132:460–468. doi:10.1016/j.apcatb.2012.12.017

Guo MY, Ng AMC, Liu FZ, Djurisic AB, Chan WK, Su HM, Wong KS (2011) Effect of native defects on photocatalytic properties of ZnO. J Phys Chem C 115(22):11095–11101. doi:10.1021/jp200926uv

Harraz FA, Abdel-Salam OE, Mostafa AA, Mohamed RM, Hanafy M (2013) Rapid synthesis of titania–silica nanoparticles photocatalyst by a modified sol–gel method for cyanide degradation and heavy metals removal. J Alloys Compd 551:1–7. doi:10.1016/j.jallcom.2012.10.004

Kabra K, Chaudhary R, Sawhney RL (2009) Application of solar photocatalytic treatment to industrial wastewater from a chrome plating unit. Int J Green Energy 6(1):83–91. doi:10.1080/15435070802701868

Karunakaran C, Gomathisankar P (2013) Solvothermal synthesis of CeO_2-TiO_2 nanocomposite for visible light photocatalytic detoxification of cyanide. ACS Sustain Chem Eng 130917082519009. doi: 10.1021/sc400195n

Karunakaran C, Gomathisankar P, Manikandan G (2010) Preparation and characterization of antimicrobial Ce-doped ZnO nanoparticles for photocatalytic detoxification of cyanide. Mater Chem Phys 123(2–3):585–594. doi:10.1016/j.matchemphys.2010.05.019

Karunakaran C, Abiramasundari G, Gomathisankar P, Manikandan G, Anandi V (2011a) Preparation and characterization of ZnO-TiO_2 nanocomposite for photocatalytic disinfection of bacteria and detoxification of cyanide under visible light. Mater Res Bull 46(10):1586–1592. doi:10.1016/j.materresbull.2011.06.019

Karunakaran C, Gomathisankar P, Manikandan G (2011b) Solar photocatalytic detoxification of cyanide by different forms of TiO_2. Korean J Chem Eng 28(5):1214–1220. doi:10.1007/s11814-010-0503-1

Kumar SG, Devi LG (2011) Review on modified TiO_2 photocatalysis under UV/visible Light: selected results and related mechanisms on interfacial charge carrier transfer dynamics. J Phys Chem A 115(46):13211–13241. doi:10.1021/jp204364a

Kuyucak N, Akcil A (2013) Cyanide and removal options from effluents in gold mining and metallurgical processes. Miner Eng 50–51:13–29. doi:10.1016/j.mineng.2013.05.027

Li Q, Easter NJ, Shang JK (2009) As(III) removal by palladium-modified nitrogen-doped titanium oxide nanoparticle photocatalyst. Environ Sci Technol 43(5):1534–1539. doi:10.1021/es8025837

Litter MI (2009) Treatment of chromium, mercury, lead, uranium, and arsenic in water by heterogeneous photocatalysis. 36. doi: 10.1016/s0065-2377(09)00402-5

Liu X, Lv T, Pan L, Sun Z, Sun CQ (2012a) Microwave-assisted synthesis of ZnO for photocatalytic reduction of Cr(VI) in aqueous solution. Desalin Water Treat 42 (1–3):216–221. doi:10.5004/dwt.2012.2825

Liu X, Pan L, Zhao Q, Lv T, Zhu G, Chen T, Lu T, Sun Z, Sun C (2012b) UV-assisted photocatalytic synthesis of ZnO-reduced graphene oxide composites with enhanced photocatalytic activity in reduction of Cr(VI). Chem Eng J 183:238–243. doi:10.1016/j.cej.2011.12.068

Liu X, Lv T, Liu Y, Pan L, Sun Z (2013a) TiO_2-Au composite for efficient UV photocatalytic reduction of Cr(VI). Desalin Water Treat 51(19–21):3889–3895. doi:10.1080/19443994.2013.781739

Liu X, Pan L, Lv T, Sun Z (2013b) Investigation of photocatalytic activities over ZnO-TiO_2-reduced graphene oxide composites synthesized via microwave-assisted reaction. J Colloid Interface Sci 394:441–444. doi:10.1016/j.jcis.2012.11.047

López-Muñoz M-J, Aguado J, van Grieken R, Marugán J (2009) Simultaneous photocatalytic reduction of silver and oxidation of cyanide from dicyanoargentate solutions. Appl Catal B Environ 86(1–2):53–62. doi:10.1016/j.apcatb.2008.07.022

Marugan J, van Grieken R, Cassano AE, Alfano OM (2008) Intrinsic kinetic modeling with explicit radiation absorption effects of the photocatalytic oxidation of cyanide with TiO_2 and silica-supported TiO_2 suspensions. Appl Catal B-Environ 85(1–2):48–60. doi:10.1016/j.apcatb.2008.06.026

Meichtry JM, Rivera V, Di Iorio Y, Rodriguez HB, San Roman E, Grela MA, Litter MI (2009) Photoreduction of Cr(VI) using hydroxoaluminiumtricarboxymonoamide phthalocyanine adsorbed on TiO_2. Photochem Photobiol Sci 8(5):604–612. doi:10.1039/b816441j

Mohamed RM, Baeissa ES (2013) Preparation and characterisation of Pd–TiO_2–hydroxyapatite nanoparticles for the photocatalytic degradation of cyanide under visible light. Appl Catal A Gen 464–465:218–224. doi:10.1016/j.apcata.2013.05.043

Mukherjee K, Saha R, Ghosh A, Saha B (2012) Chromium removal technologies. Res Chem Intermed 39(6):2267–2286. doi:10.1007/s11164-012-0779-3

Osathaphan K, Chucherdwatanasak B, Rachdawong P, Sharma VK (2008) Photocatalytic oxidation of cyanide in aqueous titanium dioxide suspensions: effect of ethylenediaminetetraacetate. Sol Energy 82(11):1031–1036. doi:10.1016/j.solener.2008.04.007

Osathaphan K, Ruengruehan K, Yngard RA, Sharma VK (2013) Photocatalytic Degradation of Ni (II)-Cyano and Co(III)-Cyano Complexes. Water Air Soil Pollution 224 (8). doi:1647 10.1007/s11270-013-1647-5

Pifferi V, Spadavecchia F, Cappelletti G, Paoli EA, Bianchi CL, Falciola L (2013) Electrodeposited nano-titania films for photocatalytic Cr(VI) reduction. Catal Today 209:8–12. doi:10.1016/j.cattod.2012.08.031

Pineda Arellano CA, Silva Martinez S (2010) Effects of pH on the degradation of aqueous ferricyanide by photolysis and photocatalysis under solar radiation. Sol Energy Mater Sol Cells 94(2):327–332. doi:10.1016/j.solmat.2009.10.008

Qamar M, Gondal MA, Yamani ZH (2011) Laser-induced efficient reduction of Cr(VI) catalyzed by ZnO nanoparticles. J Hazard Mater 187(1–3):258–263. doi:10.1016/j.jhazmat.2011.01.007

Qiu R, Zhang D, Diao Z, Huang X, He C, Morel J-L, Xiong Y (2012) Visible light induced photocatalytic reduction of Cr(VI) over polymer-sensitized TiO_2 and its synergism with phenol oxidation. Water Res 46(7):2299–2306. doi:10.1016/j.watres.2012.01.046

Rivera-Reyna N, Hinojosa-Reyes L, Guzmán-Mar JL, Cai Y, O'Shea K, Hernández-Ramírez A (2013) Photocatalytical removal of inorganic and organic arsenic species from aqueous solution using zinc oxide semiconductor. Photochem Photobiol Sci 12(4):653. doi:10.1039/c2pp25231g

Saleh TA, Gondal MA, Drmosh QA (2010) Preparation of a MWCNT/ZnO nanocomposite and its photocatalytic activity for the removal of cyanide from water using a laser. Nanotechnology 21 (49):495705. doi:10.1088/0957-4484/21/49/495705

Salinas-Guzmán RR, Guzmán-Mar JL, Hinojosa-Reyes L, Peralta-Hernández JM, Hernández-Ramírez A (2010) Enhancement of cyanide photocatalytic degradation using sol–gel ZnO sensitized with cobalt phthalocyanine. J Sol-Gel Sci Technol 54(1):1–7. doi:10.1007/s10971-009-2145-5

Sharma VK, Sohn M (2009) Aquatic arsenic: toxicity, speciation, transformations, and remediation. Environ Int 35(4):743–759. doi:10.1016/j.envint.2009.01.005

Siboni MS, Samadi MT, Yang JK, Lee SM (2011a) Photocatalytic reduction of Cr(VI) and Ni (II) in aqueous solution by synthesized nanoparticle ZnO under ultraviolet light irradiation: a kinetic study. Environ Technol 32(14):1573–1579. doi:10.1080/09593330.2010.543933

Siboni MS, Samarghandi MR, Yang JK, Lee SM (2011b) Photocatalytic removal of cyanide with illuminated TiO_2. Water Sci Technol 64(7):1383. doi:10.2166/wst.2011.738

Tian Q, Zhuang J, Wang J, Xie L, Liu P (2012) Novel photocatalyst, $Bi_2Sn_2O_7$, for photooxidation of As(III) under visible-light irradiation. Appl Catal A Gen 425–426:74–78. doi:10.1016/j.apcata.2012.03.005

Tsimas ES, Tyrovola K, Xekoukoulotakis NP, Nikolaidis NP, Diamadopoulos E, Mantzavinos D (2009) Simultaneous photocatalytic oxidation of As(III) and humic acid in aqueous TiO_2 suspensions. J Hazard Mater 169(1–3):376–385. doi:10.1016/j.jhazmat.2009.03.107

Vignesh K, Priyanka R, Rajarajan M, Suganthi A (2013) Photoreduction of Cr(VI) in water using Bi_2O_3-ZrO_2 nanocomposite under visible light irradiation. Mat Sci Eng B-Adv Functional Solid-State Mat 178(2):149–157. doi:10.1016/j.mseb.2012.10.035

Waldmann NS, Paz Y (2010) Photocatalytic reduction of Cr(VI) by titanium dioxide coupled to functionalized CNTs: an example of counterproductive charge separation. J Phys Chem C 114 (44):18946–18952. doi:10.1021/jp105925g

Wang L, Wang N, Zhu L, Yu H, Tang H (2008) Photocatalytic reduction of Cr(VI) over different TiO_2 photocatalysts and the effects of dissolved organic species. J Hazard Mater 152(1):93–99. doi:10.1016/j.jhazmat.2007.06.063

Wang N, Zhu L, Deng K, She Y, Yu Y, Tang H (2010) Visible light photocatalytic reduction of Cr (VI) on TiO_2 in situ modified with small molecular weight organic acids. Appl Catal B-Environ 95(3–4):400–407. doi:10.1016/j.apcatb.2010.01.019

Xu Z, Long Y, Kang S-Z, Mu J (2008) Application of the composite of TiO_2 nanoparticles and carbon nanotubes to the photo-reduction of Cr(VI) in water. J Dispers Sci Technol 29 (8):1150–1152. doi:10.1080/01932690701817982

Xu S, Zhang Y, Wang S, Xu J, Ding H, Li G (2013) Structure-enhanced photocatalytic removal of Cr(VI) by a TiO_2 superstructure with ultrathin rutile nanorods and abundant {110} faces. Eur J Inorg Chem 14:2601–2607. doi:10.1002/ejic.201201475

Yang L, Xiao Y, Liu S, Li Y, Cai Q, Luo S, Zeng G (2010) Photocatalytic reduction of Cr(VI) on WO_3 doped long TiO_2 nanotube arrays in the presence of citric acid. Appl Catal B-Environ 94 (1–2):142–149. doi:10.1016/j.apcatb.2009.11.002

Yang Q-L, Kang S-Z, Chen H, Bu W, Mu J (2011) $La_2Ti_2O_7$: An efficient and stable photocatalyst for the photoreduction of Cr(VI) ions in water. Desalination 266(1–3):149–153. doi:10.1016/j.desal.2010.08.018

Yang JK, Lee SM, Farrokhi M, Giahi O, Siboni MS (2012a) Photocatalytic removal of Cr(VI) with illuminated TiO_2. Desalin Water Treat 46(1–3):375–380. doi:10.1080/19443994.2012.677564

Yang JK, Lee SM, Siboni MS (2012b) Effect of different types of organic compounds on the photocatalytic reduction of Cr(VI). Environ Technol 33(17):2027–2032. doi:10.1080/09593330.2012.655325

Yang J, Dai J, Li J (2013) Visible-light-induced photocatalytic reduction of Cr(VI) with coupled Bi_2O_3/TiO_2 photocatalyst and the synergistic bisphenol A oxidation. Environ Sci Pollut Res 20 (4):2435–2447. doi:10.1007/s11356-012-1131-6

Yao S, Jia Y, Shi Z, Zhao S (2010) Photocatalytic oxidation of arsenite by a composite of titanium dioxide and activated carbon fiber. Photochem Photobiol 86(6):1215–1221. doi:10.1111/j.1751-1097.2010.00813.x

Yao SH, Jia YF, Zhao SL (2012) Photocatalytic oxidation and removal of arsenite by titanium dioxide supported on granular activated carbon. Environ Technol 33(9):983–988. doi:10.1080/09593330.2011.604857

Zhang YC, Li J, Xu HY (2012) One-step in situ solvothermal synthesis of SnS_2/TiO_2 nanocomposites with high performance in visible light-driven photocatalytic reduction of aqueous Cr(VI). Appl Catal B-Environ 123:18–26. doi:10.1016/j.apcatb.2012.04.018

Zheng S, Cai Y, O'Shea KE (2010) TiO_2 photocatalytic degradation of phenylarsonic acid. J Photochem Photobiol A Chem 210(1):61–68. doi:10.1016/j.jphotochem.2009.12.004

Zhou W, Fu H, Pan K, Tian C, Qu Y, Lu P, Sun C-C (2008) Mesoporous $TiO_2/\alpha\text{-}Fe_2O_3$: bifunctional composites for effective elimination of arsenite contamination through simultaneous photocatalytic oxidation and adsorption. J Phys Chem C 112(49):19584–19589. doi:10.1021/jp806594m

Chapter 8
Photocatalytic Materials in Water Disinfection

Erick R. Bandala and Erika Bustos

Abstract It has been estimated that 1.2 billion people lack access to safe drinking water around the world resulting not only on the prevalence of waterborne diseases but also threatening the economic development. Water resources management is critical, mainly in regions with low rainfall and growing population.

The lack of access to safe drinking water has become a main concern usually related with poverty since 1.5 million children die yearly as a result of waterborne diseases. In Mexico, waterborne diseases affect 6 % of the total population with rural communities as the most affected anytime; 22 % of the population does not has access to piped water (CNA 2011). This situation, however, is not limited to Mexico but common in other developing countries.

In order to deal with the generation of safe drinking water, several different water disinfection processes have been developed. Free chlorine and its derivatives are currently the most widely used disinfecting reagent capable to inactivate pathogenic species in drinking water. However, it is also well known that free chlorine reacts with organic matter present in surface or underground water to generate disinfection by-products (DBPs) reported as human carcinogens. Within the variety of disinfection technologies, advanced oxidation processes (AOPs) have emerged as efficient and cost-effective for pathogen inactivation in water. Specifically, heterogeneous photocatalysis (HPC) is the AOP with the widest variety of technological applications resulting in its ability to inactivate nuisance microorganisms. The use of metal-based semiconductors as photocatalyst for water disinfection has been widely reported demonstrating being effective for the inactivation of several microorganisms and carcinogen cells.

E.R. Bandala (✉)
Grupo de Investigación en Energía y Ambiente, Universidad de las Américas,
Puebla, Sta. Catarina Mártir, Cholula 72810, Puebla, Mexico

E. Bustos
Grupo de Electroquímica Ambiental, Centro de Investigación y Desarrollo
Tecnológico en Electroquímica, S. C., Parque Tecnológico Querétaro s/n,
Sanfandila, Pedro Escobedo 76703, Querétaro, Qro., México

A. Hernández-Ramírez, I. Medina-Ramírez (eds.), *Photocatalytic Semiconductors*,
DOI 10.1007/978-3-319-10999-2_8

From the economic point of view, the possibility of using solar energy to drive HPC processes may be considered as an interesting alternative for use in developing countries to assure access to safe drinking water. The application of these technologies to water disinfection using solar radiation, coined as enhanced photocatalytic solar disinfection, has allowed the efficient inactivation of highly resistant microorganisms. The aim of this chapter is reviewing the state of the art in the application of HPC processes with TiO_2 and some new semiconductor materials for inactivation of waterborne pathogens, discussing its advantages and potential limitations as well as analyzing challenges and opportunities for its application at real scale in regions with lack of access to safe drinking water.

8.1 Process for Water Disinfection

The primary purpose of water disinfection is the prevention of waterborne diseases. Of particular concern is that in a sizeable fraction of the disease outbreaks recorded, the etiologic agent was not identified. However, in many of these cases, enteric viruses are believed to be the causative agents. The mechanisms for microbial inactivation include the following (Rajeshwar and Ibañez 1997):

- Laceration of the cell wall
- Modification of cell permeability
- Modification of the nature of the protoplasm
- Alteration of nucleic acids
- Disruption of protein synthesis
- Induction of abnormal redox process
- Inhibition of enzyme activity

The promulgation of a number of new regulations for the control of microbiological and chemical pollutants in drinking water has prompted the search for suitable, cost-effective alternative methods for primary disinfection. Of particular concerns are the disinfection by-products (DBPs) of chlorination and that groundwater high in natural organic matter (NOM) may be incompatible with the more traditional chemical disinfectants. Even the alternatives currently considered such as chloramines may be inappropriate because they are weak virucides and would be unlikely to meet primary disinfection requirements (Rajeshwar and Ibañez 1997).

A variety of chemical agents can be used to inactivate microorganisms (Metcalf and Eddy 1991). These include halogens and derivatives (Cl_2, Br_2, I_2, HOCl, OCl^-, ClO_2, HOBr, HOI polyiodide anion exchange resins), oxygenated and highly oxidizing compounds (ozone, hydrogen peroxide, phenols, alcohols, persulfate and percarbonate, peracetic acid, potassium permanganate), metal ions (Ag^+, Cu^{2+}, etc.), dyes, quaternary ammonium compounds, strong acids and bases, and enzymes. There are also physical processes included in the disinfection as the electromagnetic radiation (ultrasonic waves, heat, visible light, UV light, gamma radiation, X-rays), particle radiation (electron beam), and electrical current.

A variety of factors influence the disinfection efficiency, including the contact time, chemical nature, and concentration of the disinfecting agent as well as the initial mixing mode and point of injection, nature and intensity of the physical agents, temperature, type, concentration and age of the microorganisms, and the nature of the liquid carrier (Rajeshwar and Ibañez 1997). Common attributes of the important chemical disinfectants (Cl_2, OCl^-, ClO_2, and O_3) are the following (Metcalf and Eddy 1991):

- Highly potent microorganism inactivation and relatively high toxicity to humans and animals
- Active interaction (normally oxidation or addition) with organic matter and with inorganic reducing agents
- Sufficient solubility in aqueous media (except the dihalogens due to their nonpolar nature)
- Penetration capability through surfaces and cell membranes
- Moderate to good deodorizing ability

The relative stabilities of these chemicals follow the order $Cl_2 > OCl^- > ClO_2$, O_3, and their relative costs are $O_3 > ClO_2$, $OCl^- > Cl_2$. Undesirable characteristics include DBP production, corrosivity to metallic materials, membrane attack, and discoloration of dyes and tints (Allison 1993). In fact, it has been stated that it would be ideal to separate the oxidation and disinfection functions in the water treatment system (Miller 1993; Rajeshwar and Ibañez 1997).

8.1.1 Electrochemical Disinfection Processes

Microorganism can be electrochemically inactivated either directly or via the generation of an oxidant or reductant agent such as $^{\bullet}OH$, where a third route involves the electrosorption of bacteria and the like on the electrode surface and their subsequent inactivation.

Electrode materials vary widely, depending on the disinfecting agent desired. Cathode materials include stainless steel, copper, graphite, carbon cloth, and reticulated vitreous carbon (RVC). Anode materials include platinized titanium or niobium, tantalum, graphite, carbon, metal oxides, silver, copper, nickel, Monel, dimensionally stable anodes, and combinations thereof. Electrocatalytic materials can be incorporated into electrodes, for example, in the form of coatings, or incorporated into cell separators. Three-dimensional electrodes have also been successfully used (Hinden et al. 1985; Rajeshwar and Ibañez 1997).

Narrow gap cell technology involving the use of a solid polymer electrolyte (SPE) has been applied to electrochemical water disinfection to decrease cell resistance and avoid the need for adding supporting electrolytes (Rajeshwar and Ibañez 1997). Direct and low- and high-frequency alternating current (AC) have been used for disinfection purposes. Deposit formation may occur on the cathodes, particularly in the case of hard water. This problem has been prevented by periodic current reversal and producing oscillations of the electrode (Stoner et al. 1982; Stoner 1973).

In the case of the electrosorption of microorganisms and direct electron transfer, bacteria show a tendency to adsorb onto surfaces such as activated carbon, fibrous carbon, or ion exchange resins. This tendency is driven mainly by electrostatic forces between charged groups on the cell wall (i.e., amino and carboxylic groups) and on the adsorbant. The potential-induced adsorption of solutes onto the surface of an electrode is called electrosorption, and its effectiveness depends on the potential of zero charge of the adsorbate (Oren et al. 1983; Golub et al. 1987).

An interesting electrochemical disinfection process has been described based on AC perturbation of the electrode/electrolyte interface. In the anodic part of the cycle, the pathogen adsorbed on the electrode is oxidized. In the cathodic portion, the oxidation products are reductively removed from the electrode surface, and a clean electrode surface is regenerated for subsequent disinfection cycles. This system is reported to work not only on bacteria but on larger organism such a protozoa (Stoner et al. 1982; Stoner 1973).

On the other hand, in situ electrogeneration of disinfection agents can be possible with the chlorine, hypochlorous acid, hypochlorite, chlorine dioxide, ozone, hydrogen peroxide, and chemical oxidation of MnO_2 to MnO_4^{2-} in presence or not of Ag^+ and Cu^{2+} (Yahya and Gerba 1992; Rajeshwar and Ibañez 1997).

8.1.2 Photocatalytic Disinfection Processes

The current definition of photocatalytic processes and the different steps involved within have been detailed previously in this book. HPC is among the most successful applications of the AOP's as suggested by the wide variety of research groups, facilities, scientific reports, and patents around the use of this technology for the elimination of toxic substances in water (Malato 1999; Bandala and Estrada 2007). The use of HPC for water disinfection, however, is an emerging field of research which has given rise to the design and synthesis of novel nanostructured photocatalysts with interesting properties for more efficient environmental applications.

As occurs with photocatalytic processes for organic chemical removal, mass transference is an important part in the inactivation using HPC processes. Initial practical approaches for a quantitative description of HPC kinetics have been commonly carried out using a Langmuir–Hinshelwood (L–H) kinetics model (Al-Ekabi and Serpone 1988). This mathematical model assumes that the reaction occurs on the catalyst surface. According to the L–H model, the reaction rate (r) is proportional to the fraction of particle surface covered by the pollutant (θ_x). Mathematically,

$$r = -\frac{\mathrm{d}C}{\mathrm{d}t} = k_r\theta_x = \frac{k_r KC}{1 + KC + K_s C_s} \tag{8.1}$$

where k_r is the reaction rate constant, K is the pollutant adsorption constant, C is the pollutant concentration at any time, K_s is the solvent adsorption constant, and C_s is its concentration. Many authors presented their data using the L–H kinetic approach. Nevertheless, despite that the L–H approach fits properly the experimental data, it does not consider the role of the radiation field on the mechanism (Bandala et al. 2004; Arancibia-Bulnes et al. 2002).

Other kinetic studies on HPC suggest that the reaction rate increases with catalyst concentration to get a maximum value concentration depending on the compound and the reactor used. Over these concentrations, reaction rate remains unchanged or decreases with further increments of catalyst concentration (Arancibia-Bulnes et al. 2002). An interesting problem associated is the relation between catalyst concentration, reaction rate, radiation absorption, and process improvement. Considering this, many different models have been proposed. Studies have suggested relationships aiming to estimate the radiation absorbed by the catalyst (Bandala et al. 2004; Arancibia-Bulnes et al. 2002). From these results, several models, most of them based on complex mathematical or statistical computational approaches, have been developed. These models are able to predict radiation absorption and scattering as function of catalyst concentration, optical path and catalyst type, and its relation to pseudo-kinetic constants obtained experimentally (Bandala et al. 2004; Arancibia-Bulnes et al. 2002; Curco et al. 2002). Based on the radiation absorbed by the catalyst, Alfano's group as well as other authors have focused on the a priori design of photochemical reactors, the improvement of HPC reactions, and the generation of intrinsic reaction kinetic that may lead to process scaling-up (Alfano et al. 2000; Cassano and Alfano 2000; Brandi et al. 2000).

Another widely used modeling approach commonly used for HPC inactivation of microorganisms is the Chick–Watson kinetics model (Younas et al. 2014) mathematically described as Eq. (8.2):

$$\log\frac{N}{N_0} = -kC^n t \tag{8.2}$$

where N and N_0 are the microorganism concentration at time t and at the starting point, respectively, k is the kinetic rate constant, C is the concentration of the disinfectant, and n is the reaction order. In recent work series, Bandala et al. (2009) have proposed modifications to the conventional Chick–Watson model for its application to the HPC processes for microorganism inactivation in water. The initial modification consisted in replacing the $C \times t$ factor (the product of disinfectant concentration and the reaction time) with the accumulate energy (dose, Q_n) entering to the photocatalytic system. The first-order reaction kinetic model used for fitting the experimental results for HPC processes for water disinfection is described in Eq. 8.3 (Bandala et al. 2009, 2011)

$$\ln\left(\frac{N}{N_0}\right) = -kQ_n \tag{8.3}$$

The use of Q_n as an estimation of the radiation dose has been proposed in the past, and it is estimated in agreement with Eq. (8.4):

$$Q_n = Q_{n-1} + \Delta t G_m\left(\frac{A}{V}\right) \tag{8.4}$$

where Q_n is the accumulate energy, Δt is the time between radiation measurements, G_n is the adjusted global radiation measured, A is the reactor area submitted to radiation, and V is the water volume treated. The models proposed in Eq. (8.3) have been demonstrated working properly for the case of inactivation of bacteria and other low-resistant microorganisms (Castillo-Ledezma et al. 2011). Nevertheless, when highly resistant microorganisms are tested, first-order kinetics may not describe properly the photocatalytic inactivation process. For these cases, the use of the delayed Chick–Watson model has been proposed (Corona-Vasquez et al. 2012):

$$\frac{N}{N_0} = \left\{ \begin{array}{l} \frac{N}{N_0} \text{ if } Q_n \leq Q_{n\text{lag}} = \frac{1}{k}\ln\left\{\left(\frac{N_1}{N_0}\right)\left(\frac{N_0}{N}\right)_c\right\} \\ \frac{N_1}{N_0}e^{-k_1 Q_n} \text{ if } Q_{n\text{lag}} \leq Q_n \leq Q_{n2} \\ \frac{N_2}{N_0}e^{-k_2 Q_n} \text{ if } Q_n \geq Q_{n2} = \frac{1}{k_2 - k_1}\ln\left(\frac{N_2}{N_1}\right) \end{array} \right\} \tag{8.5}$$

where k_1 and k_2 are the rate constants of the faster and slower pseudo-first-order kinetics (in L kJ^{-1}), N_1/N_0 and N_2/N_0 are the intercepts with the ordinate axis resulting from extrapolation of the two pseudo-first-order lines, $Q_{n\text{lag}}$ is the lag phase, Q_n (in kJ L^{-1}), Q_{n2} is the Q_n value corresponding to the intercept of the two pseudo-first-order lines (in kJ L^{-1}), and $(N/N_0)_c$ is the viability of the control.

Besides reactor design, heterogeneous photocatalytic degradation reaction can be enhanced by the use of higher active catalyst or inorganic oxidizing species. In the first case, activation of TiO_2 under visible light is a desirable technological approach. In order to utilize visible light for TiO_2 excitation, dye-sensitized and ion-doped TiO_2 has been developed in recent years showing promising results for the photocatalyzed inactivation of different microorganisms (Castillo-Ledezma et al. 2011, 2013; Bandala and Raichle 2013).

The main mechanism involved in the inactivation of pathogenic microorganisms has been suggested to be related with the cell damage produced by the so-called reactive oxygen species (ROS), mainly hydroxyl ($^{\bullet}OH$) and superoxide (O_2) radicals. According to different studies, these ROS are able to modify and eventually destroy the structure of the cell membrane (Malato et al. 2009; Alrousan et al. 2009) mainly as the result of lipid peroxidation (Dunlop et al. 2008; Alrousan et al. 2009). The initial damage is produced in the outer lipopolysaccharide and peptidoglycan walls, followed by lipid peroxidation and protein and polysaccharides oxidation (Malato et al. 2009; Dalrymple et al. 2010) affecting the regulatory function of the cell membrane for the internal and external interchange. The produced damage will further generate failure in the cell's respiratory activity and increase its permeability, allowing the attack of inner components leading to its death. During cellular metabolism, some ROS are produced, such as peroxide, hydroxyl and hydrogen peroxide, as a result of cell respiration. These highly oxidizing species are, however, in equilibrium with the immune system defense mechanism through antioxidizing enzyme production related to superoxide dismutase (SODs), catalase (CAT), and glutathione peroxidase (GPX) families (Castillo-Ledezma et al. 2011).

When microorganisms are exposed to major oxidative stress, i.e., ROS produced during HPC processes, enzyme production is no longer capable to eliminate excess radicals' production allowing cell damage. In the same way, ROS may produce additional oxidative stress in the cells generating damage in all the cell components including proteins, lipids, and DNA. In case of DNA damage produced by pyrimidine dimer formation by the generation of covalent bonds among the bases in the same DNA chain (Sichel et al. 2009), it generates mutations that may lead to loss of functional capability and cell death (Malato et al. 2009). At the same time, when microorganisms are exposed to ultraviolet radiation (UV, $\lambda < 400$ nm) during the photocatalytic reaction, DNA damage may occur directly through the radiation absorption by cell chromophores, which absorb radiation and produce heat. This interaction leads to an increase in ATP and RNA synthesis, jointly with the increase of ROS production. Microorganisms receiving a sublethal dose of UV radiation may become resistant to induced oxidative stress, partially recover their defense mechanisms, and adapt to oxidative stress generated by exposure to UV radiation alone (tanning effect) (Bandala et al. 2012).

8.2 Photocatalytic Reactor: Configurations for Water Disinfection

8.2.1 State of the Photocatalyst

Generation of catalyst sludges as side products is one of the main disadvantages of HPC processes in water disinfection. At pilot-plant level, HPC uses suspended TiO_2 in photoreactors where the semiconductor is recovered after the treatment. According to various lab scale research reports, the use of TiO_2 in suspensions is more efficient than their immobilized forms. Nevertheless, this latter form possesses specific advantages, such as cost reduction, material losses decrease, and skipping recovery steps in the process, which make desirable the generation of immobilized photocatalyst with higher efficiency as compared with those reported to date (Balasubramanian et al. 2004, Gelover et al. 2004).

Several supporting materials, from sand to quartz optical fiber, have been reported so far for TiO_2 immobilization. During the last years, the in situ generation of catalyst seems to be the most promising technology for catalyst immobilization (Gelover et al. 2004). Other authors (Guillard et al. 2003) have demonstrated that by using this method, the generated fixed form of titanium dioxide presents equal efficiency as Degussa P-25 (considered as the most efficient form of titanium dioxide) suspended catalyst for bacteria inactivation. However, more research is needed about the development of this promising idea before it can be considered for future design of efficient photocatalytic processes.

Nanostructured TiO_2 with different sizes, shapes, and morphologies (nanoparticles, nanofibers, nanowires, nanorods, nanoporous materials, and nanotubes) have been prepared, and their unique properties tested for water disinfection applications (Chuangchote et al. 2009). The preparation method plays a key role on the variety of structures formed and is well described in previous chapters of this book (Wang et al. 2009; Hahn et al. 2007; Kamat 2012). Very recent works have been reported for the inactivation of different bacterial strains using TiO_2 nanofibers (Liu et al. 2012), nanocomposites (Liu et al. 2013), and nanorods (Hassan et al. 2012).

One of the most promising TiO_2 nanostructure for environmental applications is titanium dioxide nanotubes (TDNs). Literature on TDN is the second most abundant only after carbon nanotubes. TDNs prepared from metal plates anodization have attracted interest because they can improve performance on well-known titania applications such as water photoelectrolysis, photocatalysis, heterojunction solar cells, gas sensing, and environmental purification (Zhang et al. 2006; Sun et al. 2008) since they provide a highly ordered TiO_2 nanotubular array with a large specific surface area. Two important structural characteristics of these TDN arrays are their inner diameter of approximately 100 nm and an average length of 3 μm. The highly ordered structure can improve the transport of photogenerated electrons in the TiO_2 film by providing a unidirectional electric channel and reducing grain boundaries in photocatalysis and photoelectrochemical applications, while the large surface area can allow more efficient light harvesting (Liu et al. 2008; Zhang et al. 2007). Despite limited, water disinfection using TDNs has been reported previously for bismuth vanadate-doped TDNs for *E. coli* inactivation (Wang et al. 2012), Ag/AgBr/TiO_2 nanotubes for *E. coli* inactivation (Hou et al. 2012), and the inactivation of *Mycobacterium* species using Ti/TiO_2 nanotubes (Brugnera et al. 2013; Chan et al. 2013).

Besides titania, other semiconducting materials have been used to promote or enhance microorganism inactivation using radiation. One of the most widely reported in the last two years is zinc oxide. Nanostructured zinc oxide with different sizes and morphologies has been used for water disinfection purposes (Rodríguez et al. 2010; Alarcón et al. 2011; Baruah et al. 2012) and the same for composites of grapheme (Gao et al. 2013).

8.2.2 Visible Light-Absorbing Semiconductors

Production of charge carriers is a fundamental step in degradation processes using HPC. Once generated, these species may lead to hydroxyl radical generation (and the subsequent organic matter degradation) or can recombine to generate the initial state and energy emission. This latter reaction, known as recombination, is a practical problem when using TiO_2 catalyst, and it is extremely efficient (reaction rate $= 10^{-9}$ s) when no proper electron acceptor is present in the reaction media (Hoffmann et al. 1995; Martin et al. 1995; Minero et al. 1996).

Another main disadvantage for the use of HPC is related with the wavelength for activating the photocatalysts. The HPC process using titanium dioxide occurs only when it is irradiated with ultraviolet (UV, $\lambda < 400$ nm). This specific characteristic limits the photocatalyst sensitivity when solar radiation is intended to be used since only a small part of the solar spectrum (about 5 %) falls within the UV radiation (Castillo-Ledezma et al. 2011; Srinivasan and Somasundaram 2003). In order to avoid this restriction, several modifications have been attempted into the TiO_2 photocatalytic structure in order to make it active under visible spectral irradiation, improve its photosensitivity and quantum yield, as well as reduce its band gap energy requirements for photocatalytic activation (Popa et al. 2009; Pelaez et al. 2009; Sato 1986; Asahi et al. 2001; Sato et al. 2005; Morikawa et al. 2001; Premkumar 2004).

Nitrogen is one of the most widely reported dopants for shifting TiO_2 spectral absorption into the visible range. Substitution of N in the titania lattice has been proposed to contribute to a narrow band gap through merging nitrogen and oxygen 2p states on the top of the valence band as nitride (Ti–N) or oxynitride (Ti–O–N). Another theory is the formation of oxyanion species at interstitial lattice sites, where N is bonded to one or more oxygen atoms creating localized intergap states. Both arrangements are considered to change the energy photothreshold responsible of the redshift of the optical absorption toward the visible spectra region (Emeline et al. 2008; Wang et al. 2005; Miyauchi et al. 2004; Burda et al. 2003; Gole et al. 2004; Kisch and Macyk 2002; Torres et al. 2004; Li et al. 2010). The complete and detailed information related with different doping agents is included in Chap. 1.

Among the different doped materials reported, silver, vanadium, iron, and palladium-modified nitrogen-doped TiO_2 have been tested to enhance photocatalytic performance of titania for gram-negative bacteria such as *Escherichia coli*, *Pseudomonas aeruginosa*, and *Prevotella intermedia* inactivation in water (Li et al. 2010; Wu et al. 2008; Sun et al. 2010; Ubonchonlakate et al. 2012; Nair et al. 2012) as well as some gram-positive bacteria, i.e., MRSA, *Staphylococcus epidermidis*, *S. saprophyticus*, *S. pyogenes*, *Streptococcus aureus*, and *Saccharomyces cerevisiae* (Mo et al. 2007; Sheel et al. 2008; Chen et al. 2008; Lu et al. 2006; Erkan et al. 2006), and a few protozoa such as *Tetraselmis suecica*, *Amphidinium carterae*, and *Chlorella vulgaris* (Rodríguez-González et al. 2010; Foster et al. 2011).

Nonmetallic ion doping has also been shown to be effective to induce modifications of the electronic structure of TiO_2 by creating surface oxygen vacancies due to charge compensation between the nonmetallic ion and Ti^{4+} but without producing a significant change in the optical absorption of the photocatalyst (Rengifo-Herrera and Pulgarin 2010). Moreover, co-doping of TiO_2 with nitrogen and fluorine has demonstrated high photocatalytic activity in the visible region with beneficial effects induced by both dopants (Li et al. 2005; Huang et al. 2006; Xie et al. 2007). Huang et al. confirmed strong visible light absorption and high photocatalytic activity of N–FTiO_2 for *p*-chlorophenol and Rhodamine B degradation under visible light irradiation (Huang et al. 2006). Xie et al. effectively decomposed methyl orange with visible light-activated N–F–TiO_2 photocatalyst. Both attributed their findings to the synergistic effect of nitrogen and fluorine doping. More recently, Dionysiou's group has shown the effectiveness of N–TiO_2 to remove cyanobacterial toxins in water by using visible radiation (Choi et al. 2007). Castillo-Ledezma et al. (2011) have recently demonstrated the capability of N–F–TiO_2 for *E. coli* inactivation using solar radiation under several different reaction conditions, including pH value and radiation wavelengths (visible and UV + visible), and showed N–F–TiO_2 possessing higher efficiency compared with regular TiO_2 and solar disinfection processes (Castillo-Ledezma et al. 2011). Similar results were found by Wong et al. for nitrogen-doped TiO_2 in the inactivation of *E. coli* and several other pathogenic microorganisms (Wong et al. 2006). They also found that proteins and light-absorbing contaminants reduce the bacterial activity of the photocatalyst as result of their light-shielding effects. Liu et al. found that extracellular polymeric substances (EPS) generated by some heterotrophic bacteria play an important role in controlling the kinetics of the solar-induced photocatalytic process by generating a protective layer against the presence of photogenerated reactive oxygen species (ROS) during the photocatalyst activation (Liu et al. 2007).

In a very interesting synergistic approach, Wu et al. developed and tested a metal-/nonmetal-doped TiO_2 catalyst for the inactivation of gram-negative (*E. coli* and *P. aeruginosa*) and gram-positive (*S. aureus*) cells (Wu et al. 2009). They found that by combining both metal and nonmetal dopants, photocatalytic activity was much higher than using the dopants separately under visible light illumination.

8.2.3 Reactor Design and Engineering of Photocatalytic Units

While some of the physical–chemical principles of photocatalytic processes are relatively well understood, reactor design and reactor engineering of photocatalytic units still require consideration. This is particularly true in the context of scaled reactors processing large volumes of water and using high levels of irradiation. Several aspects of design, optimization, and operation of photochemical reactors that are not usually considered in the design of conventional chemical reactors should be taken into account (Cassano et al. 1995). Photocatalytic reactors for water treatment can be classified according to their design characteristics: (a) state of the photocatalyst, (b) type of illumination, and (c) position of the irradiation source (Mukherjee and Ray 1999).

The photocatalyst can be either suspended or attached to a support, as:

- *Photocatalytic slurry reactors:* In slurry reactors, the catalyst particles are freely dispersed in the fluid phase (water), and, consequently, the photocatalyst is fully integrated in the liquid mobile phase.
- *Photocatalytic reactors with immobilized photocatalyst:* The immobilized catalyst reactor design features a catalyst anchored to a fixed support, dispersed on the stationary phase (the catalyst–support system).

The type of irradiation is a major design issue for photocatalytic reactors. Reactors can be irradiated using UV polychromatic lamps or solar light. Two subcategories branch off from solar-illuminated reactors:

- *Non-concentrating solar-irradiated reactors:* These systems employ intensities equal or lesser than natural solar irradiation.
- *Concentrating reactors:* These reactors use irradiation intensities that surpass irradiations equivalent to one sun.

The position of the lamp or source of irradiation is a distinguishing feature of a photocatalytic reactor. The lamp position determines different configurations:

- *Reactors with an immersed light source:* In immersed source reactors, the lamp is placed inside the unit.
- *Reactors with an external light source:* In external source photocatalytic reactors have lamps located outside the reactor vessel.
- *Reactors with distributed light sources:* In distributed reactors, irradiation is transported from the source to the reactor by optical means such as reflectors or light guides.

Butterfield et al. (1997) reported a photochemical and photoelectrochemical degradation of aqueous organic solutions related with disinfection by anodic, thermal, and sol–gel TiO_2 films on titanium supports up to 45 cm in diameter and for anodic films up to 30 cm in diameter, as catalysts for the photoelectrochemical detoxification of water. Photochemical falling film reactor (Fig. 8.1a) and a photochemical/photoelectrochemical vortex reactor (Fig. 8.1b) are classified as photocatalytic reactor with immobilized photocatalyst and immersed light source as it was mentioned before. The vortex design is more commonly employed as a fluidic in-line mixer, and one of the major advantages of this design is that it has no moving parts, and was readily adapted into a photoelectrochemical system. The former showed the best mass transport characteristics and most efficient light usage, while the latter reactor clearly showed the efficacy of the electric field enhancement effect. The authors indicate that the degradation of model organic compounds is increased by the application of a potential bias to the n-TiO_2 working electrode;

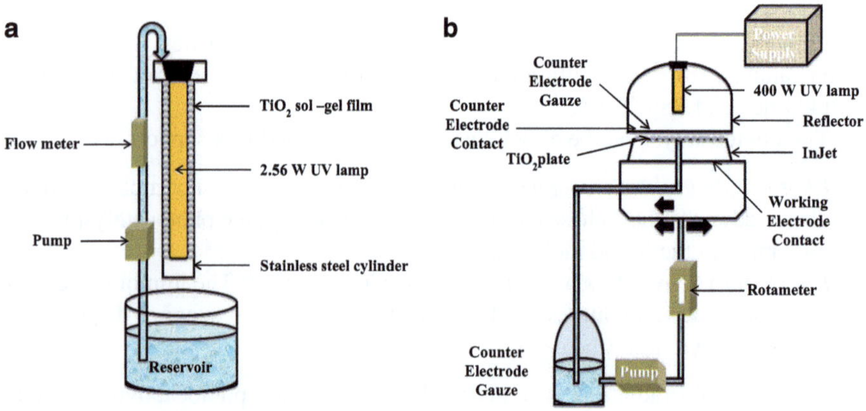

Fig. 8.1 Schematic diagram of the falling film reactor (**a**) and the vortex reactor (**b**) (Figures adapted from Butterfield et al. 1997)

however, improvements in the efficiency of the detoxification process are effectively restricted by the poor mass transport characteristics of the current reactor design. The results obtained in the absence of electrolyte are extremely surprising, suggesting that the efficiency of the photodegradation process is significantly enhanced in the presence of a positive potential bias on the semiconductor, even though no internal electric field is expected to be present in the TiO_2 under these conditions. Furthermore, there is no apparent induction period during which species may have been generated that might enhance the conductivity of the electrolyte.

Malato et al. (2003) reported a photocatalytic detoxification degradation or disinfection of four water-soluble pesticides (diuron, imidacloprid, formetanate, and methomyl) at pilot scale in two well-defined systems which are of special interest because natural solar UV light can be used: heterogeneous photocatalysis with titanium dioxide and homogeneous photocatalysis by photo-Fenton. These types of reactors are characterized like photocatalytic slurry reactors with non-concentrating solar-irradiated reactors, with an external light source as they were mentioned before (Fig. 8.2). The pilot plant is made up of compound parabolic collectors specially designed for solar photocatalytic applications related with disinfection using TiO_2 and iron as catalyzers. The total disappearance of the parent compounds, 90 % mineralization, and toxicity reduction below the threshold (EC50) have been attained with all pesticides tested using photo-Fenton. The process would be suitable for pretreatment of highly toxic and bio-recalcitrant compounds before biological treatment. Knowledge of toxicity of the wastewater evaluated by a battery of different bioassays is always necessary in order to reduce treatment costs. Once photocatalysis is complete, the catalyst separation from solution must be undertaken. Two different procedures have been proposed for this, separating and recycling TiO_2 and Fe, respectively.

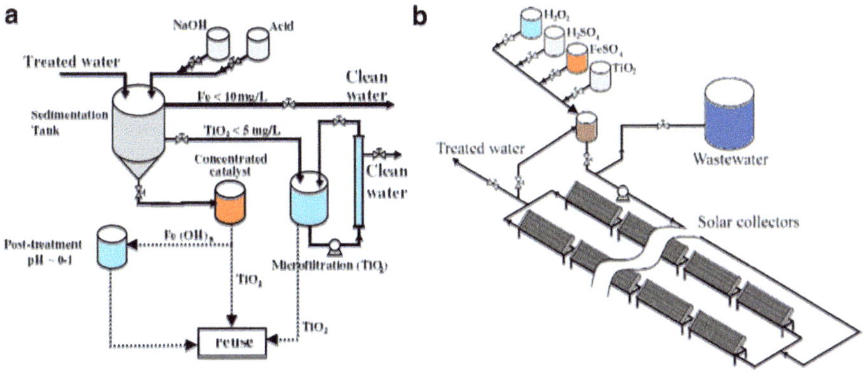

Fig. 8.2 Diagram of catalyst (TiO_2 or Fe) recycling (**a**) and simplified diagram of a photocatalytic treatment plant (**b**) (Figures adapted from Malato et al. 2003)

8.3 Efficiency of Photocatalytic Materials

8.3.1 *Photocatalytic Thermodynamic Efficiency Factor (PTEF) for Oxidation–Reduction*

The photocatalytic thermodynamic efficiency factor (PTEF) represents the used energy in the photochemical transformation and the photon energy absorbed by the photocatalyst. The PTEF is of general applicability with its application not being restricted to a specific chemical species, reaction order, reactor geometry, or reactor type (i.e., homogeneous or heterogeneous).

The PTEF can also describe the energy efficiency in a process where there is simultaneous oxidation and reduction of water contaminants as disinfection. In such process, the Q_{used} as charge of the PTEF represents the additive contribution of the enthalpy requirements to sustain both the reduction and oxidation processes. Consistently with view, the numerator of the PTEF shall include the addition of energy used for the oxidation and reduction processes:

$$\eta = \text{PTEF} = \frac{[r_{\bullet\text{OH}}(\Delta H_{\bullet\text{OH}}) - r_{\text{M+}}(\Delta H_{\text{M+}})]W_{\text{irr}}}{Q_a} \tag{8.6}$$

Or

$$\eta = \text{PTEF} = \frac{\left[r'_{\bullet\text{OH}}(\Delta H_{\bullet\text{OH}}) - r'_{\text{M+}}(\Delta H_{\text{M+}})\right]A_{\text{irr}}}{Q_a} \tag{8.7}$$

where $r_{\bullet\text{OH}}$ and $r'_{\bullet\text{OH}}$ represent the rates of formation of $^{\bullet}$OH radicals, $r_{\text{M+}}$ and $r'_{\text{M+}}$ the rates of consumption of the metal cation M^+, $\Delta H_{\bullet\text{OH}}$ the enthalpy of formation of $^{\bullet}$OH radicals (J mol e^{-1}), $\Delta H_{\text{M+}}$ the enthalpy of reduction of metal cations

(J mol e^{-1}), Q_a the energy absorbed by the photocatalyst (J s^{-1}), W_{irr} the weight of irradiated catalyst, and A_{irr} (m^2) the area of irradiated semiconductor with the next equation of PTEF in simultaneous oxidation–reduction photocatalysis that leads to:

$$\eta = \text{PTEF} = \frac{\left[r''_{\bullet OH}(\Delta H_{\bullet OH}) - r''_{M+}(\Delta H_{M+})\right] A_{irr}}{Q_a} \tag{8.8}$$

Or

$$\eta = \text{PTEF} = \frac{\left[r''_{\bullet OH}(\Delta H_{\bullet OH}) - r''_{M+}(\Delta H_{M+})\right] V}{Q_a} \tag{8.9}$$

where $r''_{\bullet OH}$ and $r'''_{\bullet OH}$ represent the rates of formation of $^{\bullet}OH$ radicals, r''_{M+} and r'''_{M+} the rates of consumption of the metal cation M^+, V the reactor volume, and V_{irr} the irradiated volume section. Considering that

$$\eta_{\bullet OH} = \frac{\Delta H_{\bullet OH}}{\alpha\left[\frac{N_A hc}{\lambda}\right]} \tag{8.10}$$

$$\eta_{M+} = \frac{\Delta H_{M+}}{\alpha\left[\frac{N_A hc}{\lambda}\right]} \tag{8.11}$$

$\eta_{\bullet OH}$ represent the fraction of the photon energy used to form $^{\bullet}OH$ groups, and η_{M+} the fraction of the photon energy employed to reduce a metal cation, one can express the PTEF parameter as follows:

$$\eta = \text{PTEF} = \left(\varphi_1 \eta_{\bullet OH} + \varphi_2 \eta_{M+}\right) \tag{8.12}$$

with φ_1 and φ_2 representing the quantum efficiencies for the oxidation and reduction process, respectively. Furthermore, in the ideal case of a photocatalytic reactor with a quantum yield of $\varphi_1 = 1$ and $\varphi_2 = 1$ (only primary processes involved), one can expect a PTEF_{max} for combined oxidation–reduction processes equal to the following:

$$\eta = \text{PTEF} = \left(\eta_{\bullet OH} + \eta_{M+}\right)_{max} \geq \eta = \text{PTEF} \tag{8.13}$$

8.3.2 Inactivation Apparent Quantum Yield (IQY)

Photocatalytic processes can benefit from the definition and use of efficiency parameters, which are based on fundamental thermodynamic and kinetic properties. Such parameters permit the direct comparison of results obtained from different experimental systems and conditions. An inactivation quantum yield (IQY) parameter can be defined as the ratio of the number of carbon atoms inactivated related with microorganisms over the number of photons absorbed with P_a representing the rate of absorbed photons and $(\mathrm{d}N_c/\mathrm{d}t)_{in}$ the rate of photoinactivation of carbon atoms at initial conditions (atoms of carbon inactivated s^{-1}) (Stuart et al. 2003):

$$\mathrm{IQY} = \varphi_{\mathrm{inact,in}} = \frac{\left[\frac{\mathrm{d}N_c}{\mathrm{d}t}\right]_{\mathrm{in}}}{P_a} \quad (8.14)$$

8.4 Based Materials Used for Water Disinfection

Metallic oxides and sulfurs are among the most used semiconductor materials available for photocatalytic purposes. Nowadays, titanium dioxide (TiO_2) has been frequently used as a semiconductor for HPC processes and reported to be among the most active (Blanco-Galvez et al. 2007). Nanocrystalline TiO_2 powder is synthesized using a novel reactive plasma process in which the precursor TiH_2 powder is oxidized through thermal plasma in-flight route to generate nanocrystalline TiO_2 powder. The synthesized powder consists of nano-sized TiO_2 particles, both anatase and rutile phases, in which anatase is the predominant phase. An additional feature of the plasma-synthesized TiO_2 powder is the higher surface concentration of Ti^{3+} state and hydroxyl group that enhance its photocatalytic activity. The photocatalytic inactivation of gram-positive *Enterococcus* and gram-negative *Klebsiella* bacteria is studied using the plasma-synthesized TiO_2 nanopowder with 365 nm ultraviolet (UV) light. The plasma-synthesized TiO_2 nanopowder catalyst is found effective in killing *Enterococcus* and *Klebsiella*. The results corroborated that the plasma-synthesized TiO_2 powder can be used for wastewater treatment and water purification (Vijay et al. 2013). Contrary to TiO_2 P25, the high surface area of TiO_2 nanotubes allows the simultaneous degradation of formic acid and the inactivation of pathogen fungus showing the interest of such materials for the treatment of wastewater (Turki et al. 2012).

Photocatalytic and photolytic disinfection of *Escherichia coli* in water were studied under natural sunlight using different types of photocatalyst (TiO_2 P-25, PC500, Ruana, and Bi_2WO_6) at different concentrations. The solar photoinactivation yielded complete inactivation results, which varied with the solar light intensity. Meanwhile, dark control samples in the lab (temperature constant at 25 °C) remained at constant concentration, and dark samples outside the laboratory showed a decrease due to the mild solar heating that occurred during the experiments. The adding of any kind of photocatalyst to the water accelerated the bactericidal action of solar irradiation and led to complete disinfection (until detection limit). The photocatalytic disinfection efficiency was not enhanced by the increase of catalyst concentration above 0.5 g L^{-1} for P-25, PC500, and Bi_2WO_6, where about 106 CFU mL^{-1} were completely inactivated within 5 min, 30 min, and more than 150 min of solar exposure under clear sky, respectively. An increase of the concentration to 1 g L^{-1} slightly decreased the total inactivation time. Rutile (Ruana) catalyst behaves differently; optimal concentration was lower than for the other titania materials; agglomeration of particles occurred as the concentration of catalyst increases (Helali et al. 2013).

TiO_2 has been used combined with another atom to develop photocatalytic disinfection. In this sense, Xiong et al. (2011) reported that mesoporous anatase (TiO_2) was modified with silver (Ag) nanoparticles using a photoreduction method. Performance of the resulting TiO_2–Ag nanocomposites for water purification was evaluated using degradation of Rhodamine B (RhB) and disinfection of *Escherichia coli (E. coli)* under ultraviolet (UV) irradiation. The composites with different Ag loadings were characterized using physical adsorption of nitrogen, X-ray diffraction, X-ray photoelectron spectroscopy, and UV–visible diffuse reflectance spectroscopic techniques. The results showed that metallic Ag nanoparticles were firmly immobilized on the TiO_2 surface, which improved electron–hole separation by forming the Schottky barrier at the TiO_2–Ag interface. Photocatalytic degradation of RhB and inactivation of *E. coli* effectively occurred in an analogical trend. The deposited Ag slightly decreased adsorption of target pollutants, but greatly increased adsorption of molecular oxygen with the latter enhancing production of reactive oxygen species (ROSs) with concomitant increase in contaminant photodegradation. The optimal Ag loadings for RhB degradation and *E. coli* disinfection were 0.25 wt % and 2.0 wt %, respectively. The composite photocatalysts were stable and could be used repeatedly under UV irradiation.

The photocatalytic disinfection of water containing pathogenic bacteria was investigated systematically by Hu et al. (2007) using AgI/TiO_2 under visible light ($\lambda > 420$ nm) irradiation. The catalyst was found to be highly effective in killing *Escherichia coli* and *Staphylococcus aureus*. The adsorbed $^{\bullet}OH$ and h_{VB}^+ on the surface of the catalyst were proposed to be the main active oxygen species by study of electron spin resonance and the effect of radical scavengers. The process of destruction of the cell wall and the cell membrane was verified by TEM, potassium ion leakage, lipid peroxidation, and FT-IR measurements. Some products from photocatalytic degradation of bacteria such as aldehydes, ketones, and carboxylic acids were identified by FT-IR spectroscopy. These results suggested that the photocatalytic degradation of the cell structure caused the cell death. The electrostatic force interaction of the bacteria catalyst significantly affected the efficiency of disinfection on the basis of the *E. coli* inactivation under different conditions. AgI/TiO_2 is highly effective for the killing of pathogenic bacteria under visible light irradiation. The destruction of *E. coli* and *S. aureus* cells was followed by TEM, FT-IR, and the formation and degradation of MDA by FT-IR measurement; it was found that the constituents of the cell wall membranes disappeared and that some intermediates such as aldehydes, ketones, and carboxylic acids were formed. The results provide solid evidence for the photocatalytic degradation of pathogenic bacteria under visible light irradiation. According to the results of electron spin resonance and the effect of radical scavengers, H_2O_2, free $^{\bullet}OH$, adsorbed $^{\bullet}OH$, and h_{VB}^+ on the surface of the catalyst, reactive active oxygen species were involved in the photocatalytic reaction (Fig. 8.3). The bactericidal activity obtained under various experimental conditions indicated that the electrostatic force interaction of the bacteria with the catalyst is crucial for high bactericidal efficiency.

Recently, Venieri et al. (2014) studied the photocatalytic inactivation of *Escherichia coli* and *Klebsiella pneumoniae* in water using novel Mn-, Co-, and

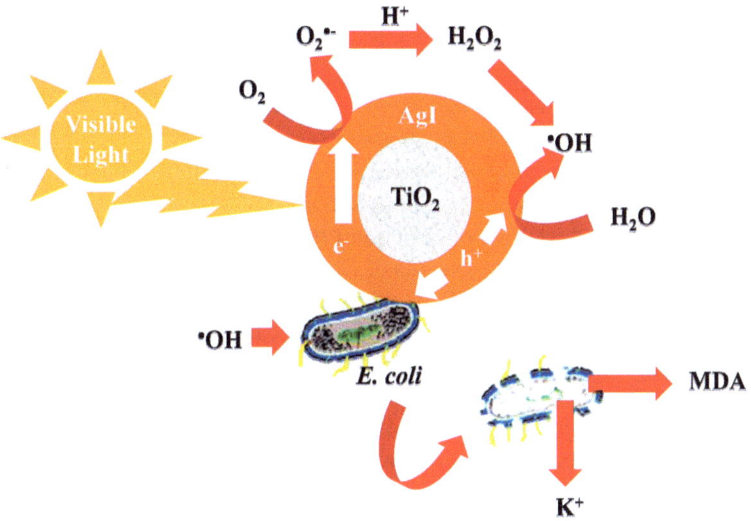

Fig. 8.3 Representation of photocatalytic degradation of *Escherichia coli* with AgI/TiO_2 under visible light irradiation (Figure adapted from Hu et al. 2007)

Mn/Co-doped TiO_2 catalysts under simulated solar irradiation. Doping shifted catalyst absorption to the visible region (up to 600 nm) decreasing the band gap energy and improving the activity of the Mn-, Co-, and binary Mn/Co-doped TiO_2; their activity was superior to the respective Degussa-TiO_2 P25 achieving the bacteria inactivation after 10 min of irradiation. Elimination of the bacteria was attributed to the oxidative degradation of their cells and increase of their cell permeability and not to the potential toxicity of the metal-doped semiconductors.

On the other hand, Kim et al. (2013) reported the degradation of humic acids (HA) which are known as the precursors of carcinogenic compounds formed by the disinfection of drinking water. Conventional treatments are inefficient for HA removal in drinking water; in this context, advanced oxidation processes have been proven to have a significant effect in the treatment of HA. The degradation of HA was investigated using nano-sized zinc oxide (ZnO)/laponite composite (NZLC). The reactions occurred in a UVC reactor by considering the following variables: pH, initial HA concentration, catalyst loading, addition of hydrogen peroxide (H_2O_2), and catalyst reuse. Water samples containing HA were analyzed by ultraviolet–visible spectrophotometer and high-performance size-exclusion chromatography. Initial HA concentrations were tested by the Langmuir–Hinshelwood model with k and K_{ads} values, determined to be 0.126 mg L^{-1} min^{-1} and 0.0257 L mg^{-1}, respectively. The change in pH affected the HA degradation efficiency by the photocatalytic activity where it was higher under acidic conditions rather than the alkaline ones. Optimal catalyst loading was proved to be a constrained factor in influencing the photocatalytic efficiency: the increase of catalyst concentration enhanced the HA decomposition efficiency up to an optimum value of 20 g L^{-1}, where there was no further degradation with excess loading. The addition of H_2O_2 was investigated through homogeneous and heterogeneous photocatalysis, and heterogeneous photocatalysis showed higher removal efficiency due to the combined effect of both catalysts and H_2O_2. Finally, NZLC was effective for reuse and exhibited an excellent stability after six times of usage.

Graphene-based semiconductor photocatalysts have attracted extensive attention because of their usefulness in environmental applications such as air cleanup, water disinfection, hazardous waste remediation, and water purification. Rani et al. (2013) prepared a series of zirconium oxide (ZrO_2, zirconia) and graphene (Gr) composites with different contents of Gr (5.7, 7.3, 8.3 %). The used precursors were zirconium oxychloride ($ZrOCl_2 \cdot 8H_2O$) and graphene oxide as the starting materials. The synthesized ZrO_2/graphene photocatalysts were characterized by X-ray diffraction, TGA, Raman spectroscopy, and UV–visible spectroscopy. The photocatalytic activities of the graphene composites were evaluated for the degradation of methyl orange dye. The rate of decolorization was recorded with respect to the change in intensity of absorption peaks for methyl orange. The absorption peaks diminished and finally disappeared during reaction, indicating that the dye had been degraded. The photocatalytic activity is strongly affected by the concentration of graphene in the ZrO_2. Finally, it was concluded that graphene when employed as catalytic support for ZrO_2 increases its photocatalytic efficiency.

On the other hand, the growth of ZnO nanorods on a flat substrate containing γ-irradiated seeds and their ability to photocatalytically eliminate bacteria in water were studied. The seed layer was obtained, by the spray pyrolysis technique, from zinc acetate solutions γ-irradiated within the range from 0 to 100 kGy. Subsequently, to grow the rods, the seeds were immersed in a basic solution of zinc nitrate maintained at 90 °C. The rate of crystal growth on the seed layer during the thermal bath treatment was kept constant. The resulting materials were characterized morphologically by scanning electron and atomic force microscopies; X-ray diffraction was used to study their morphology and structure and ultraviolet–visible spectroscopy to determine their absorbance. The obtained seed films were morphologically dependent on the radiation dose, and this was correlated with the ZnO nanorod films which presented a texture in the (0 0 2) direction perpendicular to the substrate. The rods have a hexagonal mean cross section between 20 and 140 nm. Using these rods, the photocatalytic degradation of *Escherichia coli* bacteria in water was studied; a positive influence of the crystalline texture on the degradation rate was observed (Alarcón et al. 2011).

In the case of nano-palladium loaded on nano-tungsten trioxide (n-Pd/n-WO_3), with 10 % wt, Pd loading was prepared by the impregnation evaporation method. The n-WO_3 support was prepared by dehydration of tungstic acid (H_2WO_4). This material was tested as a photocatalyst for inactivation and killing of coliform bacteria, by applying 355 nm pulsed UV laser radiations, generated from the third harmonic of Nd:YAG laser, to a model water sample, prepared using bacteria strains of *Escherichia coli*. The killing effect of n-Pd/n-WO_3 on coliform bacteria was characterized by means of selective culture media. The photocatalysis process did result in a very high irreversible injury (99 %) under investigated conditions. This process is cost-effective because no bacteria regrowth was recorded under optimum environment conditions. The disinfection rate of water was estimated by exponential decay. The conventional titania (TiO_2) semiconductor and commercially available WO_3 display a lower decay rate than that for n-Pd/n-WO_3 (Bagabas et al. 2010).

Concluding Remarks
Heterogeneous photocatalysis is the advanced oxidation process with the widest variety of technological applications resulting in its ability to inactivate nuisance microorganisms. The use of metal-based semiconductors as photocatalyst for water disinfection has been widely reported demonstrating being effective for the inactivation of several microorganisms and carcinogen cells to safe drinking water. The application of these technologies to water disinfection using solar radiation, coined as enhanced photocatalytic solar disinfection, has allowed the efficient inactivation of highly resistant microorganisms as *Escherichia coli* and *Staphylococcus aureus* using semiconductor materials as titanium, zinc, cadmium, iron, and wolframite, among others, which have some physical properties at room temperature. These materials have been used in photocatalytic slurry and with immobilized photocatalyst, with non-concentrating solar-irradiated or concentrating reactors, with an immersed, external, or distributed light source.

References

Alarcón J, Ponce S, Paraguay-Delgado F, Rodríguez J (2011) Effect of γ-irradiation on the growth of ZnO nanorods films for photocatalytic disinfection of contaminated water. J Colloid Interface Sci 364(1):49–55

Al-Ekabi H, Serpone N (1988) Kinetic studies in heterogeneous photocatalysis. Photocatalytic degradation of chlorinated phenols in aerated aqueous solutions over TiO_2 supported on a glass matrix. J Phys Chem 92:5726–5731

Alfano OM, Bahnemann D, Cassano AE, Dillert R, Goslich R (2000) Photocatalysis in water environments using artificial and solar light. Catal Today 58:199–230

Allison RP (1993) Electrodialysis membrane performance characteristics. In: Abstract of the American Water Works Association, Denver, CO. Chem Abstr 121:116850

Alrousan DMA, Dunlop PSM, McMurray TA, Byrne A (2009) Photocatalytic inactivation of *E. coli* in surface water using immobilized nanoparticle TiO_2 films. Water Res 43:47–54

Arancibia-Bulnes CA, Bandala ER, Estrada CA (2002) Radiation absorption and rate constants for carbaryl photocatalytic degradation in a solar collector. Catal Today 76:149–159

Asahi R, Morikawa T, Ohwaki T, Aoki K, Taga Y (2001) Visible-light photocatalyst in nitrogen-doped titanium dioxides. Science 293:269–271

Bagabas A, Gondal M, Khalil A, Dastageer A, Yamani Z, Ashameri M (2010) Laser-induced photocatalytic inactivation of coliform bacteria from water using pd-loaded nano-WO_3. Stud Surf Sci Catal 175:279–282

Balasubramanian G, Dionysiou DD, Suidan MT, Baudin I, Laine JM (2004) Evaluating the activities of immobilized TiO_2 powder films for the photocatalytic degradation of organic contaminants in water. Appl Catal B Environ 47:73–84

Bandala ER, Estrada C (2007) Comparison of solar collection geometries for application to photocatalytic degradation of organic contaminants. J Sol Energy Eng 129:22–26

Bandala ER, Raichle BW (2013) Solar driven advanced oxidation processes for water decontamination and disinfection. In: Enteria N, Akbarzadeh A (eds.) Solar Energy Sciences and Engineering Applications. CRC Press, EH Leiden, The Netherlands. Print ISBN: 978-1-138-00013-1

Bandala ER, Arancibia-Bulnes CA, Orozco SL, Estrada CA (2004) Solar photoreactors comparison based on oxalic acid photocatalytic degradation. Sol Energy 77:503–512

Bandala ER, Corona-Vasquez B, Guisar R, Uscanga M (2009) Deactivation of highly resistant microorganisms in water using solar driven photocatalytic processes. Int J Chem Reactor Eng 7:A7

Bandala ER, González L, de la Hoz F, Pelaez MA, Dionysiou DD, Dunlop PSM, Byrne JA, Sanchez-Salas JL (2011) Application of azo dyes as dosimetric indicators for enhanced photocatalytic solar disinfection (ENPHOSODIS). J Photochem Photobiol A Chem 218:185–191

Bandala ER, Gonzalez L, Sanchez-Salas JL, Castillo-Ledezma JH (2012) Inactivation of Ascaris eggs in water using sequential solar driven photo-Fenton and free chlorine. J Water Health 10 (1):20–30

Baruah S, Jaisai M, Dutta J (2012) Development of a visible light active photocatalytic portable water purification unit using ZnO nanorods. Catal Sci Technol 2:918–921

Blanco-Galvez J, Fernández-Ibáñez P, Malato-Rodríguez S (2007) Solar photocatalytic detoxification and disinfection of water: Recent overview. J Sol Energy Eng 129:4–15

Brandi RJ, Alfano OM, Cassano AE (2000) Evaluation of radiation absorption in slurry photocatalytic reactors 1. Assessment of methods in use and new proposals. Environ Sci Technol 34:2623–2630

Brugnera MF, Miyata M, Zocolo GJ, Leite CQF, Zanoni MVB (2013) Ti/TiO_2 nanotubes enhance *Mycobacterium fortuitum*. Mycobacterium chelonae and Mycobacterium abscessus inactivation in water. J Chem Technol Biotechnol. doi:10.1002/jctb.4243

Burda C, Lou Y, Chen X, Samia ACS, Stout J, Gole JL (2003) Enhanced nitrogen doping in TiO_2 nanoparticles. Nano Lett 3:1049–1051

Butterfield IM, Christensen PA, Hamnett A, Shaw KE, Walker GM, Walker SA, Howarth CR (1997) Applied studies on immobilized titanium dioxide films as catalysts for the photoelectrochemical detoxification of water. J Appl Electrochem 27:385–395

Cassano AE, Alfano OM (2000) Reaction engineering of suspended solid heterogeneous photocatalytic reactors. Catal Today 58:167–197

Cassano AE, Martin CA, Brandi RJ, Alfano OM (1995) Photoreactor analysis and design: fundamentals and applications. Ind Eng Chem Res 34:2155–2201

Castillo-Ledezma JH, Sánchez-Salas JL, López-Malo A, Bandala ER (2011) Effect of pH, solar irradiation and semiconductor concentration on the photocatalytic disinfection of *Escherichia coli* in water using Nitrogen-doped TiO_2. Eur Food Res Technol 233:825–834

Castillo-Ledezma JH, Bueno A, Pelaez MA, Sanchez-Salas JL, Dionysiou DD, Bandala ER (2013) Solar water disinfection using NF-codoped TiO_2 photocatalysis: Estimation of scaling-up parameters. Int J Chem Reactor Eng 11(2):1–8

Chan CMN, Ng AMC, Fung MK, Cheng HS, Guo MY, Djirisic AB, Leung FCC, Chan WK (2013) Antibacterial and photocatalytic actives TiO_2 nanotubes. J Exp Nanosci 8(6):859–867

Chen WJ, Tsai PJ, Chen YC (2008) Functional Fe_3O_4/TiO_2 core/shell magnetic nanoparticles as photokilling agents for pathogenic bacteria. Small 4:485–491

Choi H, Antoniou MG, Pelaez M, de la Cruz AA, Shoemaker JA, Dionysiou DD (2007) Mesoporous nitrogen-doped TiO_2 for the photocatalytic destruction of the cyanobacterial toxin microcystin-LR under visible light. Environ Sci Technol 41:7530–7535

Chuangchote S, Jitputti J, Sagawa T, Yoshikawa S (2009) Photocatalytic activity for hydrogen evolution of electrospun TiO_2 nanofibers. ACS Appl Mater Interfaces 5:1140–1143
Comisión Nacional del Agua (CNA) (2011) Estadísticas del Agua en México. Ed. CNA: México
Corona-Vasquez B, Aurioles V, Bandala ER (2012) Safe drinking water generation by solar-driven Fenton-like processes. In: Babatunde EB (ed.) Solar radiation. InTech Press, Rijeka, Croatia (ISBN: 978-953-51-0384-4)
Curco D, Giménez J, Addardak A, Cervera-March S, Esplugas S (2002) Effects of radiation absorption and catalyst concentration on the photocatalytic degradation of pollutants. Catal Today 75:177–188
Dalrymple OK, Stefanakos E, Trotz MA, Goswami DY (2010) A review of the mechanisms and modeling of photocatalytic disinfection. Appl Catal B Environ 98:27–38
Dunlop PSM, McMurray TA, Hamilton JWJ, Byrne JA (2008) Photocatalytic-inactivation of *Clostridium perfringens* spores on TiO_2 electrodes. J Photochem Photobiol A Chem 196:113–119
Metcalf and Eddy, Inc. (1991) Wastewater engineering: Treatment, disposal and reuse, (3 ed.), McGraw-Hill:New York
Emeline AV, Kuznetsov VN, Rybchuk VK, Serpone N (2008) Visible-light active titania photocatalyst: The case of N-doped TiO_2s - properties and some fundamental issues. Int J Photoenergy 2008:1–19
Erkan A, Bakir U, Karakas G (2006) Photocatalytic microbial inactivation over Pd doped SnO_2 and TiO_2 thin films. J Photochem Photobiol A Chem 184:313–321
Foster HA, Ditta IB, Varghese S, Steele A (2011) Photocatalytic disinfection using titanium dioxide: Spectrum and mechanism of antimicrobial activity. Appl Microbiol Biotechnol 90:1847–1868
Gao P, Liu J, Sun DD, Ng W (2013) Graphene oxide-CdS composite with high photocatalytic degradation and disinfection activities under visible light irradiation. J Hazard Mater 250–251:412–420
Gelover S, Mondragón P, Jiménez A (2004) Titanium dioxide sol-gel deposited over glass and its applications as photocatalyst for water decontamination. J Photochem Photobiol A Chem 165:241–246
Gole JL, Stout JD, Burda C, Lou Y, Chen X (2004) Highly efficient formation of visible light tunable $TiO_{2-x}N_x$ photocatalysts and their transformation at the nanoscale. J Phys Chem B 108:1230–1240
Golub D, Ben-Hur E, Oren Y, Soffer A (1987) Electroadsorption of bacteria on porous carbon and graphite electrodes. Bioelectrochem Bioenerg 17:175–182
Guillard C, Disdier J, Monnet C, Dussaud J, Malato S, Blanco J, Maldonado MI, Herrmann JM (2003) Solar efficiency of a new deposited titania catalyst: chlorophenol, pesticide and dye removal applications. Appl Catal B Environ 46:319–332
Hahn R, Macak JM, Schmuki P (2007) Rapid anodic growth of TiO_2 and WO_3 nanotubes in fluoride free electrolytes. Electrochem Commun 9:947–952
Hassan MS, Amna T, Misha A, Yun SI, Kim HC, Kim HY, Khil MS (2012) Fabrication, characterization and antibacterial effect of novel electrospun TiO_2 nanorods on a panel of pathogenic bacteria. J Biomed Nanotechnol 8(3):394–404
Helali S, Polo-López MI, Fernández-Ibáñez P, Ohtani B, Amano F, Malato S, Guillard C (2013) Solar photocatalysis: A green technology for E. coli contaminated water disinfection. Effect of concentration and different types of suspended catalyst. J Photochem Photobiol A Chem 276:31–40
Hinden JM, Ernes LM, Visel PE (1985) Electrocatalytic Electrode, U.S.A. Patent US4517068 A.
Hoffmann MR, Martin ST, Choi W, Bahnemann DW (1995) Environmental applications of semiconductor photocatalysis. Chem Rev 95:69–96
Hou Y, Li X, Zhao Q, Chen G, Raston CL (2012) Role of hydroxyl radicals and mechanisms of *Escherichia coli* inactivation on Ag/AgBr/TiO_2 nanotube array electrode under visible light irradiation. Environ Sci Technol 46(7):4042–4050

Hu C, Guo J, Qu J, Hu X (2007) Photocatalytic degradation of pathogenic bacteria with AgI/TiO_2 under visible light irradiation. Langmuir 23:4982–4987

Huang DG, Liao SJ, Liu JM, Dang Z, Petrik L (2006) Preparation of visible-light responsive N-F codoped TiO_2 photocatalyst by a sol-gel-solvothermal method. J Photochem Photobiol A Chem 184:282–288

Kamat PV (2012) TiO_2 Nanostructures: Recent Physical Chemistry Advances. J Phys Chem C 116:11849–11851

Kim JK, Alajmy J, Borges AC, Joo JC, Ahn H, Campos LC (2013) Degradation of humic acid by photocatalytic reaction using nano-sized ZnO/laponite composite (NZLC). Water Air Soil Pollut 224:1–10

Kisch H, Macyk W (2002) Visible-light photocatalysis by modified titania. Chem Phys Chem 3:399–400

Li D, Ohashi N, Hishita S, Kolodiazhnyi T, Haneda H (2005) Origin of visible-light-driven photocatalysis: A comparative study on N/F doped and N-F codoped TiO_2 powders by means of experimental characterizations and theoretical calculations. J Solid State Chem 178:3293–3302

Li Q, Wu P, Xie R, Shang JK (2010) Enhanced photocatalytic disinfection of micro-organisms by transition metal ion modification of nitrogen doped titanium oxide. J Mater Res 25:167–176

Liu Y, Li J, Qiu X, Burda C (2007) Bactericidal activity of nitrogen-doped metal oxide nanocatalysts and the influence of bacterial extracellular polymeric substances (EPS). J Photochem Photobiol A Chem 190:94–100

Liu Z, Zhang X, Nishimoto S, Jing M, Tryk D, Murakami T, Fujishima A (2008) Highly ordered TiO_2 nanotube arrays with controllable length for photoelectrocatalytic degradation of phenol. J Phys Chem C 112:253–259

Liu L, Liu Z, Bai H, Sun D (2012) Concurrent filtration and solar photocatalytic disinfection/ degradation using high performance Ag/TiO_2 nanofiber membrane. Water Res 46(4):1101–1112

Liu L, Bai H, Liu J, Sun D (2013) Multifunctional graphene oxide-TiO_2-Ag nanocomposites for high performance water disinfection and decontamination under solar irradiation. J Hazard Mater 261:214–223

Lu JW, Li FB, Guo T, Lin LW, Hou MF, Liu TX (2006) TiO_2 photocatalytic antifungal technique for crop diseases control. J Environ Sci-China 18:397–401

Malato S (1999) Solar photocatalytic decomposition of pentachlorophenol dissolved in water. CIEMAT (Ed), Madrid, Spain, ISBN: 8478343369

Malato S, Blanco J, Vidal A, Alarcón D, Maldonado MI, Cáceres J, Gernjak W (2003) Applied studies in solar photocatalysis detoxification: an overview. Sol Energy 75:329–336

Malato S, Fernández-Ibáñez P, Maldonado MI, Blanco J, Gernjak W (2009) Decontamination and disinfection of water by solar photocatalysis: Recent overview and trends. Catal Today 147 (1):1–59

Marius G (2006) The physics of semiconductors: An introduction including devices and nanophysics. Springer, Berlin

Martin ST, Lee AT, Hoffmann MR (1995) Photocatalytic degradation of pesticide-acaricides in aqueous suspensions of TiO_2. Environ Sci Technol 29:2567–2573

Miller S (1993) Disinfection products in water treatment. Environ Sci Technol 27:2292

Minero C, Pelizzetti E, Malato S, Blanco J (1996) Large solar plant photocatalytic water decontamination: Effect of operational parameters. Sol Energy 56:421–428

Miyauchi M, Ikezawa A, Tobimatsu H, Irie H, Hashimoto K (2004) Zeta potential and photocatalytic activity of nitrogen doped TiO_2 thin films. Phys Chem Chem Phys 6:865–870

Mo AC, Xu W, Xian S, Li Y, Bai S (2007) Antibacterial activity of silver-hydroxyapatite/titania nanocomposite coating on titanium against oral bacteria. Key Eng Mater 330(332):455–458

Morikawa T, Asahi R, Ohwaki T, Aoki K, Taga Y (2001) Band-gap narrowing of titanium dioxide by nitrogen doping. Jpn J Appl Phys 40:L561–L563

Mukherjee PS, Ray AK (1999) Major challenges in the design of a large scale photocatalytic reactor for water treatment. Chem Eng Tech 22(3):253–260

Nair RG, Roy J, Samdarshi SK, Mukherjee AK (2012) Mixed phase v doped titania shows high photoactivity for disinfection of Escherichia coli and detoxification of phenol. Sol Energy Mater Sol Cells 105:103–108

Oren Y, Tobias H, Soffer A (1983) Removal of bacteria from water by electroadsorption on porous carbon electrodes. Bioelectrochem Bioenerg 11:347–351

Pelaez MA, de la Cruz AA, Stathatos E, Falaras P, Dionysiou DD (2009) Visible light activated N-F-codoped TiO_2 nanoparticles for the photocatalytic degradation of microcystin-LR in water. Catal Today 144:19–25

Popa M, Diamandescu L, Vasiliu F, Teodorescu CM, Cosoceanu V, Baia M, Feder M, Baia L, Danciu V (2009) Synthesis, structural characterization and photocatalytic properties of iron-doped TiO_2 aerogels. J Mater Sci 44:358–364

Premkumar J (2004) Development of super-hydrophilicity on nitrogen-doped TiO_2 thin film surface by photoelectrochemical method under visible light. Chem Mater 16:3980–3981

Rajeshwar K, Ibañez J (1997) Electrochemistry: Fundamentals and applications in pollution abatement. Academic, San Diego

Rani S, Kumar M, Sharma S, Kumar D, Tyagi S (2013) Effect of graphene in enhancing the photocatalytic activity of zirconium oxide. Catal Lett 1–7

Rengifo-Herrera JA, Pulgarin C (2010) Photocatalytic activity of N, S co-doped and N-doped commercial anatase TiO_2 powders towards phenol oxidation and *E. coli* inactivation under simulated solar light irradiation. Sol Energy 84:37–43

Rodríguez J, Paraguay F, López A, Alarcon J, Estrada W (2010) Synthesis and characterization of ZnO nanorods films for photocatalytic disinfection of contaminated water. Thin Solid Films 519(2):729–735

Rodríguez-González V, Alfaro SO, Torrez-Martínez LM, Cho SH, Lee SW (2010) Silver-TiO_2 nanocomposites: Synthesis and harmful algae bloom UV photoelimination. Appl Catal B Environ 98:229–234

Sato S (1986) Photocatalytic Activity of NOx-Doped TiO_2 in the Visible Light region. Chem Phys Lett 123:126–128

Sato S, Nakamura R, Abe S (2005) Visible-light sensitization of TiO_2 photocatalysts by wet-method N doping. App Catal A: Gen 284:131–137

Sheel DW, Brook LA, Ditta IB, Evans P, Foster HA, Steele A, Yates HM (2008) Biocidal silver and silver/titania composite films grown by chemical vapor deposition. Int J Photoenergy 11, doi:10.1155/2008/168185

Sichel C, Fernández-Ibáñez P, de Cara M, Tello J (2009) Lethal synergy of solar UV-radiation and H_2O_2 on wild *Fusarium solani* spores in distilled and natural water. Water Res 34:1841–1850

Srinivasan C, Somasundaram N (2003) Bactericidal and detoxification effects of irradiated semiconductor catalyst, TiO_2. Curr Sci India 85:1431–1438

Stoner GE (1973) Electrochemical inactivation of pathogens, US 3725226 A

Stoner GE, Cahen GL, Sachyani M, Gilead E (1982) The mechanism of low frequency AC electrochemical disinfection. Bioelectrochem Bioenerg 9:229

Stuart J, Salaices M, Valvano M, Lasa H (2003) Photocatalytic inactivation of MS2 bacteriophage and *E. Coli*. Kinetics modeling and quantum efficiency. In: Abstracts of the 18th North American Catalysis Symp. pp 247

Sun W, Yu Y, Pan H, Gao X, Chen Q, Peng L (2008) CdS quantum dots sensitized TiO_2 nanotube-array photoelectrodes. J Am Ceram Soc 130:1124–1125

Sun C, Li Q, Gao S, Cao L, Shang JK (2010) Enhanced photocatalytic disinfection of *Escherichia coli* bacteria by silver modification of nitrogen-doped titanium oxide nanoparticle photocatalyst under visible-light illumination. J Am Ceram Soc 93:3880–3885

Torres GR, Lindgren T, Lu J, Granqvist CG, Lindquist SE (2004) Photoelectrochemical study of nitrogen-doped titanium dioxide for water oxidation. J Phys Chem B 108:5995–6003

Turki A, Ibáñez PF, Ghorbel A, Kochkar H, Guillard C, Berhault G (2012) Synthesis design of TiO_2 nanotubes and nanowires and photocatalytic applications in the degradation of organic pollutants in the presence or not of microorganisms. Mater Res Soc Symp Proc 1442:13–18

Ubonchonlakate K, Siakong L, Saito F (2012) Photocatalytic disinfection of *P. aeruginosa* bacterial Ag-doped TiO_2 film. Procedia Eng 32:656–662

Venieri D, Fraggedaki A, Kostadima M, Chatzisymeon E, Binas V, Zachopoulos A, Kiriakidis G, Mantzavinos D (2014) Solar light and metal-doped TiO_2 to eliminate water-transmitted bacterial pathogens: Photocatalyst characterization and disinfection performance. Appl Catal B Environ 154–155:93–101

Vijay M, Ramachandran K, Ananthapadmanabhan PV, Nalini B, Pillai BC, Bondioli F, Manivannan A, Narendhirakannan RT (2013) Photocatalytic inactivation of Gram–positive and Gram–negative bacteria by reactive plasma processed nanocrystalline TiO_2 powder. Curr Appl Phys 13(3):510–516

Wang ZP, Cai WM, Hong XT, Zhao XL, Xu F, Cai CG (2005) Photocatalytic degradation of phenol in aqueous nitrogen-doped TiO_2 suspensions with various light sources. Appl Catal B Environ 57:223–231

Wang J, Zhao L, Lin V, Lin Z (2009) Formation of various TiO_2 nanostructures form electrochemically anodized titanium. J Mater Chem 19:3682–3687

Wang W, Yu Y, An T, Li G, Yip H, Yu J, Wong P (2012) Visible-light driven photocatalytic inactivation of *E. coli* K-12 by bismuth vanadate nanotubes: Bactericidal performance and mechanism. Environ Sci Technol 46:4599–4606

Wong MS, Chu WC, Sun DS, Huang HS, Chen JH, Tsai PJ, Lin NT, Yu MS, Hsu SF, Wang SL, Chang HH (2006) Visible-light-induced bactericidal activity of a nitrogen-doped titanium photocatalyst against human pathogens. Appl Environ Microbiol 72:611–6116

Wu P, Xie R, Shang JK (2008) Enhanced visible-light photocatalytic disinfection of bacteria spores by palladium-modified nitrogen-doped titanium oxide. J Am Ceram Soc 91:2957–2962

Wu P, Xie R, Imlay JA, Shang JK (2009) Visible light induced photocatalytic inactivation of bacteria by composite photocatalysts of palladium oxide and nitrogen-doped titanium oxide. Appl Catal B Environ 2009(88):576–581

Xie Y, Li Y, Zhao X (2007) Low temperature preparation and visible light induced catalytic activity of anatase N-F codoped TiO_2. J Mol Catal A: Chem 277:119–126

Xiong Z, Ma J, Ng WJ, Waite TD, Zhao XS (2011) Silver– modified mesoporous TiO_2 photocatalyst for water purification. Water Res 45:2095–2103

Yahya MT, Gerba CP (1992) Water disinfection system and method. Chem Abstr 118:87342

Younas H, Qazi I, Hashmi I, Awan M, Mahmood A, Qayyum HA (2014) Visible light photocatalytic water disinfection and its kinetics using Ag-doped titania nanoparticles. Environ Sci Pollut Res Int 21:740–752

Zhang H, Quan X, Chen S, Zhao H (2006) Fabrication and characterization of silica/Titania nanotubes composite membrane with photocatalytic capability. Environ Sci Technol 40:6104–6109

Zhang Z, Yuan Y, Shi G, Fang Y, Liang L, Ding H, Jin L (2007) Photoelectrocatalytic activity of highly ordered TiO_2 nanotube arrays electrode for azo dye degradation. Environ Sci Technol 41:6259–6263

Chapter 9
Future and Perspectives for Photocatalytic Materials in Environmental Photocatalysis

Aracely Hernández-Ramírez and Iliana Medina-Ramírez

Abstract Nowadays the number of publications regarding the synthesis, characterization, and applications of photocatalytic materials for environmental remediation has increased exponentially, since the use of these materials might lead to the development of sustainable technology that allows cleaning the environment in a safe and effective way. Of particular importance is the development of technologies that will allow access to clean and affordable water. This premise is considered as one of the greatest global challenges of this century due to water scarcity, water pollution, and population exponential growth. Advanced oxidation processes (AOPs) based on semiconducting photocatalytic materials have shown to be suitable to aid in this matter. At first glance the photocatalytic process seems to be simple, but the whole process demands a fine tuning of several parameters that depend not only on the photocatalytic material to be used but also in the physicochemical properties of the effluent and characteristics of the reactor. Comparison of experimental results is not systematic and can be meaningless, since: (a) the preparation methods for photocatalytic materials are diverse and consequently the properties of the resulting catalysts (morphology, size, size distribution, crystalline phase, composition) can be quite different, which limits the comparison, and (b) the methods to evaluate photocatalytic activity vary, with no standard test being employed. Thus, most of the studies are limited to laboratory investigations while the number of practical applications is still limited. The area of environmental remediation assisted by photocatalytic technologies is an exciting subject of research that will be under investigation for quite some time.

A. Hernández-Ramírez (✉)
Universidad Autónoma de Nuevo León, Facultad de Ciencias Químicas, Cd. Universitaria, San Nicolás de los Garza, NL, Mexico
e-mail: aracely.hernandezrm@uanl.edu.mx

I. Medina-Ramírez
Departamento de Química, Universidad Autónoma de Aguascalientes, Av. Universidad 940, Ciudad Universitaria, 20131 Aguascalientes, Ags., Mexico
e-mail: iemedina@correo.uaa.mx

A. Hernández-Ramírez, I. Medina-Ramírez (eds.), *Photocatalytic Semiconductors*,
DOI 10.1007/978-3-319-10999-2_9

9.1 Economic Aspects in the Production and Application of Photocatalytic Materials

As with any industrial process, the implementation of AOPs based on photocatalytic materials will depend on their cost and efficiency, themes that can be controlled by a careful selection of the preparation method and composition of the photocatalytic semiconductor. It has been illustrated through the different chapters of this book that there are many options with regard to the composition and production of visible light active (VLA) semiconducting catalyst; however, still bare and modified TiO_2 materials are situated as a number one option for several researchers. The primary reasons for this preference rely on the outstanding physicochemical properties of the material, low cost, and green nature (biocompatible, nontoxic, do not produce hazardous wastes, do not require chemical additives).

Despite the numerous efforts for the development of efficient VLA photocatalytic materials, there are very few examples of industrial applications. From the economical point of view, broad implementation of heterogeneous photocatalytic processes in water treatment and/or environmental remediation will require overcoming the associated high cost of the photocatalysts which might be achieved by ensuring their reuse. Although this problem can be solved by immobilizing the catalyst on an inert surface, such as glass, quartz, concrete, or ceramics, this will limit the activity of the catalyst, and this, in turn, decreases the efficiency of the operation. Thus, to guarantee the reuse of the material, efficient supported materials should be produced. For practical purposes, the support should have specific characteristics, such as good adsorbent (to enhance the adsorption capability of TiO_2), chemically stable, photochemically inert, nontoxic, abundant, and low cost. Without a doubt the development of nanoscience and nanotechnology has highly influenced the field of heterogeneous photocatalysis. In this regard, nano-magnetite and nano-TiO_2, for instance, can be combined in core-shell structures, rendering hybrid materials in which the shell provides desired functionality while the magnetic core allows easy particle separation (Qu et al. 2013).

Due to the heterogeneous nature of the photocatalytic process, the adsorption of the contaminants on the surface of the catalysts is a variable that influences the efficiency and the kinetics of the system. Taking this into consideration, nanostructured TiO_2 has much higher photoactivity due to its large surface area, lower volume for e^-/h^+ recombination, and faster interfacial charge transfer. It has been demonstrated that TiO_2 nanotubes (NTs) have distinct surface features that can lead to lower charge carrier recombination (owing to the short carrier diffusion paths), better adsorption of contaminant, and high photoactivity (Qu et al. 2013).

In order to afford reliable, cost-effective water treatment processes, VLA photocatalysts should be employed, since these materials can be activated by sunlight or artificial visible light. The visible light-induced photosensitization approach based on TiO_2 (and other oxide semiconductors) is particularly amenable to use in developing country scenarios and in tropical or desert locales where solar insolation is plentiful. Even with the implementation of reactor setups using

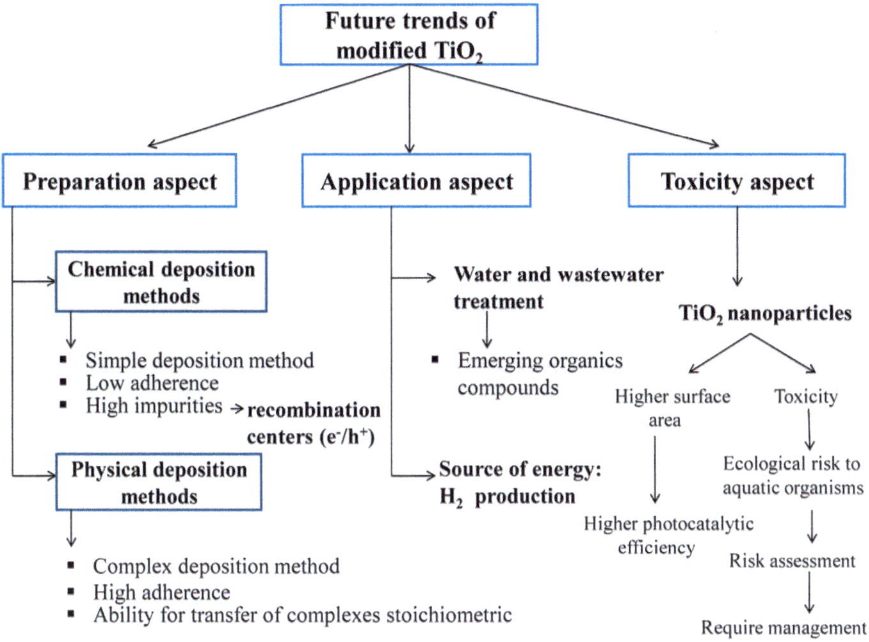

Fig. 9.1 Future trends of modified TiO_2 for environmental applications. With kind permission from Daghrir et al. 2013

artificial light, the cost for wastewater treatment will be lower compared to a process activated by UV light (Malato et al. 2009). Despite the many advantages offered by VLA photocatalysts and the several advances that numerous researchers have attained in order to improve the performance and stability of these materials, still, several challenges (toxicity and preparation aspects) have to be considered for practical environmental applications as illustrated in Fig. 9.1 (Daghrir et al. 2013).

Keeping in mind the economic aspects for the large-scale implementation of this technology, it is still very difficult to determine which fabrication technique should afford the most reliable, cost-effective method for the preparation of these photocatalytic materials. The choice of one method over another should take into consideration the cost, the effectiveness (removal rates and extent of mineralization), reaction intermediates/pathways, toxicity issues, and the stability of the catalyst for operation at large scale. For instance, only a few studies have focused on the identification of formed intermediates and have proposed possible degradation pathways during photocatalytic reactions.

Ideally AOPs will lead to the mineralization of recalcitrant pollutants; however, some systems only achieve degradation of the parent compound releasing into solution metabolites that can be more toxic to the environment or living organisms; therefore it is imperative to evaluate also the toxicity of the treated effluent. A rapid approach to evaluate the impact of treated effluents can be conducted by bioassays, using zooplankton as organisms' model. These ecotoxicological tests are considered sensitive, quick, and reliable. Nowadays, commercial kits (Daphtoxkit F™ and Microtox™; with *Daphnia magna* and *Vibrio fischeri* as tested microorganisms, respectively) are available. These kits can be used to evaluate the impact of soluble substances in water (Czech et al. 2014). Thus, extensive research is still needed in order to achieve this goal (Daghrir et al. 2013).

On the other hand, for heterogeneous photocatalytic water purification processes, the generation of photons driven by UV and solar light for catalyst activation is a significant part of the total cost for the operation of the system. The evaluation of the treatment costs is, at this time, one of the aspects that needs more attention. In order to select a purification system for wastewater reuse, a number of important factors including economics, economy of scale, regulations, effluent quality goals, safe operation, and robustness should be taken into consideration (Ahmed et al. 2011). Nevertheless, the applications of photocatalysis in this field are evidently determined by the quality and quantity (or flow rate) of the water to be treated and the requirements regarding the treated water (Agrios and Pichat 2005). Additionally, optimizing the degradation parameters is crucial from the perspective of efficient design and the application of the photocatalytic oxidation process to ensure sustainable operation. The application of heterogeneous photocatalysis for real wastewaters requires further investigation to achieve pollutant removal through the optimization of process parameters. This would make a significant impact on the potential commercial application of this technique to industrial systems (Ahmed et al. 2011).

Alternatively, coupling of semiconductor photocatalysis with other AOPs has proven their efficiency in water purification (Mohajerani et al. 2009). Some reports showed that AOP hybrid techniques appear to be promising for the elimination as well as improved mineralization compared with individual processes (Malato et al. 2009; Garza-Campos et al. 2014). The possible combination with other decontamination methods (in particular biological methods) has also shown promising results. In this context, biological treatment has been coupled with photocatalytic degradation method for the removal of phenol to reduce operational cost on single photocatalytic process. The mineralization time was reduced compared to the single biological treatment and the electrical cost saved compared to the application of the photocatalytic treatment alone (Gaya and Abdullah 2008).

9.2 Toxicological and Environmental Impacts of the Use of Photocatalytic Materials

The production and application of numerous nanomaterials in many products and/or technologies are rapidly increasing; however, there are scarce numbers of studies regarding the potential risk these materials might pose to the environment and human health (Aschberger et al. 2011). Many of the characteristics that make nanomaterials (NMs) attractive for specific applications might represent a risk to human health and for the environment. For example, it is desirable that the NMs produced for water detoxification should be able to adsorb and degrade a large quantity of ions or molecules in solution; however, this same activity inside a human being may result in negative implications such as the adsorption of proteins and their denaturation (Auffan et al. 2009). It has been stated in previous chapters of this book that many of the physicochemical properties of NMs can be controlled by the variables employed during its fabrication; thus, NMs of identical composition but different forms (spheres, NTs, films, etc.) or size-dependent properties will exert different negative effects toward living organisms making it difficult and meaningless to compare their toxicity. Also, NMs exhibit properties associated with solutes and separate particle phases, a characteristic that complicates methodologies for studying these materials. Analysis of published data regarding NMs toxicity indicates that many studies lack a proper characterization of the material under study.

The general toxic mechanisms associated with NMs can be enumerated as: (1) solubility properties—the release of metal ions in solution is considered as one of the principal mechanisms of toxicity regarding metallic and metallic oxide nanoparticles (NPs). For instance, CdS is considered toxic as it undergoes photocorrosion liberating Cd^{2+} ions in solution, whereas TiO_2, CeO_2, and ZrO_2 remain stable during the photocatalytic process. (2) Catalytic properties—nanostructured catalyst exhibits large surface/volume ratios that enhance their catalytic activity for several reactions; some of them can be detrimental to biological molecules. (3) Redox properties—some metallic or metallic oxide NMs can exhibit redox properties in biological media. Electron transfer can occur between metallic ions and biological molecules inducing an oxidation of proteins or a generation of reactive oxygen species (ROS) and an oxidative stress toward cellular organisms.

The field of nano-toxicology is still emerging; however, evidence for toxic effects of engineered NPs is increasing. Recent studies intend to correlate the physicochemical properties of NMs (as solids and in environmental conditions) to their toxicity. When assessing the toxicity of these materials, it is critical to evaluate aggregate size and structure, along with a host of other environmental conditions and potential ROS sinks (Jassby et al. 2012). In the case of photoactive semiconductor materials, it is necessary to evaluate phototoxicity and illumination-independent toxicity. It is still believed that more photoactive materials are prone to exert more negative effects toward biological molecules or cellular membranes; however, Tong et al. (2013) have demonstrated that this assumption is not properly correct. Although TiO_2 nanotubes (NTs) and nanosheets (NSs) have shown to be more photoactive than rod- and sphere-structured TiO_2, former materials (NTs and NSs) were found to be less phototoxic than their rod- and sphere-shaped counterparts.

From all the above mentioned, it is evident that interdisciplinary research is now required to reveal the full picture of the toxicity mechanisms and consequences of the impact of the nanostructured photocatalytic materials on the functioning and health of living beings and the environment.

Concluding Remarks

Crystalline TiO_2 (bare, doped, and modified) is the most used and studied as an efficient and environmentally benign catalyst for the mineralization of recalcitrant pollutants and disinfection of pathogen microorganisms. Alternatively, many studies have been focused in the research of other binary, ternary, or quaternary oxides to be used as photocatalytic materials intended to be active in visible light region.

This research in the field of heterogeneous photocatalysis has contributed to solve some of the intrinsic drawbacks encountered with this technology. For instance, there are numerous studies regarding the effects (roles) of dopants into the TiO_2 matrix and elucidation of the mechanisms for photocatalytic activity enhancement; however, it should be noted that some disadvantages have been associated with TiO_2-mediated photocatalysis and have not been properly addressed. For example, it is necessary to produce materials that present reduced recombination rates of the photoexcited electrons and holes; it is also important to optimize the preparation procedures to develop supported materials (that facilitate the separation process and materials reuse) with high photocatalytic activity. Implementation of the photocatalytic process at large-scale applications is also necessary. In order to attain this goal, economical and toxicological aspects of the effluent and systematic characterization of the risk hazards of nanomaterials should be considered.

The development of heterogeneous photocatalysis as a green technology at industrial scale must be accompanied by environmental health and safety research to alleviate unintended consequences and contribute toward the development of sustainable technology for environmental remediation.

References

Agrios AG, Pichat P (2005) State of the art and perspectives on materials and applications of photocatalysis over TiO_2. J Appl Electrochem 35:655–663

Ahmed S, Rasul MG, Brown R, Hashib MA (2011) Influence of parameters on the heterogeneous photocatalytic degradation of pesticides and phenolic contaminants in wastewater: a short review. J Environ Manage 92:311–330

Aschberger K, Micheletti C, Sokull-Klüttgen B, Christensen FM (2011) Analysis of currently available data for characterising the risk of engineered nanomaterials to the environment and human health—lessons learned from four case studies. Environ Int 37:1143–1156

Auffan M, Rose J, Wiesner M, Bottero JY (2009) Chemical stability of metallic nanoparticles: a parameter controlling their potential cellular toxicity in vitro. Environ Pollut 157:1127–1133

Czech B, Josko I, Oleszczuk P (2014) Ecotoxicological evaluation of selected pharmaceuticals to *Vibrio fischeri* and *Daphnia magna* before and after photooxidation process. Ecotoxicol Environ Safe 104:247–253

Daghrir R, Drogui P, Didier R (2013) Modified TiO_2 for environmental photocatalytic applications: a review. Ind Eng Chem Res 52:3581–3599

Garza-Campos B, Guzmán-Mar JL, Hinojosa-Reyes L, Brillas E, Hernández-Ramírez A, Ruiz-Ruiz E (2014) Coupling of solar photoelectro-Fenton with a BDD anode and solar heterogeneous photocatalysis for the mineralization of the herbicide atrazine. Chemosphere 97:26–33

Gaya UI, Abdullah AH (2008) Heterogeneous photocatalytic degradation of organic contaminants over titanium dioxide: a review of fundamentals, progress and problems. J Photochem Photobiol C 9:1–12

Jassby D, Budarz JF, Wiesner M (2012) Impact of aggregate size and structure on the photocatalytic properties of TiO_2 and ZnO nanoparticles. Environ Sci Technol 46:6934–6941

Malato S, Fernández-Ibáñez P, Maldonado MI, Blanco J, Gernjak W (2009) Decontamination and disinfection of water by solar photocatalysis: recent overview and trends. Catal Today 147:1–59

Mohajerani M, Mehrvar M, Ein-Mozaffari F (2009) An overview of the integration of advanced oxidation technologies and other processes for water and wastewater treatment. Int J Eng 3:120–146

Qu X, Brame J, Li Q, Alvarez PJJ (2013) Nanotechnology for a safe and sustainable water supply: enabling integrated water treatment and reuse. Acc Chem Res 46(3):834–843

Tong T, Shereef A, Wu J, Binh C, Kelly JJ, Gaillard J-F, Gray KA (2013) Effects of material morphology on the phototoxicity of nano-TiO_2 to bacteria. Environ Sci Technol 47:12486–12495

Index

A. Hernández-Ramírez, I. Medina-Ramírez (eds.), *Photocatalytic Semiconductors*,
DOI 10.1007/978-3-319-10999-2

MIX
Papier aus verantwortungsvollen Quellen
Paper from responsible sources
FSC® C105338

If you have any concerns about our products,
you can contact us on
ProductSafety@springernature.com

In case Publisher is established outside the EU,
the EU authorized representative is:
Springer Nature Customer Service Center GmbH
Europaplatz 3, 69115 Heidelberg, Germany

Printed by Libri Plureos GmbH
in Hamburg, Germany